# Apples and the Art of Detection

John Bunker and Francis Fenton: photo by Russell French

THE STARKEY APPLE.

from: Annual Report of the Agricultural Societies of Maine for 1877-8

# Apples and the Art of Detection

## Tracking Down, Identifying and Preserving Rare Apples


### Written and Illustrated by John Bunker

At present I am, as you know, fairly busy, but I propose to devote my declining years to the composition of a textbook, which shall focus the whole art of detection into one volume.

Sherlock Holmes, *The Adventure of the Abbey Grange*, p. 286

Cammy Watts with a small Charette

"Why not give ourselves up to the unrestrained enjoyment of the present?"
Sherlock Holmes: *The Adventure of the Mazarin Stone*, p. 1018



ISBN13: 978-0-578-50755-2

To purchase copies of *Apples and the Art of Detection or* learn more about our apple work, write to

John Bunker
PO Box 12
Palermo Maine 04354

or visit

outonalimbapples.com

# Dedication

To Francis Fenton, who may have been the oldest orchardist—and the most generous—in the world.

To Dan Bussey and Kent Whealy. Together they created the apple book the rest of us dreamed about.

And to Nate Dorpalen, who died way too young but whose grafted trees will outlive us all.

How often have I said to you that when you have eliminated the impossible, whatever remains, *however improbable*, must be the truth?
Sherlock Holmes, *The Sign of Four*, p. 111

# TABLE OF CONTENTS

# COMMON GROUND COUNTRY FAIR

CELEBRATE RURAL LIVING WITH MOFGA

SEPTEMBER 25, 26 & 27, 2009

UNITY, MAINE

GATES OPEN AT 9:00 AM

Maine Organic Farmers and Gardeners Association • P.O. Box 170 • Unity, Maine 04988 • 207-568-4142
www.mofga.org

# Introduction

Finding myself with some free time in the winter of 2008, I decided to paint a portrait of sixteen apples, each originating in one of Maine's sixteen counties. For years I'd been entranced by the colors and shapes of the old varieties. As luck would have it, the painting became the poster for MOFGA's Common Ground Country Fair. But, wonderful as the visuals are, the backstories of the apples are what I find to be particularly irresistible. For forty-five years I've been pulling over by the side of the road to marvel at ancient apple trees. I've knocked on a thousand strangers' doors, picked my way through old books and diaries, stared at apples balancing in my truck's cupholder, hung out with old-timers, grafted new trees and made lots of pies and sauce and cider. Searching for clues to the identity of the almost forgotten and the possibly extinct has made me feel a bit like a Sherlock Holmes of apples. Where did these lumpy, ribbed, striped, dotted and russeted apples come from? Who named them? What *are* their names? With their odd textures and flavors, what good are they? What can they contribute to the future? Anything? What would Holmes say? What would Holmes do? *What should I do?* He called his vocation the "art of detection." With apologies to the ultimate master, here, then, is my take on *apples* and the art of detection.

1. Red St. Lawrence, Penobscot County
2. Briggs Auburn, Androscoggin County
3. Marlboro, Hancock County
4. Cole's Quince, York County
5. Starkey, Kennebec County
6. Fletcher Sweet, Waldo County
7. Dudley Winter, Aroostook County
8. Cherryfield, Washington County
9. Cora's Grand Greening, Knox County
10. Deane, Franklin County
11. Givens, Sagadahoc County
12. Rolfe, Piscataquis County
13. Kavanagh, Lincoln County
14. Black Oxford, Oxford County
15. Blake, Cumberland County
16. Somerset of Maine, Somerset County

1. Androscoggin County
2. Aroostook County
3. Cumberland County
4. Franklin County
5. Hancock County
6. Kennebec County
7. Knox County
8. Lincoln County
9. Oxford County
10. Penobscot County
11. Piscataquis County
12. Sagadahoc County
13. Somerset County
14. Waldo County
15. Washington County
16. York County

The author channelling his inner Sherlock, by John Alsop, 2018

## A brief word about Sherlock Holmes

Throughout the text I have quoted freely from *The Complete Sherlock Holmes* by Sir Arthur Conan Doyle, 1930 edition, copyright by Doubleday & Company, Inc, Garden City, New York. It has been a pleasure to have the opportunity to get to know Holmes and Watson. They have both taught me a great deal. I can think of no one better than Holmes to be my number one guiding light in the art and science of fruit exploration.

Black Oxford

# One

# Appleton Ridge

Had he discovered a gold mine, greater delight could not have shone upon his features.
Sherlock Holmes, *A Study in Scarlet*, p. 17

Black Oxford

I'm cruising east down Route 3 in my white '71 VW truck, recently repainted green with a paint brush. It's the one that looks like a bus with the back chopped off. But it was built that way. It has sides that flap down on hinges turning it into a flatbed, with a tool compartment and the engine underneath. Perfect for loading on half a dozen logs to take to Bob's mill, or a large pile of firewood, or two wooden barrels and eighteen grain sacks stuffed with apples bound for the press. On either side of the road the remnants of ancient orchards look on as I zip by. I climb over one last rise and descend towards Belfast. Before me is Penobscot Bay, the rising sun and a hundred coastal islands. I leave the old apple trees behind and a few minutes later I'm parked in back of an old set of brick buildings. The door is open. In I go.

In the fall of 1979 I was employed as one of three managers at the Belfast Co-op Store in the small, quiet, mid-coast Waldo County town of Belfast, Maine. The Co-op was located in a storefront about the size of a large living room, just up Main Street from the harbor, a couple of doors down from the ornate offices of the Republican Journal and two doors up from what is now Chase's Daily restaurant. It would be years before the Co-op moved down the street and then eventually to its current location with its large staff, bakery, restaurant, wine and beer section, and very own parking lot. Back then it was just a hole in the wall. Waldo County still had only one traffic light, the one at the intersection just down the street. The brewpubs, the Green Store, the cheese shop and the assortment of other boutiques were decades away. The poultry factories were still humming along, although it would only be a couple of years before they would suddenly close shop and the chickens would all head south forever. Upstairs from the Co-op was Suzette's used clothing store. I shopped there during my breaks. "Trux" was our landlord, and he was in and out every day.

Most of what we sold in the Co-op Store was what you'd imagine if you were around back then. There were bins of grains and flour on wooden racks, large tins of various vegetable oils, five gallon plastic pails of nut butters, and tubs of tofu floating in murky water. We had a decent selection of cheese in big pieces that we cut up on demand. I'm sure we had the most extensive spice and herb rack in the county: five or six dozen gallon jars on their sides, all with labels on the lids. Every herb you could possibly want. I don't remember having anything in the store that was prepackaged except for ak-mak crackers and Stoned Wheat Thins.

The store had a friendly, sociable atmosphere. We had a small, dedicated bunch of customers, most of them back-to-the-landers who dribbled in and out throughout the day. It was a "slow food" kind of a place thirty years before Slow Food was invented. I got to know most of our shoppers by name, although there was the occasional exception. One dark, late-fall afternoon when we were cashing out shortly after closing time, a half dozen men in turbans knocked on the window. We let them in. They were sailors on a freighter from India docked for the night in Belfast Harbor. Belfast was not what anyone would call a bustling port, but evidently sometimes ships sailed in from India. Or at least one time one did. The men were Sikhs, and they were desperate for spices. They had seen our rack of gallon jars through the window. We sold them pounds of curry, cardamom, cumin, turmeric, and cayenne that night. They were ecstatic. They were our best spice customers ever. Another time, an old fellow came in and asked for fifty pounds of hulled sunflower seeds. Being a nosy guy, I asked him what they were for. Decades ago, he told me, his next door neighbors in Connecticut passed away and bequeathed him their parrot. He had since moved to Maine, and still had the bird who was now about seventy-five.

Ira Proctor 1974: courtesy Wendy and John Krueger

Though the "local food" movement was still a long way away, at the Belfast Co-op Store we were doing our small part to support the few local farmers in the area. It was the height of the back-to-the-land movement. Most of our customers had their own gardens. Lots of food was being grown throughout the county, but there wasn't much available wholesale. There were no farmers' markets or CSAs back then. We didn't consider ourselves to be farmers. We were something else. Homesteaders, maybe. No one I knew had a farm name, although a lot of us named our cars. We didn't even have a vegetable cooler in the store. But there were a few small commercial mixed-vegetable farms nearby, and we would occasionally offer their produce on consignment. In season we usually had a small selection of vegetables and fruit set in baskets on the floor or out on the sidewalk.

I was twenty-eight when I took the job, and I stayed there about two years. Not a long time, but one of the most important events in my life happened one day while I was there. It was about an apple.

I had been living in Waldo County for a half dozen years, and during that time I fell in love with apples. It was a love affair I never expected. Like many love affairs, it just happened. Perhaps it was all those free edibles that rained down in the fall on every lawn in town. Knock on the door. Of course you can have all you want. Fill your cardboard boxes and paper bags and plastic buckets. Free food for a young scavenger. Though I had yet to plant an apple or a pear, I was already making friends with the old hollow trees on our road and around town, and I'd begun to extend my explorations into a few other pockets around the county. I noticed that there were many different types of apples. Red and green and big and small. These must be different varieties. What a revelation! I was becoming obsessed with the shapes and sizes and colors. I assumed they must have names, but I only knew a few. I wanted to know them all.

One late-October afternoon that fall, a stocky, soft-spoken, older man with a large, gentle smile appeared at the Co-op Store with two bushels of apples. He asked us if we'd be willing to sell them on consignment. His name was Ira Proctor, and he was from Appleton Ridge in Appleton. He had come to the store other years with apples to sell. Judi and Anna, the other two store managers, knew him well, but I had never met him. The apples he brought with him that day were dark, purply black. He called them Black Oxfords. I had never seen a black apple before. I never set them out. I bought them all and took them home with me that night. I had discovered a goldmine. No, it was better than a goldmine.

Seven years after the treaty of Paris and thirty-three hundred miles away in what would become Paris, Maine, John Valentine was making nails one by one in what was then called Plantation Number Four, in Oxford County. William Lapham and Silas Maxim describe Valentine, the nail maker, in their 1884 *History of Paris, Maine:* "He used a treadle and clamp for holding the nail while he headed it with a hammer.

The motion thus acquired by using his foot on the treadle while striking with the hammer, became noticeable in his walk, and adhered to him through his life."

One day that fall of 1790 when the country was still so new, one of Valentine's neighbors, Nate Haskell, noticed a wild seedling apple tree with black apples on Valentine's land. Apple seedlings were as common throughout Maine back then as they are today. Apparently, Haskell was someone on the lookout who did more than see. He noticed. He was an observer. He probably saw that young black-fruiting apple tree when he was coming over to buy a few pounds of handmade nails. The apples had not dropped. He knew that meant they wouldn't ripen fully until they'd spent a month or two in the root cellar. This was good news; you wanted fruit in the winter. And who had ever seen a black apple before?

Black Oxford's defining characteristic is its rich, dark purple skin. Sometimes the purple looks almost black, especially when infected with sooty blotch, an innocuous, tasteless, fungal disease common in Maine during the summer. Sooty blotch looks just like you might imagine. It's a smokey, dark, blackish smear. On a yellow apple it looks unappetizing. On a Black Oxford it blends in with the purple, giving the apples an even darker and blacker color profile. Occasionally there may be patches where the purple is striped and the yellow-green ground color peeks through. Ground color is the term for the base coloring beneath the stripes and blush. It's the color of the canvas before the artist applies the paint. Around a

Black Oxford stem you might find a light, pinkish-purple splash of color. Numerous scattered light purple dots decorate the entire surface of the fruit. A tree full of Black Oxfords looks like a tree full of plums. Everyone should see this some time before they die. I need to see it every fall.

In apple lingo, the Black Oxford fruit is medium-sized and roundish in shape. The stem is thin in diameter and medium to long in length. The cavity is medium deep and acute. The basin is often shallow and usually slightly wavy and wrinkled. The flesh is white, or slightly green, sometimes with a bit of red bleeding just below the surface. The calyx is closed or sometimes open and the calyx tube is conical. The stamens are median or nearly basal. The core lines are meeting or just barely clasping. Carpels are sometimes tufted. The core is axile and closed.

Old Black Oxford tree in Brunswick: photo by Laura Sieger

Black Oxford apples get picked in late October when the nights routinely drop below freezing and the root cellar begins to cool down. Wait to pick them as late as you dare but pick them before the temperature drops below about 27. Too cold and the fruit may be ruined. The apples are hard and not very tasty in November, but they keep well and reach their best flavor as the winter rolls on. At a time when supermarkets were non-existent, having a source of winter storage fruit on the farm was essential. For many in central and southern Maine, Black Oxford trees became that source. The variety quickly spread throughout the state thanks to the work of itinerant grafters. Ten years after its discovery, it was already growing in Hallowell nearly fifty miles away.

Black Oxford reached its peak popularity during the first half of the nineteenth century. You can find it described in several of the pomological books of the day. Its earliest listing might be in Cole's American Fruit Book where it's called "Black apple." A few years later Downing and Thomas both include brief descriptions. By 1900, however, the apple had fallen out of favor, replaced by other winter varieties such as

Bud grafting: 1: make a t-shaped incision in the rootstock 2: slice a single bud off the scion 3: pull back the flap 4: insert the bud 5: tape up the graft 6: next spring, watch it grow!

Baldwin, Ben Davis, and Northern Spy. Beach doesn't even mention it in *The Apples of New York.* The fruit was considered poor for cooking, and some people complained that it lacked flavor. The trees also tend to overbear, resulting in lots of smaller fruit. Maybe these were valid complaints, or maybe they weren't the real reasons. Maybe apple growing was becoming a commodity business, and Black Oxford didn't fit the mold. It wasn't red. It didn't bear annually. It didn't do well all over New England. It was too finicky and too local. It appeared to be doomed.

But old apple trees can live on for decades or even centuries with no care at all. Eighty years after it fell out of favor, I thought the apple was great. And I still do. We now have two large, thirty-year-old bearing trees on our farm with another one coming along. There will be more. We can't live without this apple. We keep our fruit all winter in the root cellar in wooden boxes. Contrary to some naysayers of the past, it makes an excellent pie and sauce. I'm still cutting them up early each morning in May and making a small sauce pan of Oxford apple sauce to eat with my oatmeal. They might be a little shriveled by then, but they are still exceptionally aromatic and very tasty. Though we never spray it for scab, the fruit rarely shows much infection. You don't even need to tag your tree since there's no other apple like it. You'll always know which one is your Black Oxford. On top of all that, the flowers in spring are light pink.

A year and a half after my introduction to Ira Proctor and the Black Oxford, in August 1981, I learned to bud-graft. A fellow I met at the Co-op Store had just learned to graft and offered to show me how. He had sent away for scionwood from the USDA collection in Geneva, New York. We sat on the ground and inserted buds under the bark of the quarter-inch diameter rootstocks. I had acquired a new skill.

I spent the next two weeks bud-grafting every young seedling tree I could find along our road and down what we called Finley Lane. I hope I grafted onto apple trees, though they may have been cherries or juneberries! I wasn't very good at identifying young trees (or anything, for that matter) back then. Was my knife even sharp? I didn't care. I just wanted to graft. In my journal I drew maps of the neighbors' fields and the grafts I made. Where are they now? I wonder if any of them survived. Maybe I grafted apples onto birch trees.

After seeing the scion sticks from Geneva, I realized I didn't have to send away for scionwood. I could go cut my own and graft whatever I wanted. I collected buds from here and there, including a small orchard of old trees on a hilltop in Benton above the Kennebec River, behind the abandoned Fitzpatrick's Dairy.

Some attribute the name Nodhead to an eccentricity of the fellow who first popularized it. Others say it was the tree's habit of waving in the breeze. Sometimes it's called Jewett's Fine Red; I like Nodhead better: photo by Abbey Verrier

I called the apple "Fitzpatrick's." I had no idea what it really was. It was a smooth-skinned, reddish-pink winter storage apple. I liked it. Years later we determined it was probably Benton Red, a local synonym of the Washington County apple, Cherryfield. We still have one of those Benton Red trees I grafted way back when. We climb around in it every year in late October.

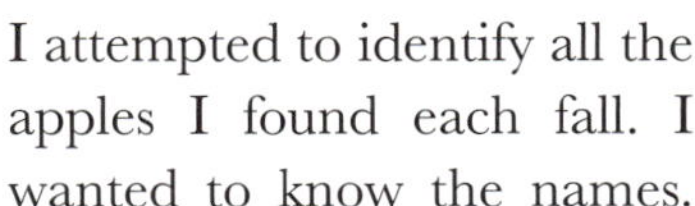

I attempted to identify all the apples I found each fall. I wanted to know the names. Black Oxfords were easy. Look for a black apple and you probably have it, especially if you're in central Maine. The others were harder. I had no one to show me how. I didn't think of it as "doing IDs"; I just wanted something to call them. I have a list I made back then with the names of several of the older trees in the neighborhood. I had them all wrong, but at least I was consistent! Actually, I did have one correct: the Nodhead tree on the lawn in front of Papa Glidden's house on the North Palermo Road. He told me. He had grafted that tree himself.

It would be years before I could actually *observe* a variety. At that time they all looked red or green or yellow or some blend of the three. I saw apples on the trees but that was about all. I did not yet recognize Northern Spy with that pink skin and ribbed torso, or brick-red Baldwin with its crown, or Stark's green that lurks through its red blush (red but not the brick-red of Baldwin), or Ben Davis with its red that appears to be black even though there's no black in it.

I had several friends back then who also wanted to learn. Snee Anthony lived a couple of towns away. She had a copy of Cole's *American Fruit Book* that I would borrow from time to time before finally getting my own. She brought me a Wolf River apple. I knew it was huge, but if I saw one again, would I know it? I didn't know yet that the two side-by-side trees up the hill on Turner Ridge Road were Tolman Sweet. The flavor was unforgettable, but what was it called? I had a path but no teachers, no apple mentors.

Four years after meeting Ira, I decided to graft Black Oxford for myself. Friends had introduced me to an old Black Oxford tree in Montville, not far from home. I could pick two or three bushels there every other year when it bore fruit. Now it was time to graft my own tree. Despite the fact that I could have cut perfectly good scionwood much closer to home, I knew the scionwood for my tree had to come from Ira. It was late afternoon on August 6, 1983.

From my journal (August 6, 1983): "I set out to find Ira Proctor. I found him two hours later at his home on Appleton Ridge. There I visited with him and collected scions from his pear tree and 3 apple trees: Blk Oxford, Northern Spy + 'Russet' The Russet tree looked very different from the Golden russets at The Apple farm and Ira also mentioned the difference. His fields were once covered with Apple trees, but many died during the terrible winter in the early 1930's when awful cold temperatures wiped out many of the trees in the state. This is a subject I find very interesting. He also spoke of keeping sheep in the grown orchard to keep down weeds + fertilize the trees. It was an enjoyable visit. I went home and grafted 5 Black Oxfords, one each Russet + No. Spy."

Black Oxford with the apex end up: the basin (depression with the calyx in the center) is shallow, somewhat wrinkled (the skin is slightly pinched around the calyx), and wavy (the rim of the basin is slightly undulating)

Not only did I now have my own Black Oxford tree, I was also getting hooked by the history and even beginning to observe differences between apples from one location to another. (More about those russets later.) There I was, standing in that small orchard perched on the side of Appleton Ridge looking east across the valley towards Penobscot Bay. The passion for the stories behind the apples had grabbed me and taken hold. Turns out, it would never let me go.

Over the course of the next few years I did begin to learn the names of the varieties. I discovered Black Oxford trees sprinkled throughout central Maine, including several in Waldo County. In 1999 I visited a Black Oxford tree in Hallowell, documented to have been planted in 1799. With an assortment of poles propping up its remaining branches, it bore seven bushels in its two-hundredth year. The tree itself was only twelve feet tall. The top had long ago died and broken off and been burned in the fireplace or wood stove. The hollow trunk had split vertically and then wrapped back around itself so that it was now inside out.

Meanwhile, the Black Oxford trees at our place have become the parents (technically the scion source) of many hundreds of other Black Oxfords. Now pushing thirty-five years old, our trees should be going strong for another hundred years or so, producing lots of scionwood and black winter keepers for people none of us will ever meet. Through the Fedco catalog, we've sold over five thousand Black Oxford trees across many states. You can find it featured in contemporary upscale apple books. A few commercial plantings are even being established. Who would have thought that one apple could inspire so many people? I'd call that better than a gold mine.

# Two

# The Art of Detection

It is of the highest importance in the art of detection to be able to recognize, out of a number of facts, which are incidental and which are vital.
Sherlock Holmes, *The Reigate Puzzle*, p. 407

Wolf River

This is a book about tracking down, discovering, identifying and preserving apple varieties. I call it fruit exploring, or as Sherlock Holmes would have it, the art of detection. Where do you find those lost heirlooms from long ago? How do you know that a particular apple tree is unusual or valuable or historic? What are their names? How do you find the names? Do all apples even have names? How do you learn about the histories behind these fruits? When you find an apple tree in the woods or in an orchard or in someone's yard, can you preserve that variety for yourself or your friends or posterity? How?

Fruit exploring is like baseball. To be good, you've got to be able to do a bunch of different things. Sometimes you hit, sometimes you throw, run, slide, scoop up grounders, catch fly balls, or steal when no one's looking. You've got to be ready for a fastball; a slider; a curveball; a changeup. Maybe even the occasional eephus pitch. In the 1960s Gaylord Perry of the San Francisco Giants was often accused of throwing spitters. You've got to be ready for spitters in this business too. Lots of apples are spitters. Sometimes you hit a single, other times a double, a triple, or a homer. More often than not, you strike out.

You could be traveling down some road you never knew was there. Or be sitting at the kitchen table with a knife and a pad of paper and some old books all afternoon while the snowflakes settle on your window sill. Or dangling from the highest branch or knocking on a stranger's door. Or spending hours alone weeding young trees and checking tags in a nursery, or bouncing around with a thousand other fanatics in a crowded convention hall drinking cider. You could swing and miss for weeks and then, all of a sudden, hit a grand slam like the one Jackie Bradley Jr. hit against Houston in Game Three of the playoffs this past fall. Wow!

Exploring for apples is like being a detective. Sometimes you have the apple before you but no name to go with it. Your job is to find the name; it's a case of lost identity. Other times you've got a name but no apple. Then it becomes a search; it's a case of missing persons. Still other times you think you have both the name and the apple. Then it's like that Gerard Depardieu film, *The Return of Martin Guerre*. You think you know it all, but are you really sure? Maybe all it needs is a quick glance. Sometimes it's a complex puzzle requiring years to untangle. Sometimes it's a mystery that forever remains a mystery.

By the mid-nineteenth century, many thousands of varieties were being grown across America. Every farm had its apples, from Maine to Georgia and west to the Mississippi. Perhaps as many as twenty thousand different varieties were being grown. Although some of these are described in old books with wonderful precision, most are not. Some appear in no books at all. Some have multiple alternate names. Some were only grown in a single town or two. There are currently over seven billion people on earth, three hundred million of whom live in the U.S. You'd think there ought to be someone out there somewhere who can identify every apple and sort this all out. Amazingly enough, this is not the case. Our grandparents and our parents got busy doing other things and didn't have time to pay attention to the names of the apples and how to use them. Along with lots of other skills, it mostly faded away. It only took a couple of generations. Tried harnessing a team lately? Not so long ago nearly everyone could. Identifying the name of an apple can be like finding the needle in the haystack or pulling the sacred sword out of the pond. Good luck.

Not only is the fruit exploring pomologist challenged by the lack of decent records, we're also dealing with sheer numbers. Birders have it easy. While there are ten thousand bird species around the world, there are only a thousand or so in North America. There may be twenty times as many apples in the U.S. alone. Not only that, but every apple seed that sprouts and becomes a tree is new and unique. These un-named impostors are foils, frauds and fakes. They look and taste and smell like apples, but they have no names. They are decoys. Not only might your apple be any of a thousand different varieties, it might not be a variety at all. It might be a seedling. It would be as though every hatchling from every robin and every blue jay and every chickadee produced a brand new bird species, different from any other that's ever been. How can anyone ever identify any apple?

Twenty Ounce can be huge, even larger than a Wolf River

What about a nifty searchable key? The skin is red, the stem is long, the basin's deep, the calyx tube is seriously conic. A few searchable apple keys have been written over the years. I'm familiar with three. All three can be useful, but all three have serious limitations. For a key to work, it must contain enough data to support a comparison. With only incomplete descriptions of thousands of historic apples, any key will be incapable of getting you very far. How do you know what it *might* be if you don't know what it *could* be? Genetic testing is making huge advances but also faces inherent challenges. Identifying a rare apple necessitates having an identified specimen with which to compare it. And with no one left who knows thousands of varieties, where would that ID'd specimen come from?

Isn't everything on the internet? Unfortunately the information we're looking for never got uploaded before all the experts passed on to orchard heaven. They all left for higher ground long ago. If you're going to identify apples, you're going to have to learn to do it yourself. You become the expert.

Tolman Sweet often has a suture line

Fret not. It doesn't have to be all or nothing. Knowing a McIntosh from a Cortland, or a Pound Sweet from a Tolman Sweet, or a Wolf River from a Twenty Ounce is an excellent place to begin. Forget the super rare ones, or at least set them aside for now, and focus on the low hanging fruit. There are many varieties you can learn by sight. You see one and bang! you know it. I had a fairly easy time learning to tell a robin from a blue jay before I was about five. Who cares about sparrows and warblers anyway? They all look the same. Learn to hit a fastball. You can work on a curveball next season, or next lifetime.

Yellow Bellflower, a.k.a. Bellefleur is mostly yellow, ribbed and late ripening

There are tricks to every trade. After he throws his fastball, how often does he throw a changeup or a curve? Good batters know these things. Wade Boggs, Ted Williams and Willie Mays knew what to look for. I'd heard about Boggs and wanted to see if he was the real deal. We got great seats. He came up to the plate. The ball lifted off his bat and rose above the second baseman's head and softly bounced into center-field. It looked like batting practice. He hit it so effortlessly. You could tell he knew what to look for.

If you've got black-capped chickadees at your feeder, you probably live in the Northeast. If you see small reddish purple birds commingling around the feeder with your chickadees, a reasonable guess would be that you've got some purple finches living nearby. If you see a jay, it's probably a blue jay. If you have white pine and you have other conifers, you don't have Sitka spruce. They might be red spruce or black spruce. They might be Canadian hemlock or balsam fir. If it's on someone's lawn, it might be blue spruce. But it's not a redwood, and it's not a Sitka spruce. So count them out. Don't even consider them. It's the context! It's the clues!

If you see a Ben Davis, look for Stark. If you see a Tolman Sweet, look for Pound Sweet. If the tree is ancient and the fruit is yellow and still hanging on in late October, think Yellow Bellflower. Who planted this tree? When did they plant it? Where did they plant it? Is it in the yard or on the stone wall or in the woods beside a creek? Putting together the right combination of bits of information can be exactly what you need in order to understand the story and make the ID. Maybe all you need to do is ask the right question to the right person at the right moment. There are people out there who still do know some of the old varieties. Find them.

Although I'm pretty familiar with both historic and modern apples, my focus is on the old ones, the heirlooms, particularly those growing in New York and New England. Preserving old varieties can become a passion. How many were there? How many of them have we lost? How many do we still have? Urbanization has not been kind to old apple trees. Many tens of thousands are gone. However, in some areas, such as much of Maine, farmland and orchards were simply abandoned. Trees have been allowed to sit undisturbed. This is good. Apple trees can survive quite well on their own despite decades of neglect. Apple trees don't mind neglect. They thrive on it. They slide comfortably into a state of undisturbed bliss. They don't eat much. They don't grow much. They don't need much. They might fruit every few years. Or not. They wait patiently for someone to come along and care. I used to assume that most of the old varieties are gone. Now I believe that most are still here. Our job is to find them.

Why bother to do this? Isn't Honeycrisp good enough? Besides, we also have Fuji and Gala, and now Jazz and Pink Lady and even SweeTango. Breeders are releasing new varieties every year. Who needs these old, discarded, odd-tasting apples?

The new varieties are selected for their potential as profitable, commercial commodities in a national or even world market. As they should be. They are the celebrities of the apple world. They all have a high glitz and glitter factor. They are dressed up and patented and trademarked. The orchards in which they grow are increasingly walled, fenced and increasingly guarded like.... Apples with names like Cosmic Crisp, Kanzi, RubyFrost, Opal, and SnapDragon will soon be waiting for you at your local Walmart or Whole Foods.

While some might argue that this pop-star apple extravaganza is ill-advised, greedy or wrong, others would say it's absolutely correct. Our work is not about quibbling with the righteousness or evils of modern agriculture. Our interest is in colors and shapes and sizes and flavors and uses. The focus here is on the nearly five-hundred year history of orcharding in America, and the stories of those who came before us; the people who selected, named and propagated thousands of varieties of apples and passed them along from generation to generation like the baton passers in a relay race. These apples are a gift to us from the past.

Years ago I met a woman at a talk I gave. She told me about her grandmother who had a small orchard of twelve apple trees, each of them different. She had gotten to know each variety over a lifetime. Each was reserved for its own special use, and she would never consider using any of them for anything else.

More and more these days, homemakers, amateur chefs and restaurant pros are using apples. They are making pies and sauce and butter and cake and bread. They are baking apples. They are drying apples. They are putting them on pizzas, cutting them up in salads and blending them in soups. On top of that, we're now in the midst of a cider revolution. Many of us are becoming like that grandmother with her twelve trees. We need apples that will do many different things. The pop-star apples of today are designed to be eaten fresh and make lots of money; they do both pretty well. But all these modern apples are worthless for those of us who cook or process or press our fruit.

Next time you go to your favorite supermarket, ask the produce manager to suggest good pie apples that ripen in August, September, October, November and December. Explain that you'd like a different variety for each month, for as we all know, they ripen in a progression and you want the best each week. "And, yes, I'd like a couple of good baking apples and a good quick sauce apple and what do you recommend for apple cake?" Let me know how he or she replies.

While a few universities and high-powered multinational conglomerates busily work and scheme away to create the next Honeycrisp, as far as I know, there's not a single person left on earth who is currently breeding pie apples. But all is not lost. Those apples are still here. Our farming ancestors carefully selected and named apples for dozens of different uses. They were great observers. They were resourceful. The best pie apples are still here. The best sauce apples are still here. Even the best apples for molasses are still here. They are waiting patiently behind some half-collapsed barn, in an old orchard hidden in the forest, or maybe even behind the local gas station where long ago some other grandmother harvested Roxbury Russets and Maiden Blush. They are waiting for us.

Fruit exploring is an art. It's like baseball…birding…fishing. You might hit a homer or see a pileated woodpecker. You might catch a king or half a dozen cohos. You also might troll for hours and snag one old soggy tennis shoe. It's always a risk. If not the big one, at least we'll catch some sun. We'll get out on the water. Maybe we'll see a humpback. Hope springs eternal. It's apple baseball, apple birding, apple fishing. It's one big mystery waiting to be solved. Today is the day you might bump into something wonderful.

Look for old orchards by the side of the road like this one in Washington, Maine

# Three

# Searching for Starkey

The nets are all in place, and the drag is about to begin. We'll know before the day is out whether we have caught our big, lean-jawed pike, or whether he has got through the meshes.
Sherlock Holmes, *The Hound of the Baskervilles*, p. 750

Starkey

If the fall isn't too warm; if December isn't too cold too early and the mid-winter fluctuations aren't too drastic; if it doesn't get too warm too fast in the spring; if the sun shines at least some of the time during bloom; if we get some rain, but not too much, and if the hail stays away, then we could have another really good apple year like the one we had in 1997 or 2015 when every branch is loaded with fruit and the trees themselves bow down to everyone who passes by. In years like that we pick and pick and pick some more. We climb our pointed ladders. Thirty bushels fill the truck. We pick at night with headlamps. We eat apple pies from August until deep into winter. We eat apple pizza and apple brownies. We eat apple sauce with our oatmeal and pancakes. We press cider until every container we can muster up is full. We sell apples. We give them away. When November comes we fill the root cellar. What a great year.

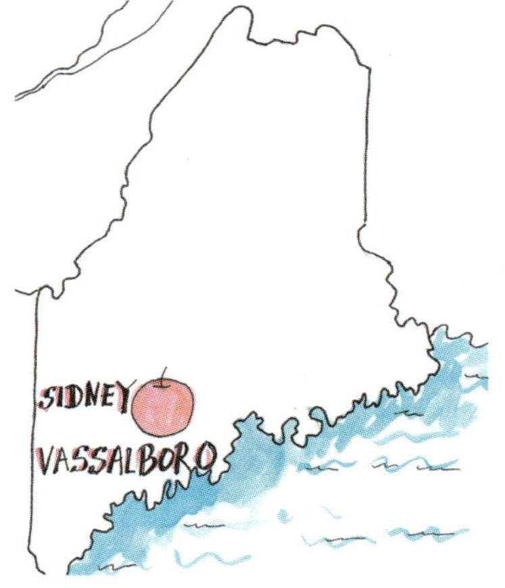


We might like to think these apples have been blessing New Englanders with this ritual for a thousand years. Or maybe ten thousand. But it's not so. Like all Americans, even "Native" Americans, apples are from away. They are from half way round the world, in the Tien Shan Mountains in Kazakstan, near the city of Almaty, which fittingly means "Father of Apples." There are still forests of apples in the Tien Shan Mountains, a place with the same latitude as much of New England. Not orchards, but forests.

Long before the great shipping fleets of Europe sailed and traded around the world, a vast network of trails and roads (the "Silk Road") stretched over five thousand miles east to west crisscrossing Asia and connecting all the great Middle Eastern cites of antiquity with one another. It joined China and India in the East with Europe and Russia and North Africa in the West. One arm passed right through the apple forests of the Tien Shan Mountains. From there the apple hitched a ride on the backs and in the guts of camels, horses, tradesmen and travelers. Off it went to Afghanistan, Iran, Iraq, Syria, Jordan, Palestine, Egypt, Turkey, Russia, Greece, Italy, Spain, and northward. Over time, with the ongoing annual assistance of tens of thousands of crusaders and religious pilgrims who traipsed back and forth across the continent, the apple made all of Europe its own.

Then inevitably, five hundred years ago, packed in wooden barrels on the decks of Basque fishing ships hunting right whales off the coast of Newfoundland, the apple most likely made its first journey to North America. For six months each year, hundreds of whalers camped out on the shores of Red Bay, cultivated gardens, drank vast quantities of Basque sagardoa (a.k.a. cider), rendered whale oil in ovens they built, and inadvertently or maybe on purpose, planted Basque apple seeds. The descendants of the apples they brought over in their ships are quite likely still with us today.

We pick and pick and pick some more

That was the beginning of the beginning. The whalers were assisted in their efforts to populate the new world with apples by Basque and Portuguese cod fishermen, then by the English, the Dutch, the French and everyone else who followed. Once all these new American settlers hit the mainland, they planted tens of thousands of orchards with millions of trees from Maine to Georgia. At first they used the seeds they brought with them from Europe. Later they used the seeds from the cider pomace of the new American trees themselves. Think Johnny Appleseed. Everyone was doing it. Apple seeds were readily available, easy to transport and easy to plant. It soon became evident that the apple loved America. Before too long there were seedlings everywhere.

I say "seedling orchards." This is important. Nearly all apples are self-sterile. They will not pollinate themselves. They need a mate. There needs to be a mom and a dad. The pollen from one McIntosh flower will not pollinate any other McIntosh flower, even if it's on a separate tree. No pollination means no fruit. But those McIntosh flowers will accept pollen from any of the thousands of other apple varieties, wild seedling apples, or even a nearby ornamental crab. The only requirement is that the two apple parents must be different.

The fruit on a McIntosh tree will always be McIntosh no matter what variety pollinates it, but the seeds are a different story. Every seed in every McIntosh apple will be some unique combination of its two parents. If you planted an entire orchard of McIntosh seeds and let them all grow to maturity and bear fruit, every one of those new trees would produce its own unique apples. None of them would be McIntosh. Each would possess traits of both of their parents, but each would be unique. You resemble both your mother and your father but you aren't identical to either. You aren't identical to your biological siblings either, except in the rare case of identical twins and triplets. And so it was with apples all over America. Seedlings of seedlings of seedlings were merrily procreating and producing an almost infinite assortment of pomological combinations. The ease with which apples cross and recross and create new and unique seedlings played the defining role in the history of apples and orcharding. Without this fact, there would be no apple history at all.

Back in the days of the colonies it didn't matter to the subsistence farmers that their seedling apple trees weren't going to be identical to the parents. They didn't care if their orchards produced nondescript or even small bitter fruit. They were using their apples for vinegar, animal food and especially cider. Any cider maker knows that those seedling apples are often the best. The beer makers get their bitterness from their hops. The cider makers only have apples. Bitter is better when it comes to cider apples.

Inevitably, the random seedling with some enviable trait appeared along a stone wall, along the road or maybe next to a fence or outbuilding. Perhaps it kept all winter in the root cellar, or ripened in late July, or made a great pie or sauce. The special ones were selected and coveted. They were given names, and from that enormous pool of seedlings the first American apple varieties were born. They were passed around the neighborhood and sometimes beyond. But not by seed.

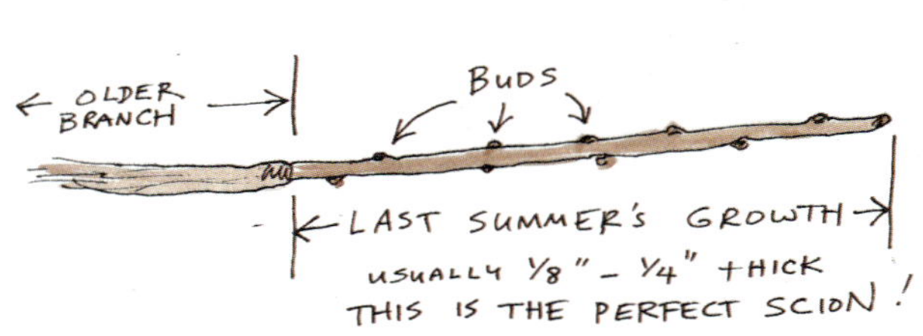


This is where grafting comes in. Grafting is a simple technique for replicating apple trees and many other woody plants. These days virtually every tree in every orchard in the world is grafted. Millions of ornamental trees are now grafted as well. It's a form of cloning that doesn't require a lab coat or chemicals. All you need is a sharp knife and some tape and some goo. With a couple of cuts you splice a small piece of a one-year-old twig of the variety you want to replicate onto a host tree. You wrap it up with tape and you're done. The splice is called the graft. The twig is called the scion (sigh-on). The rooted host tree is called the rootstock.

We gather the scions during the dormant season between New Year's and the first of spring. We want one-year wood. In other words, we're looking for last summer's growth. Theoretically you will find good scionwood at the terminal end of most branches. In reality, on many ancient trees, new growth has slowed down to a crawl and it can be hard to find. A vigorous, pruned apple tree can put on a couple of feet of new twig growth each year. Some old trees will barely grow more than a quarter inch. So-called "water sprouts" are not ideal but are perfectly usable. Some old timers say that grafts from water sprouts take longer to fruit.

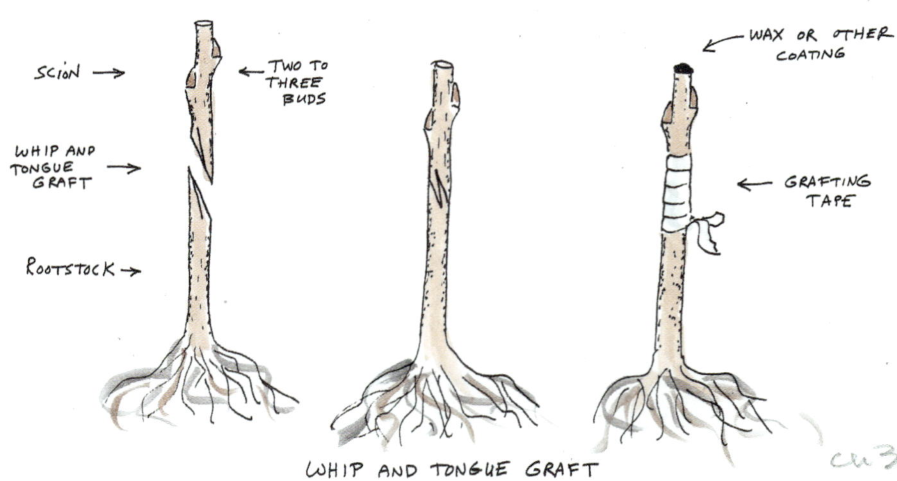


We store the scionwood up to four or five months. Triple-bagged and sealed, it stores beautifully at about 25 to 45 degrees Fahrenheit. Do not put it in your freezer. The refrigerator is fine. So is the root cellar and many unheated basements. Seal it well and it should be in perfect shape when it comes time in central Maine to bench graft in April and topwork in May.

No one knows when the first grafter did the first grafting, nor does anyone know how he or she got the idea. We think it happened about four thousand years ago. More than likely some Einstein of the distant past saw a couple of trees that had rubbed up against one another and naturally fused themselves together. It was a graft. "I can do that." They had knives, and they knew how to make them sharp. That was no problem. For the tape they used bits of cloth smeared with beeswax to stick it all together. Nowadays we use special knives, special tape and special goo all designed for grafting. But the process remains the same. You find a tree that you want to grow in your yard or orchard. You gather up some scionwood and some rootstock. You sharpen your knife. You perform plant surgery. You hope it takes.

There are many different grafting techniques. When I first learned to bud-graft, we were kneeling outdoors in August, inserting a single bud under the bark of the rootstock, which was already planted and growing in the nursery. Then I learned how to "whip and tongue" graft. In Maine you whip and tongue indoors in March and April. You splice a dormant stick of scionwood about the length and thickness of a birthday candle onto a dormant rootstock that you lay across your lap. You cover the waiting rootstocks and completed grafts with a damp rag to keep the roots from drying out. You turn on the radio, crank up the wood stove, chitchat with your grafting mates, and solve the world's problems as you create little trees by the dozen. We call this "bench grafting."

In early May we topwork. Topworking is a form of outdoor grafting onto existing trees. In the old days, most orchardists used a cleft graft when they topworked. We prefer to use a bark graft when we do our topworking. We cut off limbs or even whole trunks and insert the scions around the perimeter. This is how you switch a tree from one variety to another. Don't like Macs? Turn them into Black Oxfords. This is also how you can grow ten or twenty varieties on a single tree. You can topwork each branch with its own variety. You can turn your yard into a library of apples. Find a variety, cut the scionwood in winter, topwork in May. Now you have it saved and growing at your place.

By one form of grafting or another, the grafter can replicate McIntosh or Black Oxford or Benton Red or any other variety. Grafting made it possible to spread apple varieties around the neighborhood, county, state, and eventually, in a few cases, around the world. When you take your family to pick apples at your local U-pick orchard, every single tree you see has been grafted.

Helen and Earland Goodhue: courtesy of the family

Farmers everywhere planted apple seeds and kept an eye out for the best ones. We were a culture of apple breeders. By the time of the Civil War, many thousands of local American varieties had sprouted into being. In Maine over two hundred apples were named before 1900. A number of these native Mainers are still being grown today. Others have disappeared forever. Another unknown number have gone into hiding, living out their retirement years in abandoned orchards, along old stone walls, behind broken down barns, sometimes even tucked into well-mown suburban backyards, the yards that replaced so many of the small farms of long ago. More and more of us are searching for these remaining varieties. It's a modern day treasure hunt, knocking on doors, visiting orchards and agricultural stations, corresponding with people around the world, scouring old books, going to fairs and grange halls. We look for the few people still living who can identify this variety or that. It was in this context that I entered the world of Earland and Helen Goodhue.

Earland and Helen lived in a large, two-story clapboard house on the Goodhue Farm, on the Goodhue Road in Sidney, Maine. Behind the house, between the dairy barns, around the shop, and beyond the gardens were apple trees, about a hundred or so. Theirs was not a typical orchard. Many of the trees had been topworked to multiple varieties, some with as many as three or four on a tree. Most were unusual, many were rare, some were all but extinct.

I had heard of Earland and his collection of apples, and I knew that I wanted to meet him. I called and asked if I might stop by. A week later we were tromping through the snow together looking at dozens of

leafless fruit trees representing decades of collecting, grafting, observing, pruning and spraying.

"This one here is a Washington Sweet. We had it as kids. This one's a Brock. It was bred by my cousin Russ Bailey at Highmoor Farm, University of Maine. Over there's an Arctic. [Elwin] Meader identified it when he came to see me. Over here's a Winekist. It was given to me by Morris Towle."

We pored over Earland's orchard map at the kitchen table. Helen fed us cookies and fresh cider as she pulled dry split wood from a cupboard on the kitchen wall and stoked the cook stove. As she prepared supper, I marveled at the map, a mass of x's and o's and names on top of names, and notations on top of notations. Fortunately Earland was there to interpret. Their kids had grown up and mostly moved away, although one of their sons had taken over the family dairy operation. He lived just down the road towards the lake.

Over the next few years I returned often to visit the Goodhues. In the spring, lifted high up in the bucket of his farm tractor, I grafted new apple varieties onto old trees. In the fall I picked apples. In the winter I collected scions. Earland gave me free rein. It was there that I first discovered a number of my favorite varieties, and it was there in the fall of 1997 that I came across an old tree with a faded tag on it that said STARKEY.

By that time in my fruit exploring, I was focusing on Maine apples. You might say I was obsessed with Maine apples. "Think globally act locally." I suppose it could have been New Hampshire or Vermont apples, or Missouri or North Carolina apples. But I was here in Maine, and no one else seemed interested in Maine apples. Why not me? Starkey was one of those Maine apples. Starkey was an adventure in waiting. Let the adventure begin.

Although not much is certain about Starkey's history, it was known to have originated in Maine. It was one of those countless chance seedlings that were coming up all over the state, this one discovered in about 1800 by a Quaker farmer named Moses Starkey on his farm in North Vassalboro, across the Kennebec River from Sidney and twenty miles from our farm in Palermo. One of the apple's parents is thought to be Ribston Pippin, a famous English apple brought to Maine by Benjamin Vaughan about the time of the Revolution. It spread up and down the

Lifted high in Earland's tractor bucket

Kennebec River from farm to farm primarily through the efforts of Vaughan's head gardener, John Hesketh. Although once popular locally, the Starkey apple is virtually unknown today. If I could locate a true Starkey, I could propagate it and offer it for sale in Fedco's tree catalog, as we'd already done with a few other Maine varieties. Another apple would be saved.

I was aware that it's one thing to find an interesting apple and take scions and propagate it for your own orchard. It's another thing to offer it for sale. I wanted to be certain I had the real thing before I sold trees. As I tasted it on that cold, gray, windy September afternoon, I knew I liked it. I picked about half a peck, took them home and put them in our root cellar along with the several dozen other varieties I had picked so far that fall.

Moses Starkey was born in Attleboro, Massachusetts, in about 1770. He moved to Vassalboro before the turn of the century, during the time when many Quakers left intense persecution in southern New England for the relative safety of Maine. Quakers were not a popular bunch. They were, as the Massachusetts General Court described them in 1657, a "cursed sect of heretics." Starkey married twice. His first wife died young and childless. He and his second wife, Janette Warren, had five children, three of whom got sick and died within a month of each other in the fall of 1840. Moses himself lived into his seventies, and Janette lived on to be ninety-three.

So thee left it to thy horse, did thee?

In 2004 I received a letter from a Ralph Greene, who had written a University of Maine Master's thesis called *Development of Quaker Communities in Central Maine 1770-1800*. Mr. Greene had come across Moses Starkey during his research and that had led him to me. In a second letter, he related this story:

> An experience happened to Moses Starkey who developed a particular brand of apple which was known as the Starkey Apple and who was a successful carpenter and orchardist in Vassalboro. He had heard that the Quaker preacher, David Sands, was in the area but felt little inclination to hear him. While going home one day he came to a point where the road divided; one road going to his farm, the other to where Friends were meeting for worship. Moses decided to let the horse choose the road and let the bridle drop. The horse went to the road where the Quakers were meeting. When he arrived, Moses got off and went in while David Sands was preaching. The minister paused in his message, looked directly at Starkey and said, "So thee left it to thy horse, did thee? It would have been well if thee had left it to thy horse years ago." Needless to say, Moses Starkey became a convinced and useful Friend.

The Starkey apple is described by S. A. Beach in his 1905 *The Apples of New York.* That suggests that it was fairly well known at one time. It was also written up in the 1893 Maine Experiment Station report. Both descriptions are reasonably detailed as far as old apple descriptions go, although I could find no illustrations. As I read through Beach that night, however, I wasn't convinced that I knew what sat in the bag in our root cellar. The descriptions seemed to match the apple I had, but they also could have matched any number of others, including another apple I had found that same day.

Earlier in the afternoon Earland had introduced me to another variety, Garden Royal. Garden Royal is an old chance seedling that originated in Sudbury, Massachusetts, in the eighteenth century. Earland had topworked one branch of Garden Royal with scionwood he'd cut from an ancient tree not far from his farm. Like Starkey, Garden Royal is a somewhat roundish apple overspread with red splashes and stripes and covered with conspicuous off-white dots. Both are good fresh eating apples. Maybe these were both Garden Royal trees. Maybe neither one was a Garden Royal or a Starkey.

I decided I would look for a Starkey tree in Vassalboro that fall. Vassalboro is not far from Palermo, and it seemed like a good thing to do. I talked to Earland, and he suggested that I visit Frank Getchell, a retired orchardist who had lived in Vassalboro most of his life, and who ran a small convenience store a couple of miles from the location of the original Starkey tree. On September 16, I went to find Frank Getchell.

Frank Getchell (left with shovel) and friends planting a Starkey tree in Vassalboro c 1970: courtesy of the family

Getchell's store sat beside Route 201 at the corner of the Getchell Road, about halfway between Waterville and Augusta along the Kennebec River. The store was not much bigger than our living room. There was only one car out front when I pulled in. The store was empty, but I could see someone sitting in a back room behind a large window. Frank Getchell rose and came out to meet me, cigar in hand. He didn't appear to be all that happy to see me, but when I mentioned Starkey, his eyes lit up.

We talked for about half an hour. Of course he knew all about the apple. Twenty years back he had even grafted several trees from what he thought was the last Starkey in town. He and friends had planted one of the resulting young trees in the center of town, but some young mischief makers had done it in. None of the others had made it either. My only hope would be to locate the old tree from which he had obtained that scionwood. Unless the tree was still alive, he was afraid that Starkey was gone.

Frank gave me directions to the old tree and off I went. I should find it about five miles away along a stone wall between the southern edge of Lemieux's Orchard on Priest Hill and a field recently purchased by a Charles Burns. After introducing myself to Mr. and Mrs. Burns, I was able to locate an ancient apple tree, three feet in diameter and about twenty-five feet tall. All of its branches had rotted and broken off, but a couple of sprigs were still growing at the top. About ten apples were ripening in the September sun.

I climbed the tree and picked two apples. They were completely different from Earland's. This apple was distinctly conical and ribbed. Earland's was round and not ribbed much at all. Not only that, but this apple had no conspicuous dots. I took the apples home, leaving the rest of the fruit on the tree to ripen. I called Frank Getchell and asked if he'd be willing to look at the fruit with me in a few weeks. They'd be riper then. The color would be better, more indicative of the variety. He said he would, so my search went on hold.

Searching for Starkey was not my only mission in the fall of 1997. I did have to go to work. There were also the usual fall farm chores. It was an excellent year for apples. With so many trees fruiting for the first time in years, I went to many orchards and visited hundreds of trees.

One such orchard was a couple hours north of us behind a large, white, colonial home in downtown Orono not far from the University of Maine. At that time the property belonged to a family named Comstock. The orchard itself had been created years earlier by a professor, named Bobaleck, from the university and included many old and unusual varieties. I had long heard about Bobaleck's orchard and finally saw it for the first time myself on October 4.

A friend had a photocopy of the orchard map, but I was unable to get over to his place to borrow it before my visit. I went ahead anyway, hoping for the best. The orchard was not in good shape; many of the trees were dying or dead. All were overgrown. It was another of so many cases of a private collection lost after the owner died. Still, I was able to take some notes, collect some fruit, and identify a few varieties from the tags that remained. Over in one corner of the orchard, nearly obscured by young maples, I found another tree with a STARKEY tag.

I climbed the tree. The fruit looked nothing like the apple Frank Getchell had led me to. It did look like Earland Goodhue's Starkey and Garden Royal. They were roundish. They had the rusty red stripes and splashes. They had the whitish dots. The taste was good.

The description of Starkey in *The Apples of New York* calls its season October to January. Garden Royal, on the other hand, is said to be at its prime from late summer to early fall. I decided I would hold my samples until Christmas and see how they compared then. Meanwhile, I called Mr. Getchell and arranged to meet him at the tree in Vassalboro the next day. Sunday, October 5, was a beautiful warm fall day. It was late afternoon when Frank and I met on the Priest Hill Road. I climbed my ladder and fetched the apples. We leaned against my truck as the sun began to set. He smoked his cigar and looked at the fruit.

"No, this isn't a Starkey," he said. I showed him the samples I had picked the day before in Orono. He tasted one. "Maybe," he said.

I was disappointed. I drove home feeling dejected. I thought about giving up.

The following weekend one of our neighbors came by. "I have a friend at work who lives in Vassalboro," she explained. "She gave me this bag of apples from their old tree and asked if I could bring them to you. She's hoping you'll know what they are."

Sue Ernst and friends with her Starkey tree, 1999

It had been fifteen years since I came home from work that night at the Belfast Co-op Store with two bushels of Black Oxfords. I had acquired a reputation around town as a modern-day Johnny Appleseed. Earlier in the fall a local TV station interviewed me for the nightly news about my apple exploration adventures. I'd already received a number of calls, so I wasn't too surprised to see my neighbor that afternoon.

I opened the bag. They were roundish, rusty red, and covered with those now familiar white dots. This had to be Starkey. "I think I know what these are," I said, to my neighbor's surprise.

I called Sue Ernst, the owner of the apples, and a few days later I visited her and her tree. It was very old, large and prominently placed in a small field across from her farm house on the Nelson Road. It had been windy and rainy. The apples had dropped, but I was still able to collect a bushel of decent specimens. I was thrilled.

That night I drew diagrams of Earland Goodhue's Starkey, Professor Bobaleck's Starkey, and Sue Ernst's Starkey, cutting them in half from stem to blossom end and carefully describing them according to the nomenclature in Beach. The form (shape) of the fruit is roundish-oblate and slightly conic. The sides are obscurely ribbed, though only barely so.

The cavity is medium in size, not very deep and colored with a small greenish russet splash that you might not notice until you really look. Then you can't miss it. The calyx is large and closed, "or nearly so." Basin is medium and shallow and "somewhat wrinkled." The three apples were almost identical. I decided that they were all the same variety. They must be Starkey. Maybe this tree on the Nelson Road was the last remaining Starkey tree in Vassalboro. The only thing left to do was convince Frank Getchell.

I balked. Did I want to know? Could I stand the disappointment, especially after meeting Sue Ernst and her husband Walter and their old tree? And besides, I was falling in love with this apple. It was practically the only apple I was eating that fall, even with dozens of others from which to choose.

Of course I knew that at some point I would have to take these apples to Getchell's Store. All I could do now was delay the inevitable. I decided I would wait until December. If these were Starkeys they would be at their finest then. That would give them their best chance of convincing Frank Getchell.

The day before Christmas I headed off in my truck with seven painstakingly selected, beautiful, round-oblate, rusty-red splashed and striped, off-white spotted apples from Sue Ernst's tree on the Nelson Road in Vassalboro. I walked in. Frank was by the cash register. Without a word I took out the apples one by one and placed them on the counter. He picked one up, looked it over, and took a bite. A customer was next to me buying cigarettes.

"Try this apple, Bob," Frank said. "It's a Starkey."

Rosy red Starkey showing the basin (apex) on the left and the cavity (base) on the right

Andrew Tufts and Northern Spy tree, Frankfort ME:
photo by Jack Dodson

# Four

# Identification 101

Please arrange your thoughts and let me know, in their due sequence, exactly what those events are which have sent you out unbrushed and unkept, with dress boots and waistcoat buttoned awry, in search of advice and assistance.
Sherlock Holmes, *His Last Bow,* p. 870

Northern Spy

Each year around Labor Day, apple ID season begins. As baseball season winds down, we crank it up. Soon packages appear in the mail. First one or two a week, then one or two a day. Sometimes it seems like one or two an hour. I also give about twenty apple talks during the fall, mostly in Maine but occasionally out of state. I usually return home from each talk with another five or ten bags of apples to identify. At the Common Ground Country Fair and Great Maine Apple Day, I pick up another few crates of IDs. You get the picture. Meanwhile I'm dashing about visiting orchards, tracking down rare old varieties, sourcing fruit for our apple CSA, and so on and so forth. Fall means apple fishing season, six or seven days a week.

Of course, there's going to be those periods when not much happens. You might get lulled into complacency. But you still have to be on your toes because when a Canadian Strawberry or a Moses Wood or a Somerset of Maine nibbles, don't be dozing. They'll pull the fishing rod right out of your hands. No daydreaming allowed.

Fall fruit exploration is like being in the fire department: stretches of inaction interrupted by essential activity. Off we go to save another endangered apple tree. Unlike the Palermo Volunteer Fire Department, however, most of the time I can fend off emergencies long enough to eat. We sit down to dinner, and the phone rings. "You don't have to answer that," Cammy says as I head for the phone on the wall in the kitchen. At least it's one of those retro-phones attached to the wall by a long tangled cord so I can't sit and eat with the phone under my chin. "We just sat down for supper. Can you call back in a half hour?"

The apples continue to roll in whether I'm eating dinner or not. I open up my various apple books and my computer and my file folders of lists and drawings. I open up another box of apples. Why do I do this? And then I remember. The apples shimmer before me across the dining room table. Maybe they even remind me to "listen, listen, listen!"

The perfect bark graft, by my niece, Eliza Kuchuk

Sometimes I'm "listening" to the apples, and it's pretty loud and clear, and I know in a second that it's a Spy or a Twenty Ounce or a Tolman or a Mac. It's like music. You hear a few notes of your favorite pop song on the radio, and that's all you need. You know. Sometimes it scares me that I can hear the tiniest bit of hundreds of sixties records and I know. Why? I suppose it's because I heard all those songs a thousand times. And I listened.

When I was in college I took Classical Music Appreciation 101 with Dr. Ermano Comparetti. I thought I should learn something about classical music. During the tests the professor would plunk the needle down on the record, play a few seconds, and then quickly pick it up again. Sometimes without scratching it. And we were supposed to ID the music. Beethoven's Seventh Symphony, second movement. That sort of thing. And I thought to myself, "This is impossible. How am I ever going to do that?"

Although this was a decade or two before computers or even cassettes, I did have a reel-to-reel tape recorder. I developed a plan. I borrowed the LP's from the music library on campus and recorded the latest assignment. Then instead of the Grateful Dead, I listened to Bach, Beethoven and Brahms everyday until the next test. The classical music assignment became my default entertainment. When the big day arrived and we all sat at our desks in the auditorium, down went the needle, and I knew instantly. Most of the class looked perplexed or panicked, but not me. One second of Ravel's String Quartet in F or Mozart's Sinfonia Concertante K.364 was all I needed. Nothing like repetition!

When it comes to music, repetition is not necessarily a bad thing. The LP is tailor-made for deep listening. You put one on and sit down to pay your bills or read the paper or do a few apple IDs; 15 minutes go by, and the needle gently lifts up off the record, and "click" it shuts itself off. Since it's ten feet over to the record player, you sit there in silence for a while. That's nice. Eventually you notice the quiet, and you miss the music, so you get up. It's too much trouble to put on another record so you lift up the needle and play that side over again. And over again. And over again. A day or two later you finally flip the record over, or maybe you put on a new record; but before too long, you've listened a few dozens times to *Kind of Blue* or *Anthem of the Sun* or Beethoven's *Seventh*. Each time you listen, you go a little bit deeper. Each time you listen you hear a little more. Your brain sorts out the music—even the challenging stuff—hears it, makes sense of it. You go deep. It can take years. The best music takes years of multiple listenings.

My best friend freshman year had one of those record players with the tall spindle. What a great invention. You could slip on a stack of about fifteen records. When one was done, the needle lifted up, backed out of the way, and the next record slid down the spindle. The needle then gently lowered down, and the music began again. We would listen to all fifteen LPs a few times, then slide them all off, carefully flip the pile, slip them on the spindle upside down, and play the other sides. This allowed a little variation, but the effect was still the same. You learned the music. You went deep.

When I began to study the apples in my neighbors' orchards, I knew I'd finally gotten something out of my four years of college, or at least music class. Once someone showed me a real Northern Spy for the first time, I was ready to go. I balanced it on the dashboard in my truck. And at every stop sign, every red light, every construction delay, I stared at that Spy. I lined the kitchen counter with Northern Spys. I labeled them with sharpies. I felt like I was in Dr. Comparetti's Apple Appreciation 101. And I did the same with Baldwin, Tolman Sweet and Pound Sweet. Later when they invented cup holders, I always kept some new apple there. Was it a cup holder or an apple-display rack? I looked and looked. One by one they found their way into my brain so that whenever I saw one, I knew instantly. That's Beethoven, that's Brahms—and that's Northern Spy.

So can you tell a Mac from a Cortland every time you see one? Yes, you can, but you may have to "listen" to both of them for a while first. Then you'll never forget. It's surprising how many people bring me an apple to identify, and it's a Mac. Impossible! It should be a New England prerequisite to know a Mac when you see one, or at least when you taste one.

Sometimes I open the bag, and I know I've seen it before but what is it? There are some that get me every time. I open that bag, and I pull out the apple, and the apple says to me, "Here I am again, you bum. I

Yellow Transparent is perfect for a few days in mid-August

know I've fooled you before, and I'm back, and no clues. You're going to have to do this all by yourself!" Other times I sit at the table and open the bag, and I know I've never seen this one before. Who is this? This is a new one. I've never heard this apple before. This is awesome.

Although neighbors and others taught me the names of many varieties, it became apparent early on that some apples were going to be difficult to identify. I learned the names of a few experts around the country and occasionally sent difficult apples off to this place or that. If I could determine where an apple tree had been propagated, I'd send fruit off to that nursery. Identifying apples is very hard to do. In all the times I sent away apples, I don't recall ever receiving a correct answer back. Usually there was no answer at all. Sometimes it was a "Sorry we don't know." Often the answers were obviously incorrect. Cast no aspersions on the apple identifiers of the world! Remember that nursery people look at small trees everyday, not fruit. They can easily tell a bud 9 rootstock from an Antonovka rootstock, but they may not know a Yellow Transparent from a Yellow Bellflower.

Mistaken identity is not confined to nurseries and pomological experts. No one is exempt. Cammy and I visited an old Massachusetts orchard where Twenty Ounce was labeled as Wolf River. Outrageous! True, they are both very large apples. They both get used in pies. But they don't resemble each other. Grocery stores are also notorious mis-labelers. Even high-end produce markets. If they ask and I know, I'm happy to help. Otherwise I bite my tongue. Unsolicited advice comes across as criticism. No one likes a know-it-all. Maybe there's actually no such thing as a know-it-all anyway, only those who think they know.

They told me it was Stark but it was really Benton Red, and Benton Red may actually be Cherryfield which is also called Collins. Yikes!

One of the best ways to learn new varieties is to have someone show you an apple and tell you the name. In fact, this may be *the* best way to learn a new variety. But beware. Be grateful for the assistance but be cautious before you act too sure of yourself. One of my first favorite apples was the one I found in the small orchard behind

Fitzpatrick's Dairy, in Benton. I fell in love with the orchard of about twelve trees and stopped by every other year for a decade or more. The old couple greeted me from the door as I drove in. "We were wondering when we'd see you again," they would say as I got out of my truck. They didn't know the names of the apples but were happy to have me take all the fruit I liked. Then one day a friend identified the apple as Stark. (Not to be confused with Starkey.) Of course I believed him. Why not? Stark originated in Ohio and was planted in Maine for over fifty years as a hardier alternative to Baldwin. Old trees can still be found in central Maine. I thought I had found Stark.

Lauchlin Titus, Frank Getchell (with a Starkey) and JPB, November 18, 2004: courtesy of The Town Line

Some years after Frank Getchell and I met in his store that day, back when I was searching for Starkey, I was invited to give an apple talk to the Vassalboro Historical Society. I had begun to give apple history talks here and there, but this would be at the birth place of the Starkey apple! I wrote in my journal on November 18, 2004, "Vassalboro Historical Society. What a place. Big Crowd. It was like a dream come true to be…there and have Frank Getchell, Sue Ernst and Bob Clark all in the audience." (Bob Clark had led me to a second Vassalboro Starkey tree a few years after I discovered the tree that belonged to Sue and Walter Ernst.)

I like to bring apples with me to my talks; I brought a Starkey with me that night, of course. I also brought along a Starking Delicious (one of those surreal, glossy, poisonous-looking Red Delicious types), and a Stark. I wanted to point out that three very different apples can have deceptively similar names. It was a good decision because not only did Frank know Starkey, he also knew Stark. He even had an old Stark tree next to his garage. He set me straight for the second time: "This apple is not a Stark." Oops.

This is the real Stark. It's often confused with Baldwin…or sometimes Benton Red!

Another time a friend introduced me to "Milwaukee" growing in his small commercial orchard not far from Palermo. Our next door neighbor Emily was thrilled. Not only did she grow up in Milwaukee, this was a great apple. When I keyed it out, I included a note, "strong resemblance to Blue Pearmain." A year later I was looking at our friend's orchard map. Something was amiss. Where it said Blue Pearmain on the map, there was no tree. And yet we found Blue Pearmain in his orchard. I pulled out my books and read about Milwaukee. "Oh, that's why his Milwaukee resembles Blue Pearmain. It *is* Blue Pearmain."

Russell Libby, 2009: photo by Jean English

Many times, however, other people can be incredibly helpful. One year, Brian Milakovsky, a young fruit-exploring friend on our road, pointed out an old tree that I had never stopped to check out. "It's wonderful." So I did stop, and yes, it was wonderful. I took one of the apples to the Common Ground Country Fair and set it out in the display. On the label I put a question mark. The first day of the Fair, someone saw the fruit and the label and said to me, "This is an Opalescent." I took the pronouncement with a grain of salt. I always take those pronouncements with a grain of salt (or, at least, I should). Next day at the Fair, another person said "This is an Opalescent." Two different people, two different days, same name. I went home and carefully read the old books. They were correct. And what a delicious apple.

For nearly two decades Russell Libby was executive director of MOFGA. He travelled all over the state for his job, attending countless meetings and visiting farms and supporters. He did it all cheerfully and always seemed to have time to stop along the way. One day he stopped at a yard sale in Belgrade. On the edge of all the stuff for sale stood an ancient, broken-down apple tree. "Do you know what you have?" he asked the old-timer. "That's our Judy tree." Russell called me that night in a frenzy. "Have you heard of an apple called Judy?" I pulled out Bradford and found the page on Judy. More than likely it originated within a few miles of the yard sale. Could the old-timer have been wrong? Of course. But can you make up a name like Judy without good reason? Of course not. We collected wood the following winter, and the delicious, beautiful, red-blushed Judy was saved from extinction.

I've never counted the number of ID's I do in a season. I know that I do a lot of them every year, probably three or four hundred. The fall can be overwhelming at times, especially when I know that the clock is ticking. If I don't hop to it, many of the apples in the basement are going to melt. I shudder when I think of the postage people paid to rush a box of apples to me only to have it sit on my shelf for weeks or even months. Piles of boxes and bags line the basement shelves. Help!

Over the years I've gotten better at getting to the ID's before the apples rot. I do a lot of organizing and dating. Every week or two I shuffle the bags. Some people shuffle paper; I shuffle paper bags. Fortunately, for the last few years I've had an intern or two who organize the named varieties we collect into an A to Z of wooden boxes in the root cellar. When I'm doing an ID and I need to check a Stark or a Starkey, I know I'll find them in the box labeled "S." The interns (who we also call apprentices) go out with me collecting and visiting. They entertain me when we're on a long drive. "Name five favorite movies." "What are five ways you got in trouble growing up." "Top five favorite pie apples." Thank you, John Cusack.

The interns also help put together displays for events and come with me when I give talks. In return I teach them what I can, and make periodic detours to visit my favorite ancient trees I've discovered over the years. We have fun. One November when apple season was winding down, we were standing in the kitchen and I happened to notice a bag of ten or fifteen varieties. I dumped them out on the table and said to Emily, "What are they?" Three months earlier she couldn't have identified a single one. I looked at her. She looked at me. Then she aced it.

Emily Skrobis acing it, 2010

With the ID's, I try to do first-in-first-out. Sometimes, I confess, one of the mystery apples leaps to the head of the class when something about it strikes a chord. I get the vibe that I better do this one, now. I beg people to bring me their apples in paper bags—not plastic—because the apples stay in pretty good shape for weeks in paper. In plastic they sweat and begin to rot almost immediately. In paper they might still be in a recognizable shape when I finally open the bag.

Occasionally I get asked how many rare apples I've rediscovered, brought back from the brink, returned to cultivation, saved from extinction, and gloriously preserved for posterity. I don't count. I know there are some. Starkey is likely one of them. I know that I've saved some good apples, and I like to think that the world is a slightly better place as a result.

Apple identification can be a lonely affair. Identifiers are few and far between. Early on I feared I might be the only one. Although I got lots of practice each fall when the packages arrived, I wasn't always sure what it was I was supposed to be practicing. What is there to practice without a teacher? While it's true that Earland and Frank, and, later, Francis and Garfield taught me a huge amount, they often shrugged their shoulders when someone brought them an apple. If they couldn't do it, who could? Lessons are hard to come by for the fruit identifier. Sometimes you have to look in strange places to find them. I found a good one in the music of Sun Ra.

Sun Ra was born in Alabama in 1914. His story, however, was that he was born on Saturn. He grew up playing piano in Alabama, and formed his own jazz group here on Earth in the early nineteen-fifties. He composed, arranged and performed for several decades, primarily with his big band but also with small groups, and occasionally as a solo act. His band, the Arkestra, was unusual in the history of jazz. The slate of musicians remained relatively constant for nearly forty years. On any given evening, they might preform music spanning the entire history of jazz, from Dixieland to swing to avant-garde. Onstage, they dressed in a blend of ancient Egyptian and space costumes. Sometimes when Sun Ra spoke, you might have thought he was one of those anonymous grafters, passing along the rare apple variety through time: "I'm actually a present sent to you by your ancestors," he said.

Sun Ra was also a poet, philosopher and mathematician who talked and wrote in cosmic equations. His passions included ancient Egypt and outer space. From what I could tell, he believed above all in precision

Sun Ra leading the Arkestra 1979: photo by Jared Crawford

and discipline, two key words for the apple identifier. If you want to get it done, you have to be willing to set aside everything else, and you have to train yourself to pay close attention to the details.

Perhaps because of Sun Ra's challenging music and his interest in esoterica, few recording companies would touch his music. Never to be deterred, the Arkestra made their own records. They recorded prolifically at home and onstage. They laid out their newly-pressed discs on the living-room floor and glued on the labels with rubber cement. They inserted the discs in plain white or yellow record jackets, which they decorated with magic markers and stickers or simply left blank.

Were you lucky enough to attend a Sun Ra concert, afterwards you could purchase LPs from the band members who hung out at a table by the exit. These days it's common to purchase recordings after a concert, but back then it was rare. What was even rarer was that you went home with an album in a blank cover with a cryptic label with little or no indication as to who were the musicians or even what you were listening to. Sometimes even Sun Ra's name was nowhere to be found.

Sun Ra performed in Maine a number of times. I saw him once in Portland on January 15, 1989, at a long-gone music venue called, appropriately enough, The Tree. There were eighteen musicians and dancers that night, packed onto a very small stage. Ra himself was front and center directing the show and playing a Yamaha DX7 synthesizer. It was wild!

As far as we know, there's no record of Sun Ra's favorite apple. According to John Szwed in *Space is the Place*, Ra advised a saxophonist named Wendell Harrison that he should "forget about form and play what you feel." "Play the warmth of the sun." "Play the apple." I didn't get to met Sun Ra that night at The Tree. I should have brought the band a bushel of Golden Balls. "Golden" is a word that appears now and then in apple names. Nearly everyone knows Golden Delicious, the most popular apple in the world. Golden Russet is the most well-known russet. There's also Golden Pearmain, Golden Harvey, Grimes Golden, Golden Sweet and many others. A hundred years ago, Golden Pippin was a popular variety Down East in Washington County. Currently we've got some local fruit explorers hot on its trail. I think we'll find it within the next year or two. Although not well known outside the state, in central Maine we have the Golden Ball, a blazing yellow-orange sphere, the pomological manifestation of the fiery orb at the center of our celestial world: the perfect apple for someone named Sun Ra.

Maine's Golden Ball: a cooking apple par excellence. Like the Arkestra on a good night: it really cooks!

Towards the end of Sun Ra's sojourn on earth, a friend named Jared Crawford loaned me dozens of his Ra recordings,

The Egyptian God Ra with the sun up above (*The* Golden Ball): drawing by Paul Burgess, courtesy of the family

all of which I copied onto cassettes. A few were commercial records with decent liner notes. Others were blanks. What on earth? At first I was totally put off by these. How could I appreciate the music if I had no idea who was performing? I was walking through an apple orchard of wonderful but nameless fruit. Where are the labels? What am I eating? Someone tell me the names!

What's the big deal? For generations, popular music went mostly undocumented. The 45 rpm records we listened to as kids in the fifties and sixties were notoriously poor in that regard. Most came in blank sleeves. Even long-playing albums told you very little. Few people ever knew who performed many of pop's greatest hits until recent films like *Standing in the Shadows of Motown* and *The Wrecking Crew* helped identify the relatively small number of anonymous "session" men and women who performed on hundreds of hit records. Were they like the anonymous grafters of orchards past?

But Sun Ra was jazz, and jazz was supposed to be different. Beginning with the first 78 rpm records, the names of the musicians, the instruments they played, the date and even the location of the session often appeared in tiny print on the circular label at the center of the thick plastic disc. With the invention of the LP and the large cardboard record jacket, jazz liner notes became even more detailed. The music of Ellington, Coltrane, Monk, Mingus, Miles and hundreds of other jazz musicians is wonderfully documented. But not Sun Ra's. By the time of his death in 1993, hundreds of recordings had been made with no labels of any kind. His vast body of work was nameless, timeless, and musician-less. It was like an orchard of five thousand varieties and no tags. Would anyone ever sort this out?

Fortunately for Sun Ra fans everywhere, a Clemson University professor named Robert Campbell and a British jazz recording engineer named Christopher Trent were asking the same question. After Sun Ra's death, the two of them researched, produced and published *The Earthly Recordings of Sun Ra* (1995). Trent and Campbell identified thousands of titles, dates, locations and musicians. This was no easy feat since the band played hundreds of tunes, many of them Ra's own, sometimes bizarre, compositions. They figured out when each recording was made, where it was made, and who were the musicians. What they accomplished was astounding. All of a sudden I gained a whole new appreciation for those LPs with their blank record jackets. If they could do this with records, could I do it with apples?

In the fall 1988, I had the opportunity to try my hand at puzzling out my own Sun Ra recording when friends gave me a cassette that their daughter's friend had recorded at Bard College in the fall of 1986 while the two of them were students. No other information came with it. Over the next few years I listened to the tape now and then. Then one day, inspired by Trent and Campbell, I decided to give musical detective work a shot.

My first step was to study *The Earthly Recordings*. Not surprisingly, it had no listing of any Bard College gigs in the late eighties. Does that mean that my friends' daughter was wrong about the concert? I decided to assume "fall of 1986" was correct; if that proved impossible, I'd return to square one. The next step was to pinpoint the most plausible concert date. I pulled out my *Earthly Recordings* again.

You can practically see through the open calyx right into the center of the core of the Golden Ball; the reflexed sepals are like the wide spreading rays of the sun

Campbell and Trent list roughly forty gigs and studio dates during 1986. My goal was to reconstruct their "travel log" for the year, hoping to find the most logical date for the Arkestra to head up the river to Annandale-on-Hudson. September seemed unlikely. They were mostly on the West Coast until they returned for a gig on Saturday, October 11, in D.C. From the 13th to the 17th of October they were touring in Italy. On Friday, October 24, they played in St Louis. From the 28th to November 2 they played Sweet Basil's in New York City. Two days later, on November 4, they performed in Cambridge, Mass. Annandale-on-Hudson is a two-hour drive from Manhattan. Could they have gone from St. Louis back to Philly where they all lived and then slipped in a warm-up gig at Bard before their six days in the Big Apple? This becomes my working hypothesis. I decide the concert date was October 27, coincidently a jubilant night for the victorious New York Mets and a terribly sad one for Red Sox Nation not far away at Shea Stadium.

Now, who were the musicians, and what were the tunes that night at Bard? Presumably the personnel and the playlist can be jimmied out of the nearby dates. Although the band configuration varied from time to time, and there could be as many as two dozen musicians and dancers onstage on any given night, the core of the Arkestra remained remarkably consistent. Chances are that, from one week to the next, the lineup and set list would be similar. Fortunately, Campbell and Trent had reconstructed complete playlists and personnel for the nearby dates. I put on the cassette.

I'd listened to it many times but this would be different. A few of the tunes I know—jazz standards or iconic Ra compositions. But there are several I don't know. I go back to the book, turn to the two adjacent concerts, and go down the list of songs, scanning for the ones I'm unfamiliar with. It's like studying the varieties in an orchard map or in a list from one of those nineteenth century agricultural yearbooks. Of course, I could have looked over the complete list of Ra tunes, but I would have been totally overwhelmed. There are over fifteen hundred, many of them obscure. There's "Black Magic," "Black Mass," "Black Myth," "Black Forest Myth," and so on. The names of the tunes go on for thirty-six pages in small print. How would I ever know? I hope I'm on the right track by checking out the recent gigs. Eliminate the impossible!

I compile a list of candidates for the missing song titles. Then I listen to known versions of those tunes performed by Ra or someone else. Back and forth I go, listening to my unidentified tape and then identified versions. I feel like I'm doing an apple ID. I look at the mystery apple you gave me. Then I look at paintings and photos and descriptions of Northern Spy, Baldwin or Golden Ball. One by one I knock them off. Ra loved to play swing music by Duke Ellington and Fletcher Henderson. I put on my Henderson LPs, and I pull out my Ellington, and I listen.

The playlist is now coming together row by row. The first eighteen minutes are an improvisation. The band often started with an untitled jam. This would be analogous to the seedling apple trees you see sprinkled in the woods along the road leading up to the orchard proper, unnamed and unnameable, seemingly wild and free, but actually disciplined and precise in their own pomological seedling way. Just like Ra's chaotic-sounding improvs, those roadside seedlings have their own organization, often far too complex for any of us to see.

An improvisation of wild unnamed seedlings with an organization we humans may never understand: Lubec, 2018

Then begins the first real tune. Then another. These are the grafted trees: McIntosh, Cortland, Gravenstein. Put away your watch. Don't think about the time. Just open the next bag, pull out the apples and observe. "Prelude to a Kiss" by Ellington followed by the Henderson favorite, "Yeah Man." The Arkestra members usually don't sing but when they do, it certainly helps with the ID. It's like pulling out an apple with deep ribs or prominent white dots. "Try this apple, Bob. It's a Starkey!" James Jacson sings his hilarious version of "Mack the Knife." John Gilmore sings the Brooks Bowman composition, "East of the Sun." Mixed in with the standards are several Ra compositions including, "Discipline 27-11," "Children of the Sun," "Stardust

Sandy River Orchard is grafted, spaced and orderly; the improvisation is there if you look closely

From Tomorrow" and "History-Mystery." ("History, that's his story. Mystery, that's my story.") Then the concert ends, just as the D.C. gig did two weeks earlier, with a wild version of Henderson's "Christopher Columbus," followed by "See You Later Alligator."

Meanwhile, I sort out the musicians. Some of them are easy to spot by the long stem or the red blush. Knowing who's who onstage will also help verify the date. I can hear Marshall Allen's "harmonics" and John Gilmore's tenor. The tone of their horns is a giveaway. The swinging solo that happily doodles along for over five minutes in "Mack the Knife" is like the huge patch of russet around the stem of a Wolf River. Both are hard to miss. This has got to be Ronald Wilson, the tenor-man who played with the Arkestra for a few years beginning in 1985. His presence further helps with the ID.

What about the trumpets? Where are they? The two trumpeters of the eighties, Michael Ray and Ahmed Abdullah, don't appear on fall '86 personnel lists. I listen and listen again. No stripes on that apple? That narrows the search by about ten thousand varieties. The trumpets are missing here too. What's missing is as important as what's there. No Macs in the orchard? It probably dates from before 1920. Look for Spies and Baldwins and Ben Davis. No Ben Davis? We could be talking 1870. Look for Yellow Bellflower and Tolman Sweet. No Michael Ray and no Ahmed Abdullah? Sounds like the fall of '86, all right. Look for Ronald Wilson. I think we have the date, the musicians, and the tunes. Yes, this orchard was planted in October 1986 with some wonderful varieties.

*The Earthly Recordings of Sun Ra* is an impressive undertaking. It documents nearly eight hundred recordings, spanning nearly fifty years. The index is essentially a searchable key. You can look up any of the more than three hundred musicians and see when they appear and on what recordings. How did

As Abbott Meader's nameless seedling tree looks on, JPB ponders three blank Sun Ra albums: could they be Blue Pearmain, Black Oxford and Red Astrachan?

Campbell and Trent do this? In the introduction they explain that many of the record jackets that do list musicians and dates are actually incorrect and that Ra's personal tape collection, which could have been a valuable source of information, was purposefully stored in blank boxes. It was an orchard grafted with no labels.

As Holmes like to say, somehow the two of them got a "grip on the very little which we do know, so that when fresh facts arise we may be ready to fit them into their places." Presumably they began with bits of information from those labeled records that were correct. The contradictions and impossibilities they discovered along the way kept them on the right track. They gained access to conversations with band members, photos from known concerts and studio dates, plus the collective experience of dozens of fans and devotees who attended concerts over the decades and shared music and information with one another. In the introduction they thank seventy contributors.

Campbell and Trent's example remains an inspiration to me every time I turn on my old KLH record player and sit down to another apple ID. Although I've never communicated with either of them, I imagine their process to be like solving those wonderful cryptograms that used to appear weekly in the Saturday Review. The puzzle consisted of a long quotation made up of blank dashes except for a few letters filled in here and there. It was up to you to figure out what letters went in the blank spaces, thus filling out the quote.

The identities of many ancient apple trees are likewise obscured by blank spaces. Many are also stored in unmarked boxes. Many are mislabeled. Names have been changed over and over again. Recently Cammy and I went to visit a tree in Jonesboro, said to be a Hard Sweet. I was so excited because I'd never heard of it. Turns out, hardly anyone had heard of it. It's a rare local synonym for Tolman Sweet. Many an orchardist has left for orchard heaven, leaving us with questions unanswered. Who are the musicians, what are the tunes, when was the gig? We look at the fruit and we look at the context. We pause and then we plunge. Sometimes we're right and sometimes we're wrong. If we strike out, we proceed to plan B. As Sun Ra and the Arkestra sang on many occasions, "What do you do when you know that you know that you know that you're wrong?"

# Six

# The Givens Tree

This is an instructive case. There is neither money nor credit in it, and yet one would wish to tidy it up. When dusk comes we should find ourselves one stage advanced in our investigation.
Sherlock Holmes, *The Adventure of the Red Circle*, p. 907

Could "Jane" actually be Givens?

People sometimes ask me if I charge anything for this service. Most of the time I do not because you're sending me apples I might otherwise never see. It's true that much of the time I'm given pretty standard stuff, varieties I've seen dozens of times. But every so often someone gives me something very rare. An apple of historic significance, the one I thought was gone forever.

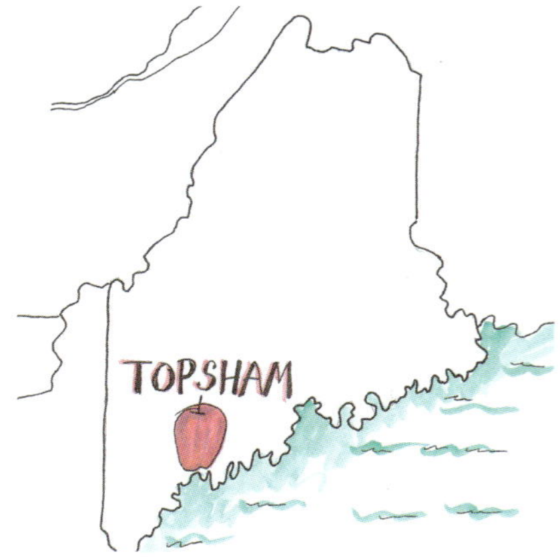


If I charged for the service, would I still receive that rare historic apple? Maybe. I'll never know if doing IDs for free actually helps to find and preserve our lost varieties. A year ago I was giving a talk near Portland. Things were winding down, and I was about to head home. A woman named Jan Bodwell came up to me and said, "Do you know of an apple called American Beauty? We have two old trees on the farm." An email followed a few days later: "The trees are located near the SW corner of the barn on the farm where I grew up. At 83 I am the last surviving sibling. I am available to meet you there almost any 'good traveling weather day.' Jan Bodwell"

A couple of months later I drove out to visit Jan in western Maine. We spent an hour together in front of her old family barn. She told me about the history of the farm and the two ancient American Beauty trees. I collected scionwood. We grafted rootstock in the spring. How can I ask people to pay me? I should be paying them.

In December 2015 I was going through a box of apples, and I pulled out a bag left for me by David Buchanan. David is a cider maker, author and fruit explorer with a good eye. He had scribbled on the bag, "two very old trees." Inside were two apples: one labeled Jane and the other Duncan with a Sharpie across the skin. People tell me I should buy stock in Sharpie. They're perfect for writing on paper bags filled with apples and especially good for writing on the apples themselves. But don't get lured into using them for tree tags. They fade quickly. Instead, we use scraps of vinyl siding marked with number two pencil.

Jan Bodwell pointing at the graft of one of her two American Beauty trees, 2017

There was something about the apple called Jane. It looked as though it came from an old tree. It also looked almost exactly like the old Massachusetts variety, Williams Favorite. But Williams Favorite is a summer apple, and this was December. Williams Favorite would be rotten by now. This is not Williams Favorite. David lives in Pownal. Could the Jane tree be growing somewhere near David's? I picked her up and held her in my hands. Could this be the long lost Givens? The one description I have says the Givens apple resembles Williams Favorite. I emailed David; he replied that the tree is exceedingly old, located at

the end of an old road in Lisbon, not far from Pownal and not far from Topsham…where Givens originated.

When I created the painting for what became the 2009 Common Ground Country Fair poster, I decided I would include sixteen apples, one that originated in each of Maine's sixteen counties. Some were no brainers. I knew I'd be using Black Oxford and Starkey. I had most of the sixteen growing in our orchard. Sagadahoc County would be tricky. Givens was the only variety I could find that originated there, and I didn't think I'd ever seen one. I had photos of what I assumed was Williams Favorite (a.k.a. Williams and William's Early), so I used one of those as a model for my Givens painting.

The only known citation for Givens was in the November 15, 1849, *Maine Farmer*. The brief description is reproduced in Bradford's *Apple Varieties in Maine*. Givens was exhibited by J. Sanford of Topsham at the Maine Pomological Society in the fall of 1849. J. Sanford would be Josiah Sanford (1785-1855). Sanford (or Sandford) lived on a farm close to the center of Topsham. He was a member of the Maine House of Representatives and a selectman of Topsham. At some point before 1849, he obtained the scionwood from Samuel Givens on whose farm the tree originated.

Samuel Given(s) was born in 1788 and died in 1876. He married Ruth Booker (Bowker), and their family farmed Lot 5 and part of Lot 6, both bordering the Cathance River in Topsham near modern day Katy Lane. Whenever possible, farms were laid out bordering the rivers. The rivers made for very decent traveling, long before anyone had time to build roads. The Cathance River flowed through much of Topsham north to Bowdoinham and then on to Merrymeeting Bay and the Kennebec River.

"Jane" Givens from Lisbon; Jane is on the two trunks to the left; a topworked Baldwin graft is still visible on the right-hand trunk as a ripple about a foot off the ground

Was it Sanford or Sandford? Given or Givens? Booker or Bowker? It wasn't just the apple names that were subject to "spelling fluidity" back then. While some of this can be attributed to handwriting, penmanship was generally a lot better than ours is today. Consistency in spelling was not a high priority. The script in those old reports and maps and lists could also be rather small and a challenge to decipher. Even on the same old town map or the same old list, you're bound to find multiple spellings of what had to be the same person or family.

Bradford says that Givens is shaped like Williams Favorite. That would mean that the fruit's "form" is roundish-conic, leaning towards conic. Red Delicious also fits that description. Most apples are closer to round than conic. Looking for an apple that resembles Red Delicious should be a huge help in narrowing the search. Unfortunately, no color is mentioned in the old description. But we'll assume it's red since Beach says that Williams Favorite is red. The description also tells us that the flesh is white, juicy and sprightly sub-acid. Its season is October and November.

Williams Favorite is a summer variety found throughout central and southern Maine; it's mostly red with a very shallow cavity

What else can we deduct about an apple from these bits of information? Can we tease out a story? Even a brief citation suggests several things. We build our case and proceed from there. We go by the facts, but we allow those facts to tell us more. We read between the lines. We use our imagination. We might be entirely right, partly right or totally wrong.

Knowing that there's only one known citation suggests that the apple is obscure. Why was nothing else written? The apple probably wasn't grown more than a town or two away from Topsham. The 1849 citation suggests that it wasn't being grown much after the war, otherwise someone else would have surely mentioned it. We can assume therefore that when we find our Givens tree it will be ancient. We have at least two possible locations: the Samuel Givens farm and the J. Sanford property. We need to find those two spots.

And so the search began. I kept a lookout for an apple on an ancient tree in the Topsham area that resembled Red Delicious, ripened in mid-late fall and tasted pretty good. At some point I knew I would have to track down the J. Sanford and Samuel Givens properties. In October 2006, a woman named Brenda Nelson sent me two fruit from her tree in Bath, only a few miles from Topsham. They were red and conic. They looked like Red Delicious. I wanted the apple to be Givens but I knew I'd have to visit the site and see the tree in person. 2006 was still a few years before everyone was emailing photos. If the tree is about a hundred and fifty years old, it could be Givens. On November 21, I headed for Bath. I pulled into the driveway. Brenda was there to greet me. "The tree is in back," she said. "Follow me." Alas, the tree was twenty-five years old, or maybe thirty. Oh well. Red Delicious.

Two years later a couple named Bill and Judith Sutter brought me an apple from their old tree in Wiscasset. An email followed a few days later:

> I think you may have succeeded in your search for the Givens apple. The apple sample that I brought to Maine Apple Day was from a tree behind my house in Wiscasset...Your mention of the variety Givens rang a bell from some distant past. I know I have heard that name before, from someone, some time in the past. I do look forward to your visit next Fall...Bill Sutter

I visited Bill and Judith in 2009 at their old farm house on the outskirts of Wiscasset, across the road from the old stone jail, perched above the Sheepscot River. A Red Astrachan, a Winter Banana and this old tree were all that remained of an old orchard in what is now their backyard. The Wiscasset Givens is a reddish version of a Yellow Bellflower. It looks like a Givens should look. The fruit is roundish-conic, ribbed at both the basin and cavity ends, striped and blushed red and peppered with distinct small gray pinpoint dots. Some of the apples are lipped or have practically no cavity at all, like a Pewaukee or a...Williams! I agonized over this Wiscasset apple for a year or more, then decided it was not Givens. Eventually I was able to identify it as Minister, another oblong-conic apple, one that originated in Rowley, Massachusetts, on Boston's north shore, not far south of the New Hampshire border. Named for the Reverend Doctor Spring of Newburyport, Minister is a classic New England heirloom. The only Minister specimen painted for the USDA watercolor collection was submitted from Union, Maine, in October 1914, only about thirty miles from Wiscasset.

Minister apples from Bill and Judith Sutter in Wiscasset

Over the next few years I visited the Sutters a number of times. They introduced me to another old tree in town a few blocks away. That one awaits identification. We've grafted their Minister and it's included in the Maine Heritage Orchard. It's quite rare, a wonderful find, but alas, still no Givens. The search returned to the back burner.

I did run into other Givens candidates from time to time, including a beautiful old tree I passed one day on the old road between Brunswick and Durham, as well as another one near Bath that turned out to be a Winesap. And there were others. Somewhere along the way had I seen a Givens tree? How many times do you see the "real" apple but you don't know what you're seeing? You see, but do you observe?

Nothing big happened in the case of the missing Givens until five years after visiting the Sutters. It was a cold bright sunny mid-January day in 2014. I was doing IDs in the dining room. My intern at the time, Abbey Verrier, was with me. She was looking through boxes of bags of apples for maps to inventory and save for future reference. We make maps of many of the sites as we travel around in the fall. People tell us we should use GPS, but I prefer the brown paper bag and Sharpie method. You draw the map on the bag and put the apples inside. Abbey was cutting out the maps and sticking them in a file folder: "Fruit exploring 2013."

I went to the root cellar and brought up one of the remaining boxes of apple mysteries and started to go through the bags. Okay, here's one from Harpswell, down the peninsula from Brunswick. I look at the apple and realize I'm off on another adventure. I can often tell at a glance if it's a modern variety or an heirloom, and if it's a seedling or a grafted tree. In this case, the apple had to be from an old heirloom grafted tree. The owner, Kara Douglas, had left the two apples at the The Great Maine Apple Day in late October. Fortunately, the fruit was still in pretty good shape. There was an ID form with her name, contact info and a brief note. The apple is sweet, and best used for eating and cider. The estimated age of the tree is one hundred. It's part of an old orchard and it ripens in late September/early October.

The apple itself was roundish, ovate and conic, like some chicken eggs. Not many apples are ovate. That's good. It means I can eliminate hundreds of possibilities right off the bat. The cavity is round and deep and acute. One of the apples has no stem. It got pulled out. The other one is medium-short and thin. The basin's worth noting. It's deep and wide and the sides are abrupt. The calyx is large and completely open. This is a distinctive apple. The coloring is right out of the nineteenth century: about a third Chinese red, a third muted but rich deep yellow, and a third patches and netting of golden russet. Sprinkled generously over the red blush are tiny white dots. This apple is beautiful. It's coloring resembles Ribston Pippin, but it's too ovate to be Ribston. This is something I've never seen before.

I write Kara an email and thank her for leaving the fruit. I ask her if I might drop by sometime this winter to cut some scionwood and do some grafting. This one should be in the Heritage Orchard even if it remains unidentified. I should cut up the apples and check out the inside, but I don't want to cut them up yet. I should photograph them first. For now they'll go back in the bag and down to the root cellar. Done for the afternoon. Abbey heads home. I try not to get too excited. Maybe this is Givens.

Next day I get an email from Kara. The old tree is one of four in their yard. They think they also have a Baldwin, and possibly a Grimes Golden. The other two, they don't know. Yes, of course, I can come visit. A day or two later I pull out the apples and look again. Then I pull out photos of Williams Favorite. I compare the form of one to the form of the other. They're identical. Then I reread the Givens description from Bradford.

"...Form of Williams Favorite; quite as handsome and some think equal in flavor. Flesh white, juicy; flavor highly sprightly sub-acid. In season between October and November...Quite worth propagating."

Well, it may be too late in the season to get a decent taste of this apple, but I will need to cut it open and check the flesh color at the very least. I hardly dare. Once you cut them open, you can't go back. So take pictures of the two apples and then decide what to do next. In goes the knife and the apple is now sliced in half. The flesh is white but not pure white. Is it white enough?

John Paul Rietz and the roadside apple from the Givens farm in Topsham

About the time I met Kara Douglas, I also met a writer from Waterville named Kate Cone, who had caught the fruit exploring bug and began to look for Givens. Kate worked for many years doing title searches and used to live near Givens country. She tracked down the site of the Givens farm on the Cathance Road. Unfortunately, much of that area has become a power-line strip, or has been divided into house lots. But an old tree did survive the onslaught of civilization over the bank by the side of the built-up road. We followed Kate's directions and found it in October 2015. There were still a few apples hanging on the tree. That in itself was a good sign. We know the fruit ripens later in the fall. We picked four or five apples. The flavor was very decent, better than Williams Favorite. Another good sign. Still, I think the tree is a seedling: no name.

When 2016 rolled around, I was on a full-scale Givens kick. David Buchanan had me all worked up about the Jane apple. Through David, I contacted Jane and Charlie Ridlon, the owners of the tree, and received

Topworking step 2: It's OK to do several grafts per tree

permission to go to their farm in Lisbon. On September 14, 2016, I made the trip. There was no one home. The old tree was right behind the house. It looked like a seedling with something else topworked onto one limb. I got out the pole picker and a bag. The graft was a large branch of Baldwin. This was not the apple that David sent me. I looked for other trees. It was getting late. Later that day I swung by David's. He wasn't positive which apple he had written Jane on or where it came from. So near and yet so far!

A month later I asked my intern, Laura Sieger, if she'd like to do a systematic search of the Cathance Road. The goal would be to find a Givens tree down that way. We set out to make a day of it on October 18. The Cathance Road is one section of the old Post Road that runs from Topsham and Brunswick up through Bowdoinham to Richmond Corners and on to Gardiner and Augusta. You know it's old the instant you turn the corner in Bowdoinham and head south towards Topsham. The houses are old, the apple trees are old, even the road itself feels old. We started at the north end and stopped at every house with any visible apple trees. The second or third stop we struck gold. There we met Nina Mendall on the front steps of her beautiful old farmhouse. No, she wasn't aware of the Givens apple but she was quite clear about where she wanted us to go next. Turn around and head for the Bowdoinham Historical Society. You need to meet Elaine Diaz and locate a copy of *The Journals of Abraham Preble*. Preble probably grafted the tree you're looking for. We found the Bowdoinham Historical Society in a cramped but sunny third floor room above the town library. They called Elaine Diaz and she came right over. Next thing we knew, Elaine, Laura and I were poring over a copy of the *The Journals of Abraham Preble*.

Abraham Preble was born in 1800 and died in 1884. He lived in Bowdoinham within a few miles of his uncle Samuel Givens' farm. He was a farmer himself, as well as a teacher, peddler, surveyor, justice of the peace and grafter. One of Preble's descendants, Ann Elden Alexander, transcribed thousands of journal entries dating from 1822 to 1854. Unfortunately, thirty years of entries are missing, apparently given away many years ago to a relative, who then disappeared. Even so, what they do have is incredible. Elaine Diaz had her own annotated version and knew just where we'd want to look. She flipped to the spring of 1846. Over the course of an eight-week period from the first week of April to the first week of June, Preble "set" 6,300 grafts. Every day but Sunday he traveled to this farm or that, topworking anywhere from a dozen to a couple hundred trees. He charged a penny a graft. On Wednesday, April 29, 1846, "I set 262 scions for Uncle Samuel Given & receive pay in full $2.62."

Three years later, on November 15, 1849, *The Maine Farmer* referenced the Givens apple. Where was Abe Preble on the 15th? Later at home, I turn to the journal entries for October and November 1849. It doesn't look as though he attended any meetings from October 1 to November 15. On Friday the 15th he was finishing up digging nursery trees for fall planting. "We have dug in all 8,446 apple trees." Had Preble grafted Givens scionwood for the Givens family that day in April 1846? Had he grafted Sanford's trees? These trees would be only be a hundred and sixty-six years old this past spring. At least some of them should still be out there. We'll need to go visit Preble's farm next year. Maybe he grafted Givens for himself.

Elaine Diaz, JPB and Betsy Steen, October 2017

In the fall of 2017 I returned to Bowdoinham and gave a talk at the historical society on Sunday afternoon, October 22. It was a packed house. Elaine was there with nearly sixty specimens she'd collected from around town. The apples lined the length of one entire wall. If the Givens is still alive, maybe it's here. I glanced over the apples but decided not to take a close look until I got home. Next day I began to go through them. Pewaukee, Northern Spy, and a number of the other usual suspects were in the mix. Some are seedlings, more than likely. Others are certainly intriguing but not what I'm looking for today. What about that Williams lookalike? Will it be in the lineup?

Then all of a sudden, one after another, I find three conic apples. All three are from the Carding Machine Road in the north part of town. Each bag contains two specimens and Elaine's handwritten notecards. The first is dark red and promising. The note read: "behind blackberry patch along north rock wall (behind house) Old 4 way split tree—largest trunk 41" circumf mid-large conical deep red apple. Yellowish flesh, softening 10/16 very tasty, sweet. Could this be Sandford-Given?" It does resemble Williams Favorite. Unfortunately, the flesh is light yellow. Givens should be white. But when does light yellow pass for white? It also seems to be past its prime. It should be perfect now.

The second apple also resembles Williams. It's striped however, not solid red. But the classic description says, *form* of Williams not *color* of Williams. Could that mean that Givens could be conic and striped? The flesh is white and firm. I take a bite. It's still good and will be for another month or six weeks. The notecard reads, "Right after red apple tree Medium large conical apple Light red striping over yellow very sweet." And then a cryptic addendum, "from same tree as the dark red sample?" Oops. Could both of these apples come from the same tree? I feel so close.

I still have more apples to look through. I pull out the third Carding Mill Road candidate. This one is smaller in size than the other two but still conic. Never let size bother you too much. Perhaps the tree just needs a good pruning. This time she wrote: "Large, split trunk tree, right next to road, N of house. Small medium apples on one trunk slightly conical red with gold spots and russet at stem. Very tasty tart firm—white flesh *One of the best apples I've tasted this year." I take a bite. Still firm, though beginning to go by.

Elaine's notes are fantastic. Who could ask for more? I open the remaining bags. No other conical Williams types but many others that have that old look to them. I know I'll need to go back to Bowdoinham, though it may not be until next summer. I call Elaine a few days later and leave her a message. During the next few weeks I find myself thinking about that striped conic apple. Then one day when I'm working on an unrelated batch of ID's, I find myself flipping through one of my folders that includes photocopies of some of the early twentieth century apple watercolors from the USDA. All of a sudden I'm staring at a conic striped apple. It looks just like Elaine's apple. The description line reads "Williams." I blink a few times. Well, I guess it's time to revisit Williams. Evidently I've missed something rather important.

In the meantime, I get a Christmas card from Elaine. She's left the state to help a friend. She'll be in touch in the new year. I pull out *The Apples of New York* and turn to the Williams entry. The color of the image is solid red and the description reads, "...a very beautiful, bright red apple..." Then I flip to Dan Bussey's new book. Williams is on page 153. The image was painted from a Kent County, Delaware, specimen submitted by A. N. Brown in August 1899. It's striped but also darkly blushed. It looks like a red apple, not a striped apple.

Williams from Prof. Henry, Connecticut, August 1921: from the USDA watercolor collection; reprinted with permission

Then I go to the USDA watercolor website. I want to see any other Williams paintings they might have. Turns out there are twelve. Looking closely I notice that five are crosses between Williams and Yellow Transparent. We can dismiss them; they don't resemble either parent. The next three "Williams" apples are from Delaware, one being the painting in Dan's book. The second of those three (also submitted by Brown) looks identical to the first, but the third is distinctly striped. Maybe Williams is sometimes striped. The next one is from Ohio and mostly red. No new information there. The next two are from Arlington, Virginia, both submitted on August 1, 1928. They look identical in form but one is solid red and the other is striped. This is getting interesting.

Then I pull up one submitted by someone identified as "Prof. Henry" in "Connecticut" on "8-10-'21." I had a loose copy of this one in my folder. It's got that Red Delicious conic shape. The coloring, however, is not solid red. It's striped—distinctly striped. It looks almost exactly like Elaine's Carding Machine Road apple. We have red Williams and striped Williams from Delaware, red and striped from Virginia and striped from Connecticut. Maybe Elaine's two apples did come from the same tree. Could they be Williams? No, Williams is a summer apple. The paintings were all done in August. The apples from Carding Machine Road are mid to late fall apples. Both resemble Williams, just like Bradford said they would. Maybe we've found Givens!

I resist the temptation to declare victory. Elaine is out of state. She doesn't do email. It's snowing outside. In fact we've got over a foot on the ground. The temperature is a few degrees below zero. The Givens is a work in progress. It will all have to wait. I'll write to Elaine and the two of us will take a ride in the spring. The tree will need to be old. Or maybe Givens isn't in Bowdoinham at all. Maybe it's the Kara Douglas tree in Harpswell. Maybe it's that apple David left me. In a more recent conversation, David said he thought the apple did come from behind the Ridlons' house. That tree I dismissed as a seedling with a Baldwin branch may have been itself a grafted tree. Maybe in my haste I overlooked the real deal. I'll return to the Ridlon tree. I'll go to Carding Mill Road.

In March of 2018 I visited another ancient Givens lookalike at the home of Bill and Judy Debray in Richmond, only fifteen miles north of Topsham. Judy had brought specimens to the Fair and I was impressed. That one turned out to be a Yellow Bellflower. I'll find the Givens tree. Meanwhile I promise I won't charge you for the apples you send. Keep them coming, especially all the conic-shaped apples from any of those hundred-and-sixty-six-year-old trees in the Topsham-Bowdoinham area. It's totally fine if they're red, or red-striped. I'll take either.

I should mention one other thing here before we move on. Although I do refrain from charging for the service, I am always happy to take bribes. This past fall a chocolate bar appeared with a note: "to inspire you while you ID our apple." Not long after, a jar of honey appeared from the bees that pollinated the tree that grew the apple that's in the bag that I'm about to reach into. Or a funny note or a poem or an old article about some orchard or ancient cider mill. Are these bribes? I hope so. Are they welcome? Of course. But a word of advice to the would-be

Bill and Judy Debray and the Richmond Yellow Bellflower tree, 2018

The Sand Hill tree in Somerville: "Graft this one!"

briber: if it's perishable, please note the bribe's expiration date on the package exterior. I want to rush you to the head of the line, and I certainly don't want to open up the bag and find the bribe has rotted along with the apple.

So many apples flow into my life each fall. Each one with a story. Everyday they appear. If they don't fit in the mailbox, the mail deliverer leaves a green postcard indicating another package at the post office. "Apples" he says cheerfully as he pushes another two or three boxes over the counter. "Please bring back the green card. We like to reuse them."

And the apples appear on the porch, or in the barn or on the roof of the car. Apples everywhere. One afternoon a few years ago, Cammy was coming home when she met a stranger driving away. "I left some apples for you on your kitchen table. You have a lovely house!"

I also don't charge because I might get it wrong, or I might not get to it at all. Maybe I won't feel so guilty if I haven't charged you. I do try. I owe you my best shot but, hey, no guarantees. Mostly people are patient and appreciative, though that's not always the case. One year I received a rather nasty email from someone who thought they waited too long in line at the Common Ground Fair where I do quite a few IDs. Apparently I ruined somebody's fair experience. It's only an apple!

Do I make mistakes? Of course. I try to make it clear when I'm uncertain. Some years ago Cammy and I had the pleasure of visiting Isabella Dalla Ragione, the Italian fruit explorer and apple identifier, at her farm north of Perugia. A couple of years later she was in Maine for a conference and came to visit me. I took her around town to see my favorite trees. She took a look at the old one at the top of Sand Hill and said, "Graft this one!" When we talked about doing apple ID's, she told me how she always begins her answer: "Your apple *most closely* resembles…"

JPB with Isabella Dalla Ragione at her home in Italy holding a beautiful mela muso di bue, November 2006

# Seven

# Len Alexander

"My dear fellow," said Sherlock Holmes as we sat on either side of the fire in his lodgings at Baker Street, "life is infinitely stranger than anything which the mind could invent."
Sherlock Holmes, *A Case of Identity*, p. 190

Aunt Penelope Winslow

"The Fair," as we call the Common Ground Country Fair, has been good to me in my search. It's the Maine Organic Farmers and Gardeners Association's annual, three-day agricultural bash in Unity, Maine. I have made some wonderful friends, found mentors and connected with many of my favorite apples, including some that otherwise would have been lost forever. It's always in late September and has been here in central Maine for forty years. The first two were held in Litchfield. It then moved to Windsor for a long stretch before settling into its current permanent home in Unity in 1998.

The Fair is billed as a "celebration of rural living." Each year about 60,000 people from all over the country converge for organic food, crafts, workshops, concerts, displays, oxen, sheep dogs, solar panels, tractors, post and beam barns, hand-spun wool, and much more, including apples.

I've been every year. Early on I had a booth and sold "live food" salads. (Not that popular.) Later I sold cider to vendors. (Way more popular.) In the early 1990's we established a Fedco booth in one of the animal stalls. I began handing out Fedco Trees catalogs and talking with whoever stopped by. It was at one of those early Fairs, years before we had a booth ourselves, that I first saw Francis Fenton. He stood beside an apple display he had created by stabbing twenty heirloom apples onto nails protruding from a large upright sheet of plywood. People and dust swirled around him. That was Francis' style. He was an old-time hawker, and this was his sideshow—only these were apples, not strippers. It was clear that here was someone who loved to talk and teach. I was transfixed but too shy to say hello, although I sensed that someday this person would become my mentor.

A year or two later Francis stopped coming to the Fair. Perhaps he was too busy with his orchard. He sold all his apples from a small farmstand on the edge of a quiet dirt road just outside his packing house. He had no crew. It was all Francis and a scattering of customers who drove up to purchase some of his Wealthys or Blue Pearmains or Dollie Delicious apples (named after his wife) and listen to him chatter away about life. Of course I didn't know any of that at the time. We still hadn't met. I just noticed that he was gone.

It was also about then that Fedco decided to rent a free-standing tent for the Fair. We needed something to fill it up, so we assembled three displays: potatoes, tomatoes and apples. I was in charge of the apples. I was going to be the new Francis. I collected varieties from here and there, forty or so the first year. Instead of a sheet of plywood with apples stabbed on nails, I took a bed sheet and drew a grid of small squares with a magic marker. It proved to be a big hit. How could it fail with all those colors, shapes and sizes? We only did displays of potatoes and tomatoes a few times. The apple display, however, never quit. We still do it today, twenty-something years later. And it's still fun. Last year we had nearly two hundred varieties spread out on the grid.

Those first displays were made up of whatever apples I could find. The only criteria was that they grew in Maine and were at least vaguely ripe at Fair time. I wanted them to have some color. I wanted them to look ripe even if they weren't. A monochromatic display of a bunch of pale green apples is not much to look at. Fortunately by mid-September in Maine the apples are beginning to show pretty decent stripes and blush.

Over the years the display gained in sophistication. It's always been a reflection of the apples I know. As I've learned more, the display expanded. Each year I would add a different twist, organizing them in some new way. One year I laid out a couple dozen McIntosh strains and sports and offspring. Another year I did

The first year of the Fedco display at the Common Ground Fair, 1993: forty-eight different varieties!

all the various apples I could find called Pumpkin Sweet. I had six different apples. One was the big yellow-green variety described in the *Apples of New York*. Another was red-orange and creased like…a pumpkin. Will the real Pumpkin Sweet please stand up? In recent years we usually create two side-by-side displays. One display is typically a mishmash of whatever we find growing around Maine that fall. The second display is the one we put more attention to. We label it "Apples You'd Find In Old Maine Orchards."

It's the Old Maine Orchard display that many people specifically come to see. They bring their own apples with them and stand and stare and hold theirs up to compare. "I found it!" Meanwhile a gang of us hang out and assist. Be patient. We'll get to you!

A long table and a bedsheet and a Sharpie. I mark out a grid on the sheet with squares about 4" x 4", although 3" x 3" will work if space is limited. Each square gets a peel-off Avery label. At first I wrote the names in Sharpie. Later I got hi-tech and printed them with the computer. Down goes the sheet, the labels, the apples, and in come the curious. The display is a magic magnet. All day long for three days people gather in front of the table to stare at the apples. To tell each other about them. To touch them. To marvel at the names. To hold them up and pretend to bite them. To tell me stories about the old tree in their backyard. To invite me to visit them. "Are they real? Do you have to collect them again every year?" And to ask me to identify the fruit that they brought with them. But back then it was stories and invitations. The ID thing hadn't taken off yet. I wasn't ready anyway; I still had a lot to learn.

Cammy checks out the display twenty-one years later in 2014

These days I still get invited to visit a lot of old trees, more than I have time for. I've learned to accept the fact that I can't go everywhere. Contemplating what I might be missing can be painful. Like that fisherman, and the one that slipped the hook. But I teach myself not to think that way. Part of going on these fruit exploring adventures—these treasure hunts—is sorting out how to measure success. Being able to go fishing at all could be the measure of success. It's the boat and the rod and the water. It's some new combination of roads and weather and people and trees and fruit. Often when I visit someone and their trees, they apologize to me as soon as I arrive. "There probably won't be anything here you've never seen before," they say. Or more often: "I don't want you to be disappointed." I think it was Ernie Banks who spent his whole career with the Cubs and never played in a World Series or even a playoff game. One of his favorite declarations was "It's a great day for a ball game; let's play two!" I was chatting with a friend recently, and she told me the following: "Expectation is premeditated disappointment."

Many times I've arrived at the house of strangers in some town I've never been to, and next thing you know, I'm in their car, and we're heading off together to find a tree on some abandoned farm. Or I'm in their house, and they tell me the story of an apple they thought no one would ever care about.

In September of 1993, I'm at the Fedco tent with my brand-new display of forty-eight apples at the Common Ground Country Fair; people come, people go, and then there appears a slender old man with a soft brimmed fishing hat and a few teeth missing and a huge grin. He stays for over an hour. He lives a few towns away in Chelsea along the Kennebec River, just south of Augusta. He loves to talk and tell stories, and he loves apples. His name is Len Alexander. He becomes my first apple mentor.

Over the next few years I visited Len spring and fall. Apples generally mean at least two visits a year: in the spring you cut the scionwood, and in the fall you pick the fruit. The next spring I was there on his farm in early April. I noted in my journal, "Went to Leonard Alexander's to collect pear and apple scions and then to the scionwood exchange." Len was not an orchardist. He and his wife had a quiet little farm where Len restored old vehicles in a long shed next to the vegetable garden set back high above the Kennebec. A dozen or so fruit trees he or his dad had grafted were placed here and there around the house and gardens. The varieties were rare, especially the pears: Russet-Bartlett, Bosc-Bartlett cross, and Washington State Pear, among others.

Len Alexander and JPB at the Fedco booth, 1994

One favorite apple tree was next to the driveway. This was a tree that Len and his father, Leonard Sr., had discovered not far from the Alexander place. They loved to explore the old abandoned farms along the river. Those farms dated back to when there were few bridges, and most travel was done on the water or on the ice. When Len and his dad found something they liked, they tagged it and returned in the winter to collect scionwood. Leonard Sr. topworked for people all around the state. One apple they particularly liked was huge and red. They grafted it and, not knowing what it was, they just named it Alexander after themselves.

The Alexanders' Alexander really was an Alexander!

The tree grew and thrived and produced beautiful, large, red-striped, roundish-conic fruit that were excellent for cooking. One year in the fall when the apples were ripe, the County Extension agent stopped by to visit and chat and maybe even give a bit of advice about this or that. Len and his father showed the fellow around. Their route took them by the big, red apple tree they had grafted. There was the fruit looking ripe and ready. "We discovered this one down along the river," they told him. "We don't know what it is." The agent looked at the fruit and said, "Why, this apple is an old Russian variety called Alexander."

Another of the apples Len introduced to me came from his family's days on North Haven Island. The Alexanders were some of the early inhabitants of the island's north shore until they left in 1922 and moved to "the old Williams place" in Dresden, a few miles south of Chelsea. Farmland was in short supply on the island, and they were farmers. Len and his three brothers grew up on a hundred acres of good Kennebec River farmland on what became known as the Alexander Road in Dresden. His brother Bruce still lives there. Later Len moved up to Chelsea with two of his brothers, where they shared 200 acres. Eventually their parents joined them.

The North Haven apple became known as Aunt Penelope Winslow. It's a variety with a story dating back to Charlemagne. Evidently, Penelope Winslow brought a small tree in a tub out to the island from her home in Marshfield, Massachusetts, most likely sometime between 1760 and 1770. Penelope and her husband were among the first group of settlers from Massachusetts to farm on the island, then known as Fox Isle or North Island. When Penelope's husband died young out on the water in 1773, she continued running the farm by herself. Perhaps it was then that she took on the title "Aunt" Penelope Winslow. She came from a titled background.

A few years ago I struck up a correspondence with Ellen Kennedy, herself a descendant of Aunt Penelope. Ellen has been a big help in sorting out who Penelope was. Apparently there were at least three and maybe even four Penelope Winslows dating from roughly 1650 to 1850. Two married into the name a hundred years apart, both to men named Josiah Winslow. All were related to one another by birth or marriage or both: "Apparently she is an ancestor on both my mother's paternal AND maternal side. Those naughty puritans!"

It's quite the genealogy. Penelope's ancestors go all the way back to Charlemagne—*the* Charlemagne, born in 747. Also included is Fulk III, also called the Black, who died in Jerusalem in June 1040, fifty years before the First Crusade. Other notable ancestors of this cheerfully red-striped Maine apple include Henrys I-III and Edwards I-III. Then there are the founders of Salem, Massachusetts, who were on both sides of the witch-burning travesty of 1692-3. Another descendant, Thomas Wentworth Higginson, founded the first two all-black regiments in the Civil War. Apparently they rode through town together with nooses hanging around their necks. After the war, Higginson became Emily Dickinson's patron and close friend. We could probably do a history of western civilization from one apple. Do we ever know what we might find when we shake that family apple tree?

Aunt Penelope Winslow tree on North Haven Island; probably the old tree Bruce Alexander remembered: courtesy of the North Haven Historical Society Lloyd Whitmore collection

Aunt Penelope Winslow, the apple, became locally famous. It never quite reached the status of Charlemagne or Emily Dickinson, but people out there loved it. It's about the size of a McIntosh. Its shape varies from roundish to somewhat conic to somewhat oblate. It's striped and blushed and has a russet area around the stem. It resembles Duchess as well as the Canadian variety, St. Lawrence. It ripens in September and cooks quickly into a slightly coarse, slightly tart, delicious, yellow apple sauce. The fruit is good all fall and keeps until about mid-December.

Len had a branch of Aunt Penelope Winslow on one of his trees and gave me scionwood one early spring when I was visiting. Over a decade later, in 2007, at the invitation of Gil Foltz, Cammy and I visited North Haven where I gave a talk and toured orchards and old trees. Gil took us to the farm of James S. Brown. The Browns have what is likely the last mature Aunt Penelope Winslow on the island. The following winter, Gil cut scionwood from the Browns' tree and sent it to me. I've since grafted many Penelopes, several of which are now planted back on North Haven and one of which is in the Maine Heritage Orchard. According to Len and Ellen, there may still be old trees in Thomaston, Maine, and Cohasset, Massachusetts. Someday there'll be an ancient one in Unity at the Maine Heritage Orchard.

In more recent years Len and I drifted apart. He stopped coming to the Fair when it moved thirty miles north to Unity. I visited Len's younger brother, Bruce, after Len had died. It was winter, and we sat together by the wood stove. Bruce told me about the ancient Aunt Penelope Winslow tree. "I can remember it over on the North Shore Road. Set a young one on the same spot. North Haven has such wonderful soil."

L to R: Summer Sweet(ing), Winekist (showing its red flesh) and Moses Wood: photo by David Grima

# Eight

# Bill Reid and Morris Towle

As Cuvier could correctly describe a whole animal by the contemplation of a single bone, so the observer who has throughly understood one link in a series of incidents should be able to accurately state all the other ones, both before and after.
Sherlock Holmes, *The Five Orange Pips*, p. 225

Winthrop Greening

Two years after meeting Len Alexander, I met Bill Reid at the Fair. He was admiring the display, and we got to talking. He had a collection of heirloom apples, and he thought I might like to visit him in New Sharon and see his orchard. He seemed like a good guy. Something told me I should go.

A few days later I drove up Route 27 through the village of Belgrade Lakes and onto Route 2 in New Sharon where I turned west and headed up his dirt road. The houses thinned out, the woods thickened up, and then I was there. I paused. A few yards from the road sat a long yellow mobile home.

As I walked in the driveway I could see that the mobile home was connected to an old red cape that sat perpendicular to the trailer. Attached to the cape on the south side was a greenhouse. Grape vines clung to the interior framework. It looked as though it had made the transition from greenhouse to storage unit some years ago. Beyond the greenhouse was a small barn. A few random cars peppered the yard. Fresh food scraps were scattered on the ground several feet below a door to nowhere in the side of mobile home.

Between the cape and the trailer was a short connecting hall with what appeared to be the front entrance. I knocked and then stepped back a few feet and waited. I like to give a stranger some space when they open the door. Ferocious barking came from somewhere inside. The door opened, and out came Bill accompanied by Cyrus, his three-legged pit bull.

Bill Reid at work in his shop: courtesy of Bill Reid

Bill Reid was a big man with a smile and a beard. He was wearing a large pair of overalls. He was alone that day. As it turned out, he was always alone when I visited and he was always wearing overalls. His wife worked as a factory manager several towns away and returned to New Sharon on weekends. Their grown-up daughter lived out of state. Bill invited me in and showed me around. The living room walls were lined with books. Everywhere books. "I throw the food scraps out for the birds," he explained.

As we talked, I learned about his life, his politics, his homemade treadle lathe, his banjo playing and his PhD in philosophy. I was visiting a large man in overalls in a trailer on a back road in New Sharon, Maine, talking about philosophy. Or rather I should say, I was being taught philosophy.

Bill's orchard was small, but hardly disappointing. There were about a dozen trees, all probably thirty years old. It was enclosed with four feet of woven wire fence. A single sheep nibbled down the grass between the trees. We walked from tree to tree. They weren't far apart. He knew all the varieties and all the stories behind them. He had grafted them himself. Some of them I knew, and several I did not. I had an idea they might be rare.

Bill had assembled his modest collection primarily with the help of an orchardist named Morris Towle, who lived nearby in Winthrop, and collected unusual apples throughout the middle years of the twentieth century. If Maine has a historical apple epicenter, it would be Winthrop. Bill read an article in the local newspaper about Towle and then went to visit him. He returned home with scionwood of several apples. Two of them, Rhode Island Greening and Black Oxford, were familiar to me. Three others were not.

The three were Moses Wood, Sweet Sal and Winthrop Greening. All three originated in Winthrop, and all three are rare. Later I discovered that Winthrop Greening is also included in the Lothrop Davenport collection at Tower Hill Botanic Gardens in Massachusetts. I've also found it elsewhere in central Maine. As far as I know, the other two appear in no other collections and would likely be gone were it not for Bill. From what he told me, it sounded as though Towle had assembled a large collection of rare apples and that the orchard might still exist. It was possible that these varieties might still be there too.

I met Bill before I met Earland Goodhue and before I met Francis. Bill was a philosopher and a tinkerer. He was friendly, good natured and well read. I enjoyed listening to him talk about the world. He chuckled. I got the feeling he wasn't fooled by anyone. He gave me lots to contemplate whenever I came over. Although he was not deeply into apples and we spent only a short time together, he became my next apple mentor. He gave me apples. He gave me scionwood. Later on he gave me an exquisite applewood jackknife he made for me. I think of Bill as my diving coach and his orchard as a diving board. Each time I entered his backyard and touched a Moses Wood or a Winthrop Greening, I felt as though I was inching closer to the edge of a very high cliff. Bill inspired me to take the plunge. It was during that time I began to think about the baton that comes to us from the past, ours to pass on to the future. We are the link. We have the opportunity to look backwards and forwards simultaneously. This is our opportunity…or is it our responsibility? Well, either way, we stand in the middle. In the past are those who planted the seeds, selected the best, and grafted them for us. We are at the balance point. Do we save the Winthrop Greening and the Moses Wood and the Black Oxford? Or do we let them go. We make a decision to take the baton—the scionwood—and pass it on. We stand in the present. We graft the past and hand it on to the future. Other anonymous grafters passed the batons along the continuum until one day we stand by

"…we're carrying that torch, like the Olympic torch, passing it hand to hand, to each generation…and that's our job."
John Lennon, 1980

Morris Towle admiring one of his trees: photo from Sally Dawson

the ancient tree and the baton is held out to us. What do we do? Is it time to sharpen the grafting knife?

Morris Towle (1911-1993) devoted much of his life to tracking down and saving old apple varieties. He had been an orchardist in Wilton, Maine, about thirty miles north of Winthrop until the terrible freeze of 1933-34 when he and thousands of other orchardists lost most of their trees and went out of business. After serving in the Second World War, he began a second orchard, this time in Winthrop south of Route 202 on Route 135. He built a small barn for the family to live in temporarily until he could build a proper house. He never did build that house. Instead he and his wife and daughter lived in the barn, and he began to collect apple varieties. His day job was working for the state, but his passion was his orchard. Eventually he had about two hundred and sixty trees on five acres of orchard.

An old Kennebec Journal article featured Towle and his mission to save rare apples. Maybe that was the article Bill read. "Towle starts his summer day at 5 or 6 o'clock and putters around until it's time to report to work. After work at night he's out in the orchard or garden again. As Towle says, 'If they pay off, all well and good. If they don't—well, I'll have had fun.' "

Topworking #3: plunge your knife all the way through the bark and give it a slight wiggle

Towle was a collector at a time when local heirloom apples were not much more than a curiosity to a few folks. He was one of a group of diehards from the Northeast that included Fred Ashworth, Henry Converse, Lothrup Davenport, Ira Glackens, and Wendall Mosher. They were fruit explorers during a time when fruit exploring was not glamorous. Somehow they found out about one another and communicated by letter. The only recognition they received was an occasional article in the local paper. Without them, many of our historic apples would be gone forever.

Not only was Towle a collector, he was also a breeder of sorts. One of his creations was an apple tree he called Sour Sal near the road in front of his house. Another one out behind the house he called Sweet Sal (both were named for his daughter). Though some of his orchard still stands today, both Sal trees are gone. The only mature Sweet Sal tree is the one in Bill's backyard. The medium-sized, roundish fruit has a short stubby stem. The skin color is mostly washed with vibrant purple-pink, overlaid with stripes of deep rusty red and covered with pink dots. It's beautiful. Sweet Sal is all sweet—a true "sweet" apple. The flesh has no acidity. Because it's never tart, it tastes about the same from August to April, although it's at its best in October. Like its seed parent, Northern Spy, Sweet Sal comes into bearing later than most other varieties. Also like Spy, it keeps extremely well. One year when we cleaned out the root cellar on June 1, the Sweet Sals were still firm and looked great. I learned that Towle sent a box of Sweet Sal apples to a large nursery to see if they might be interested in propagating and selling trees, but he was turned down. No one wanted sweet apples anymore.

Morris Towle introduced Sour Sal and this one, Sweet Sal

In the fall of 1996 I was able to meet Morris Towle's widow. Later I met his daughter, Sally Dawson. It was Sally who informed me that there had also been a Sour Sal. A year or two later I sent a young Sweet Sal tree out to her in Ohio.

Sally also sent me a map of the orchard. Unfortunately, by that time the property had been divided up, and many of the trees were gone. When I was finally able to visit the orchard, it was difficult to reconcile the map with the site. A few years ago the bulk of the property was sold, and the new owner, a blacksmith named Dereck Glaser, is working to bring back the trees. He and I communicate with one another every so often, and one of these falls I will return with the map to identify and retag everything I can.

There had also been two Winekist trees on Morris Towle's farm, although both were gone by the time I visited the orchard. Winekist is a dark red-skinned apple with electric red flesh. There are many red-fleshed apples. Most of them are small-fruited ornamental crabs with catchy names whose attraction was never intended to be the fruit, but rather the pink flowers they all have. The best time to locate red-fleshed seedlings, by the way, is in the spring. Scan the roadsides for pink flowering trees. Note their locations, and come back in the fall for the fruit. Some years ago I found one by the side of the road not far from our farm. It was clearly not a grafted tree. I came back in the fall and picked a bag of fruit. While not a dessert apple, they did make a very good pie. I collected scionwood, and we now have a couple of trees started here on the farm. Meanwhile, the original tree got swallowed up in roadside jungle and finally died. Since it was a seedling, I took the opportunity give it its own proper name. The original tree is in China, Maine. With it's red flesh, how could I resist? We call it Red China.

Winekist— it's really red! photo by David Grima

I assumed that Winekist was a third Towle introduction. It was Dan Bussey who gently straightened me out. Turns out the apple originated out west. It's true that Winekist is quite rare, but there are trees still scattered here and there around the country. I had thought that the only surviving tree was in Sidney in Earland Goodhue's orchard. Earland was a friend of Towle's and had received wood from Morris many years earlier. Earland then introduced the apple to me and gave me scionwood. Winekist is quite a good dessert apple, especially for a red-fleshed variety, most of which are too tart or too weird to be good fresh eating.

It was fitting that Morris Towle's orchard was in Winthrop. Of the couple hundred apple varieties that originated in Maine, about twenty-five originated in Winthrop. No other town comes close. Winthrop had the right soil and the right location. It's not far from Augusta and the Kennebec River. Early on the river meant accessibility. Immigrants came up the river, and products went down the river to markets in southern New England and eventually overseas. Two hundred years ago was a good time to be a farmer and an orchardist in Winthrop. Nowadays most of the orchards are gone, and the town has largely become a bedroom community for Augusta. Other early central Maine orchard meccas, such as Wilton, Monmouth and Turner, are farther away from the river and have been somewhat spared from the pressures of modern life. They've done a slightly better job holding onto their orchards.

Winthrop attracted, or maybe it created, a number of apple fanatics over the generations; Morris Towle was not the only one. One of the town's first citizens was Ichabod Howe (1731-1810). Born in Massachusetts, Howe moved to New Hampshire as a young man and then came to Pondtown (Winthrop) in 1768. With him he brought apple seeds and started the town's first nursery. From that planting he selected and named several varieties, including Lambert, Watson's Favorite and Winthrop Greening. It appears as though Winthrop Greening is the only one of Howe's introductions that's still around, now approaching its 250th birthday. It's a large, oblate, multi-colored, partially russeted cooking apple. It is absolutely beautiful and it makes a thick creamy sauce. In Maine it may also be known as Fall Jenneting. The two apples are similar.

Like most early central Maine settlers, Ichabod

Howe was into cider. David Thurston writes in his 1855 *History of Winthrop*:

> The first cider made in the town was from the orchard of Mr. Icabod How, [sic]... They had neither cider mill nor press. But thirsting for a beverage to which they were formerly accustomed as almost one of the necessities of life, but had been now for a long time without, with true Yankee ingenuity they pounded a quantity of apples in a sap trough, and extracted the juice in a cheese press. In this way they obtained a few gallons. All the neighbors (and that included a long distance) were invited to partake of it as a rare luxury.

Icabod Howe was the visionary and catalyst for what was to follow in Winthrop. By the time of his death, several of Winthrop's most prominent orchardists all lived in that same neighborhood just west of Cobbosseecontee Lake. Between 1828 and 1857, twenty more new apples were introduced in town. These guys were a bunch of apple-growing buddies all living on the same couple of roads. They were the creators of a renaissance of apples. Some of their varieties are gone forever, but others still remain.

Moses Wood is another Winthrop introduction that would have been lost forever were it not for Morris Towle and Bill Reid. Moses Wood, the person, was born in 1730. A year older than Icabod Howe, Wood's son married Howe's daughter. Moses was the first of three Winthrop residents with the same name. Either he or his son or grandson, or maybe all three of them were itinerant grafters, topworking seedlings throughout central Maine. In their travels one of them discovered a roundish-conic, light red-striped seedling that ripened in August and made an excellent pie. It also made a highly desirable summer dessert apple if you like them tart. This was the seedling apple that became known as Moses Wood.

Moses Wood always makes a really good tart pie in late summer

Morris Towle found and saved Winthrop Greening and Moses Wood. He passed the scionwood along to Bill, and Bill passed it along to me. It was still early in my career as a fruit explorer. Would I become another baton passer in the continuum? Maybe I would track down some of the other Winthrop apples. I began to learn the names. Someday I would set my sights on Brokedown, Fairbanks, Jerry Brown (not the governor), Never-Equaled, Phoenix, Winthrop Orange and Winthrop Pearmain. For now, however, it was time to leave Bill behind and go find Francis Fenton.

Sometimes the graft is less vigorous than the rootstock; Jill Piecut shows how the graft can double as a bench during a picking break at Cayfords Orchard in Skowhegan

Other times the graft is more vigorous than the rootstock; Angus Dieghan points to the graft of a Williams Favorite tree in Winterport: photo by Abbey Verrier

# Nine

# Francis Fenton

By the method of exclusion, I had arrived at this result, for no other hypothesis would meet the facts.

Sherlock Holmes, *A Study in Scarlet*, p. 84

There was a fourth apple that Bill Reid introduced to me that day behind his yellow trailer in the woods of New Sharon. Bill had learned about this one from a fellow named Glen Harris. Glen Harris lived and farmed in New Sharon, about three miles north of Bill's place, not far from the Starks line, across the Sandy River from Mercer. As Bill described him, he smoked a pipe and farmed by hand. Like all the farmers of that generation, he also had an orchard. According to Glen's great grandniece, Darlene Powers, "He was a farmer all his life. He was also sexton of the village cemetery. I've heard tell nobody dug a more perfect grave than him, almost a work of art. He was very attentive to make sure they were smooth and neat and had a professional finished look. Spoke volumes of his work ethic."

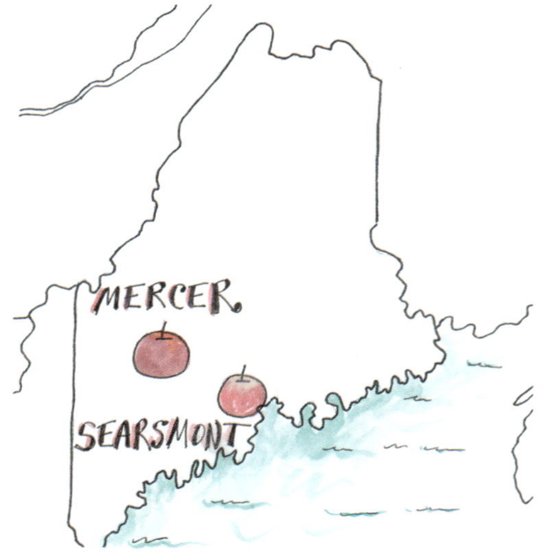


Glen Harris also dug holes for planting apple trees. One of them was a rare local selection he called Somerset of Maine, a variety that originated on the farm of someone named Thompson, a few miles east of New Sharon in Somerset County. Unfortunately I never met Harris. I wish I had. He died in 1997, six months before his ninetieth birthday. I did, however, get to know the Somerset of Maine.

On a snowy afternoon the next winter I was rummaging through some articles I'd saved and I came across one from the Waterville Sentinel about a farmer who lived in Somerset County not far from Bill Reid and Glen Harris. Francis Fenton. Of course I remembered his apple display at the Fair years earlier. Who could forget the apples stabbed onto the nails and the man who rattled on about apples and orchards and life? What caught my eye about the article that afternoon was mention of an apple variety called Thompson. "You probably won't find this variety in any catalog. You will find it, however, growing in Fenton's orchard. It originated in Mercer in the orchards of, who else but, a man named Thompson. It's an early eating apple with very white flesh that tastes real sweet. Children like them, says Fenton, who took his grafts from an old tree in Thompson's abandoned orchards."

Glen Harris: courtesy of Darlene Powers

It was time to meet Francis. I called information and got his number. There was no answer, but there was an answering machine. A gruff voice said, "I can't come to the phone right now." Click. I left a message, but no one called back.

A couple of weeks later I was back at Bill Reid's collecting scionwood. Bill was working in his shop, Cyrus by his side. He was turning tool handles on his treadle lathe. He took a break and played me a couple of tunes on his banjo. We sampled his latest drinkable concoctions, and I left with a cigar he had rolled from his own tobacco. I knew the locations of all his trees by then, and I snowshoed out into the orchard alone to collect the wood. The snow was deep, and I could walk right over the fence. Old Mrs. Sheep was not around that day. She must have been in the barn. The birds kept me company as they fed on the orange peels and bits of bread strewn across the snow below the kitchen door. I cut a bunch of sticks, thanked Bill, said goodbye and then headed out. I wasn't far from Mercer so I decided that I would look for Francis Fenton. I turned off Rte 2 and located the Mercer town office. "Francis is away in California." He'd be returning in a few weeks. I called him again, left a second message and went about the rest of my life.

He did call a few weeks later, in early April. Yes, he did have Thompson which originated in Mercer on the farm of John Thompson, one of the original settlers. He'd be happy to send me scionwood at no charge. The wood arrived in the mail, and several weeks later I topworked it onto a tree by the grape arbor.

John Thompson was a Mainer with two enviable distinctions: he may have been the originator of two apple varieties. One was the variety that bore his name; the second, apparently, was Somerset of Maine. There was not much written about either of them, and what little information I could find was confusing. Were they two varieties? Were they the same variety? Was it a Starkey "deja-vu all over again"?

Somerset of Maine as grown by Glen Harris in Starks, Maine

Mercer Maine was settled shortly after the Revolution, when many southern New Englanders were heading north to find their own "forty acres and a mule." They had fought, and they felt they deserved a piece of land for their efforts. Early settlers built cabins along the Sandy River, which today forms the border between Mercer and Starks. The settlement that became known as Industry Plantation in 1799 was incorporated as Mercer five years later, named in honor of Brigadier General Hugh Mercer.

Mercer was typical of small central Maine communities. In the center of town, Bog Stream was dammed a few miles above the Sandy River. A grist mill was built, followed later by two starch mills, a saw mill, a shingle mill, a joining mill, and a tannery. Mostly though, everybody farmed. The population was forty-one in 1800 and climbed to 1,432 in 1840. Today it's less than half that. Like many other small Maine towns, the population peaked just before the Civil War. The mills have been closed for many years, and the factory buildings have disappeared. Today it's a pretty quiet place.

One of the early settlers was John Thompson. Born in Middleton, Connecticut, in 1784, he grew up in central Massachusetts and eventually headed for Maine. "I had always entertained the idea that I should go into some of the western states where many of my old acquaintances had gone, but it was otherwise determined by an overruling Providence. I had my mind fixed on going Down East," Thompson wrote in his autobiography.

Thompson as grown by Francis Fenton, Mercer, Maine

Thompson was a nineteenth century version of a person from away. Today he'd probably be called a "flatlander" or an "out-of-stater." Had it been 2007, instead of 1807, he probably would have been coming up on weekends. Somehow he found Mercer and established his homestead. Then, despite a lack of roads, he continued to travel back and forth to Boston. This was a hundred and sixty years before the Maine Turnpike. He would grab a schooner in Boston, head for Popham Beach and then up the Kennebec River to Gardiner where he'd disembark and walk thirty-eight miles north to Mercer. Back and forth between Boston and Mercer. In the spring, he brought new plants with him, including the first lilacs and quince in the area. You could write an entire book about either plant and its historic contribution to Maine. Somewhere along the way Thompson also developed or introduced an apple or maybe two. Or maybe none.

A few weeks before Francis sent me the Thompson scionwood, I bought a copy of *The Apples of Maine,* independently published by George Stilphen in 1993. George had obtained a copy of the University of Maine thesis written by Frederick Charles Bradford. "The Bradford Thesis" was well-known to aspiring fruit explorers at the time. The book is an inventory of all the apples grown in Maine in 1911. George edited *Bradford* and then had it typeset and printed. The new version is nearly identical to the original thesis. The typesetting made it more readable, and, for the first time, Bradford's extensive research became available to the general public. As soon as I heard of it, I sent away for a copy from Stilphen himself. I've opened the book a thousand times or more over the years. There are errors and omissions. There are no graphics, and the descriptions of most of the varieties lack detail. Still, it's an excellent resource for anyone interested in the history of apples in Maine and the rest of the Northeast. Many years later I picked up a copy of Bradford's original, which is worth having as well, if you can find it. Now it's also available online.

Topworking #4: whittle the scion perfectly flat on one side

So, by the time I began to sort out Thompson and Somerset of Maine, I had my own brand-new copy of *The Apples of Maine*. I also had S.A. Beach's iconic 1905 *The Apples of New York*. I don't recall how I knew about "Beach," but once I did, I bought the two-volume set. I think I paid $80. It was the most expensive book I'd ever bought. *The Apples of New York* was, at that time, the gold standard for books about American apples. It is not an identification key, and it misses dozens of local and regional varieties no matter where you are. Nonetheless, it represents a detailed inventory of apples grown throughout most of the U.S. in 1900. The descriptions are quite good. The color plates are decent and very helpful. During that time I also obtained two other books, both from the mid-nineteenth cent-ury: Downing's *Fruits and Fruit Trees of America* (1872) and Cole's *The American Fruit Book* (1849). Both would also prove to be very useful over the years, beginning with my efforts to sort out the two Mercer apples.

Beach says that Thompson was "originated by J. S. B. Thompson, Grundy County, Ia." He continues, "It is also reported as promising in the Northern apple districts of Maine." He calls the apple "roundish or roundish oval." This was not a big help. I think he was describing two different varieties as though they were the same apple. As I said, Beach is excellent, but he's not necessarily the go-to source for local varieties.

Of Thompson, Bradford says, "Several apples were grown in Maine under this name." He explains that Thompson may be a synonym of the famous Massachusetts apple, Williams Favorite. It may also be synonymous with Somerset of Maine, which he describes as "resembling Porter in shape and Williams [Favorite] or Sopsavine in color but a much better and larger apple than either." Beach doesn't mention Somerset of Maine at all, though he does describe Somerset of New York. If Somerset of Maine resembles Williams, and Williams resembles Givens, does that mean I'll find Givens growing in Francis Fenton's backyard?

It was also about this time that I made the decision to forgo the rush to Kazakhstan and instead spend the rest of my life focusing on the apples of Maine. Why? Why not? Perhaps it was Bill Reid's inspiration. Did

Francis Fenton with Fancy in about 1998

I really need to go to Kazakhstan when endless adventure was brewing right here? What was the sense in going halfway around the world when all these ancient trees were calling out and no one else seemed to be listening. Get in the truck and head off down any road in town. They're waiting for you.

There were several well-publicized apple exploration trips to Kazakhstan in the early 1990's. These were trips to the forests of apples where it all began. At first I thought I wanted to go. Take me too! I have a nursery and a nursery catalog! Who cares if it's four sheets of paper stapled together in the upper lefthand corner. So what if I have no credentials. So what if I was an English major who took a lot of art and music classes. So what if I don't work for a university. I live on a tiny farm that's not really even a farm, but rather a small hole carved out of the woods. Take me anyway! I waited, but no invitation arrived in the mail.

I was spending a lot of time in the old orchards around Palermo. I was beginning to love those old orchards. Everywhere I looked I was noticing the old trees. The ancient trees. The crumbling trees. The neglected trees. And I thought to myself one day, "Who is caring about these old trees? Does anyone care about these old trees? What will happen if no one cares about these old trees?" These old trees that patiently wait, unpruned, unfertilized, sometimes even unpicked for decades. Have they all given up? Are they all ready to die? Or are they just hanging out with no agenda in mind at all? If no one comes, yes, they will die peacefully without so much as a murmur. But if you listen closely, can you hear them whisper? When you do show up one day, maybe you can hear them say, "You came!"

I knew that "Think Globally, Act Locally" meant something, but I thought it was something *you* should be doing, not me. Other times I thought of it as an excuse for not really doing anything at all. Hey, don't hassle me, I have a garden. I'm recycling my plastic bags. I couldn't afford a plane ticket to Kazakhstan anyway. I'm not even sure what I would have done had I gone there. I was getting to know the trees in the neighborhood. I could now tell a Northern Spy from a Wolf River! I had learned to graft. No one seemed to know the names of any of the trees up on Turner Ridge Road, a mile from the end of my driveway. But I knew a few of them, and I knew I'd learn the rest before too long. And these trees, these ancient trees that were topworked a hundred and fifty years ago by some anonymous grafter of the past, are a gift from the past to us—to me—from someone who never knew that any of us would ever be. These gifts of the past were waiting so patiently for someone to come along and care.

You know that old jazz standard "Back in Your Own Backyard." "You'll find your castles in Spain through your window pane, back in your own backyard." I thought to myself, I love apples. I love the shapes and the colors and the tastes and the trees and the picking and the pruning and the grafting and the older the better. And the stories and the people who own the trees. There are all these old trees all over Maine, and no one seems to be paying any attention to them. They are dying, and they will be gone forever while everyone's gone to Kazakstan. Maybe it's a good thing that I can't afford to go to Kazakstan. Kazakstan doesn't need me. Palermo needs me. Waldo County needs me. Maine needs me. It's all here. I could spend the rest of my life right here tracking down these apples one by one, and I'll never run out of stuff to do.

I had operated under the assumption that most of the old apple varieties were gone. Time, pavement, new houses and shopping malls had done them all in. Meanwhile, to my delight, I kept finding another one or two old Black Oxford trees every fall. One day I found myself wondering why I was locating so many old Black Oxford trees when all the old apples were supposed to be gone. While I used to think that Ira Proctor's Black Oxford was the only Black Oxford left in the state, now I was pretty sure that there were a few hundred Black Oxfords scattered around central and southern Maine.

So, if I'm finding all these Black Oxfords, why am I not finding Franklin Sweet, Fletcher Sweet, Marlboro and Briggs Auburn? Then it occurred to me—maybe they were all still here too. I just couldn't see them

220-year-old Black Oxford tree in Hallowell, planted in 1799: trunk folded to the ground with one branch forming a new trunk; still bearing every other year

yet. There is no apple in Maine like a Black Oxford. An old tree full of Black Oxfords looks like a huge plum tree or a tree filled with black croquet balls or maybe even a billboard in Times Square. Once you see one you won't miss another one. They're not super common, but they're also not extinct. At least, not yet.

A stem of Black Oxford and a stem of Ribston Pippin each topworked about four feet high in Kennebunk—wow!

If old Black Oxford trees can still be found today, might it follow that some of the other old apples are out there as well? These would be the ones that were of comparable popularity in their day to Black Oxford. Taking into consideration variations in longevity, it would make sense that a bunch of other varieties were still out in the landscape somewhere. What about the varieties that were not as popular as Black Oxford a hundred and fifty years ago? This could be another three to five hundred varieties in Maine alone. Do we assume that they're gone? Unlikely.

I've seen about twenty old Black Oxford trees over the years. You can assume you won't find them much north of Bangor; they aren't extremely hardy. So I focus on the southern half of the state. Although I've done quite a bit of fruit exploring, I've only seen a small fraction of a fraction of the state. So if I've found twenty old trees, there could still be many times that number out there somewhere. I'll be conservative and say twenty times what I've seen. That would mean there are four hundred old Black Oxford trees left.

Now let's consider the lesser known varieties, only known in a town or two. Let's say Variety X was a twentieth as popular as Black Oxford. Having determined that there are at least four hundred old Black Oxford trees still alive, we should be able to say with confidence that there are at least twenty each of these other old rare varieties still around. And, if Variety X was a tenth as popular or a fifth as popular as Black Oxford, that means that there should be even more of them still alive.

The challenge is that most of the named varieties and most of the unnamed seedlings are red, yellow, or red and yellow. Although a couple of other apples look vaguely like Black Oxford, it's pretty safe to say that if you find a black apple in Maine, it's a Black Oxford. If you find a red and yellow apple, it's a different story. The possibilities become almost infinite. While many of those other old apples are quite likely still out there, that's only half the equation. That's the good news part. We're going to have to find a way to sort through a lot of red and yellow. It's pretty easy to recognize Django Reinhardt after hearing him play guitar a few times. But can you identify Al Casey? Yes, but it takes more listening.

Those other guys have Kazakhstan covered. I've got my work cut out for me here. So I gave up all plans, hopes, desires and dreams of ever going anywhere much beyond Maine. A few years ago I got as far as the orchards of the other Somerset County, the one in England. And it was wonderful. I've also been to the orchards of Normandy and Northern Spain. But for the most part I stay in Maine and do my searching here.

I decided I was finally ready to meet Francis Fenton. I would get to know him, he would become my mentor, and I would figure out this thing about Thompson. Maybe it's not a variety at all. Maybe it's just a synonym of Williams. Maybe it's a unique variety. Maybe it's the same as Somerset of Maine. And maybe that means that Williams and Somerset of Maine are the same apple. Is this algebra or fruit exploring?

By mid-September each year, collecting apples for the Common Ground Country Fair display was in full swing. I'd wait as late as I dared to give them time to ripen, but at a certain point I had to get out there and start collecting. That was the perfect time to go meet Francis. It was one of those clear blue fall days when the maple leaves are orange and the fruit clusters on the sumacs are red. The best days of the year. I found the Sandy River Road and made the turn. A mile later I was driving through the center of an orchard. There was Francis, and there were the trees all covered with apples.

Sure, he would help me collect apples for a display. No, he would not allow me to pay him anything for the apples. Off we went, he ahead and me behind. We didn't walk slowly. We had a lot of ground to cover. He identified the varieties as I grabbed two or three of each, scribbled the names on brown paper bags with my Sharpie, and then stuffed them into my pack basket. It was all I could do to keep up with him. He was eighty, and I was forty-five.

Francis' father planted the orchard in 1906. Of the original trees, about forty remained. These included a Mac, a Pewaukee, a Ben Davis and about thirty Wealthys. Francis left Maine for California and the Navy in 1940, and when he returned home thirty-two years later, the orchard was grown up to white pine, and the old farmhouse had all but collapsed. He renovated the house and barns in the 1970's. He cut the pine out of the orchard and rescued as many of the old apple trees as he could. He then filled in the gaps in the orchard rows with young whips, many of which he grafted himself using scionwood from the ancient trees on surrounding farms. One of them was Thompson. There were also Blue Pearmain, Scott's Winter, Northern Spy, and Fallawater, or "Fallswater," as Francis called it. By the time I met him, he had four hundred apple trees of more than forty varieties as well as an assortment of pears, plums and other fruit.

Over the next twenty years Francis and I spent hundreds of hours together. One year I stopped by on my birthday, which conveniently falls at the height of apple season. He dropped everything, and we spent the rest of the day labeling every tree in the orchard. He pointed and called out each name, and I wrote them on aluminum tags as fast as I could. I would tack a tag to the tree with a shingle nail and then race on to meet him at the next. He had one of those long-handled pole pickers, the kind with the small cage at the end of a long stick. I'd always hated those things, but in Francis' hands the pole picker was like a lacrosse stick. He would reach up into the tree, and with a flick, a whisk, a flash and a twist the little basket cage would have five or six apples in it, all in perfect shape. He had easy access to apples way out on some limb that no one could ever reach from a ladder. How had I lived without this tool! I bought one immediately. Ever since then it lives in my truck from August to December. This tool is magic. Over the years I've gotten pretty good at grabbing apples from the most inaccessible branches, although I'll never be as good as Francis. I can leap from my car, grab a couple apples and be gone before you dash out of your house and yell, "Hey, those are mine!"

Did a more generous orchardist ever walk between the rows? Orchardists are, as a group, a very generous lot. But Francis was right up there at the top in that department. Many a display at the Fair has been populated with apples from his Sandy River Orchard. And many a tree grafted in central Maine was grafted with his scionwood. I visited him and his dog, Fancy, whenever I could. He loved to talk about his adventures in the Pacific during The War. He loved to play the saxophone. He loved Fancy. He attributed his spunky energy and his longevity to his daily lunch of ice cream and hot apple sauce.

When Francis was no longer confident in his own grafting ability, I grafted for him. I never needed to do any chain-sawing though. He was still cutting wood and fixing things around the farm right up to the end. "What's that sound?" Cammy asked when we were out in the orchard. "Oh that's Francis and his chain

Francis at age 95: still a master with the pole picker

saw." On May 11, 2015, he died in the room in which he was born, missing a hundred by only two months. By that time he was likely America's oldest orchardist. He also had one of Maine's largest apple collections. But he wasn't just a collector; he grew the varieties and knew them well. And he had his opinions. He would wax poetic about some and admonish others. "That apple ain't good for nothin'," he'd grumble. He was a man with a great deal of knowledge, an infectious love of life, a wonderful sense of humor, and a passion for apples and for people. And he loved to talk.

As interested as I was in his life and his orchard, he became equally interested in my various projects and activities. Soon he became a Fedco Trees fan and customer. He attended many MOFGA events, especially the Scion Exchange and the annual fall Fruit Swap, later to become the Great Maine Apple Day. For a number of years he appeared at the Fedco Tree Sale in the spring. We would give him a name tag and make him an honorary employee for the day. He would hold court and sell trees. Customers would flock to him. He never stopped planting trees either. One year, when he was about ninety-five, I saw him standing in line holding a tree he intended to purchase. (He always insisted on paying.) I heard him say to the stranger next to him, "I always wanted a Red Astrachan."

Although Francis slowed down over the years, nearly every time we visited the orchard he was out there mowing between the rows or picking fruit or conversing with another customer who stopped by the farmstand to purchase a peck of Wealthys or Wolf Rivers or his famous Dollie Delicious. If not in the orchard, he'd be puttering around the barns with his chain saw. There was Francis, late nineties, working on some project.

Another perfect Wealthy from Francis Fenton's orchard

Francis' favorite apple was Wealthy. It is an excellent apple. He always said it was the best there was. His hundred-year-old standard Wealthy trees are still putting out large crops of beautiful fruit. Not only that, but the trees themselves never seem to get too big, even on old-fashioned, standard seedling rootstock. Wealthy is exceptionally hardy and prolific,

Wealthy is all-purpose, hardy, roundish, striped, with a deep cavity and deep regular basin: photo by Abbey Verrier

truly an all-purpose fruit. It makes a fabulous pie and wonderful hot applesauce, especially with ice cream.

So on that day in 1995, Francis and I began our twenty-year adventure together. I now had at my disposal a beautiful collection of heirloom apples that I could pick and photograph and set on my counter to stare at and put in my various displays. And one of them was this apple he called Thompson.

Oddly enough, Francis did not have Somerset of Maine, though he was well aware of the apple. So I went to Bill Reid's later that day and collected them there. Bill had stopped spraying, and the lone sheep was gone. The trees were looking a little more neglected each year. I could usually find a few fruit to show, but that was about all. Still, his small orchard continued to be a great resource until my own trees began to bear. For me, it has always remained a model of a vision fulfilled.

By this time I was topworking each new rare variety I found onto my own trees. There wasn't much room in our three-acre clearing in the woods. I didn't want to be an orchardist. I was becoming a preservationist. I didn't want to use up all my time and space taking care of trees. I just wanted to save the varieties I was discovering. Many of the trees I found were neglected, exceedingly old, or both. The best way to insure their survival was to graft a branch back at our place. Before long, most of our trees had six or eight or ten varieties grafted onto them. I all but gave up the allure of the modern, highly-flavored dessert fruit. If a variety was being grown around the country and it didn't appear on the unofficial apple endangered list, I resisted the temptation to graft it. Why bother? With limited space, I made the decision to focus on the indigenous, endangered heirlooms.

Having the varieties there in our yard also enabled me to generate scionwood for grafting new trees to be sold or given away in the future. I assumed that I would get fruit at some point, but that was a low priority back then. Finding fruit elsewhere for us to use was pretty easy. Growing for production at home could wait.

It was also about that time that I first visited Chris Holt in Mercer, not far from Francis' orchard. Chris has a picturesque old farm overlooking North Pond, one of the Belgrade Lakes. He also has several ancient apple trees, one of which he called "Somerset." I met Chris through Denis Culley. Denis was a homesteader I met at one of the early Scion Exchanges.

The Scion Exchange, and its fall "sister" event, The Fruit Swap, were both creations of Jack Kertesz, a fanatic fruit explorer, forever on the lookout for anything edible in the landscape. He invented an organization he called the Maine Tree Crop Alliance. MTCA put out a newsletter for several years that Jack wrote and printed. MTCA sponsored the annual Scion Exchange first in a small state office and then in a church, both in Augusta. After that, the Scion Exchange moved for a number of years to Unity College; finally for the last fifteen it's been held at MOFGA in Unity. It's a fantastic event.

Jack Kertesz and Denis Culley at the Scion Exchange 2014

About twenty of us attended that first Scion Exchange. Now we're up to several hundred. There are workshops and demonstrations, rootstock and grafting supplies for sale, seeds of all sorts and, of course, piles and piles of scionwood laid out on long tables that stretch down the hall. Everyone is encouraged to bring scionwood and seeds to swap. It's all free. You bring what you can and take what you want. But you do have to be careful. You have to hope that what you're taking is correctly labeled!

The Scion Exchange is held on the last Sunday in March when everyone is ready to do something besides put wood in the stove. A few years ago it was snowing wildly all day. Some people assumed it was cancelled, and they didn't show. Not so—it was the biggest crowd ever. Almost every year Mark Fulford gives his now-famous grafting demonstration. We schedule him last in the day because his hour-long class has been known to stretch to two hours or more.

Shortly after meeting Denis Culley, he and I spent an afternoon in his wagon riding up Beach Hill to look for Somerset of Maine at the farm of Arthur Johnson. Denis knew many of the oldest trees in town. He also knew Francis and he knew Chris Holt. The day I went to Chris's to get apples for the Fair display, I left with what he thought were Somersets of Maine. They were very different from the Somersets I'd

The Scion Exchange in 2018; the scionwood is all free to take

picked at Bill's. As I recall, Denis himself didn't think Francis had Thompson. Another friend and apple grower, Warren Balgooyen of nearby Norridgewock, thought that the Thompsons around town were all Williams Favorite.

I decided to assume for now that Bill Reid did have Somerset of Maine. And I decided to assume that Francis, Warren and Chris all had Thompson. This would make Somerset of Maine its own variety. Perhaps Thompson was the same as Williams Favorite, while not the same as Somerset of Maine. Maybe when John Thompson sailed up the coast of Maine and walked the thirty-eight miles from Gardiner to Mercer, he brought scionwood—or maybe a small tree—of Williams Favorite. No one would have known who Williams was. Williams who? But they would've all known Thompson. Oh, of course, *his* apple. This was two centuries before patents and trademarks and club varieties. Glamor was not part of the equation. It was more useful for apples to have names like Thompson or Somerset or Starkey. Names such as Cosmic Crisp, Honeycrisp, SweeTango, and even Red Delicious were many decades away. One of our favorite apples is an old Winchester Connecticut variety that was commonly grown in central Maine a hundred years ago. It's called Hurlbut. No one's trademarked "Hurlbut" as far as I know.

Bundles of scionwood line the tables

A selectman of Searsmont named Bruce Brierley introduced me to Hurlbut, in 2008. Bruce had been fruit exploring all his life, from what I could tell. He contacted me that fall and invited me to come visit an old Hurlbut tree. We met at his farm on Route 3 and drove off to the foot of Appleton

Ridge, not far from Ira Proctor's. When I first saw the tree rising above the goldenrod, I was disappointed. This was not an old tree. It must be a seedling. But after wading through the weeds, I saw that the tree was ancient. It had split in half, laid down on the ground for about twenty feet and then had risen up like Phoenix from the fire into a fresh young tree. It was awesome. That day I began a longstanding appreciation of Hurlbut, as well as a friendship with Bruce. I visit him at least once a year. He's always got something to show me. Recently he discovered an old tree that could be the long lost Hayne's Sweet.

Bruce Brierley at the Fedco Tree Sale, 2015

You could study the names of the old apples for a lifetime. Who was Haynes? Who was Hurlbut? The names tell many stories. Some are pretty obvious. Starkey, for example. Some are tongue-in-cheek. Joseph Taylor, a stonemason and orchardist in Belgrade, selected and introduced a dozen or more varieties during the mid-nineteenth century. I've been attempting to track them down. I think I've found Childs and Zachary. Childs is a local family name. But what about Zachary? Turns out Joseph Taylor introduced the apple in 1849, the same year Zachary Taylor became president. Zachary Taylor. Joseph Taylor. Cute.

Apple names were like folk songs. One name here, another name there. One set of verses here, another set there. One tune here, another tune there. Same with apples. The most famous apple of the eighteenth and nineteenth centuries, Baldwin, had at least ten different names. When I visited a small commercial orchard a few years ago I found several hand-painted signs along one row of trees. The faded signs read "Hall-berts." Boy, did that apple look familiar. Then it occurred to me, ah yes, this is a sort of sanitized version of Hurlbut.

Fall Pippin tree in Coventry Connecticut

The name thing worked in reverse as well. Sometimes a few apples shared the same name. Fall Pippin, Golden Russet and Pumpkin Sweet all come to mind. None of this mattered much back then, since most commerce was local. But for the twenty-first century fruit explorer, it can be a little wonky. Will the real Somerset of Maine please stand up? That evening when I got home after visiting Bill and Chris, I wrote in my journal, "I think Chris has Thompson—not Somerset of Maine, but the jury is still out."

It's wonderful that John Thompson wrote an autobiography. I've collected a few of these obscure memoirs over the years. They can be extremely useful, particularly in revealing what apples were being grown where and when. Much of the work of tracking down and identifying varieties is about the context. How old is this tree? Where is it? Who planted it? Why does Thompson not mention any apples in his writing? Could it be that back then, naming an apple was not worth mentioning? We may never know.

Bradford says, "Several apples were grown in Maine under this name." We'll accept that for the moment but with a grain of salt. Sounds like conjecture. One: Thompson is a synonym of Williams Favorite. If this is true, Thompson would not only resemble Williams Favorite, they would be identical. Bill Reid's Somerset of Maine does *not* resemble Williams Favorite. Remember that.

Back to Bradford's page on Thompson. Two: Several seedlings from Minnesota called Thompson's 24, 26, and a few others were brought to Maine. This sounds irrelevant. We know there was a guy named

Thompson in Mercer growing this apple. Maybe he had Williams Favorite and called it Thompson. It's possible, though unlikely, that he was growing apples from Minnesota with his name. So we'll throw that one out.

Three: Thompson and Somerset of Maine are the same apple. A few things work against this. The spreadsheet of eighty-six recommended apples, first published by the Maine Pomological Society in the 1874 *Catalog of Fruits for the State of Maine* includes separate listings of Thompson, Somerset of Maine *and* Williams Favorite. Glen Harris, who had farmed all his life a stone's throw from Mercer, gave Bill Reid the scionwood for the Somerset of Maine that Bill gave me. That fruit bore no resemblance to the Williams Favorite of Beach and others, nor to Francis Fenton's Thompson. Not only that, but Downing does have an entry for Somerset of Maine in his *Fruits and Fruit Trees of America*, and his description matches Bill's Somerset of Maine and doesn't match Williams Favorite.

Thompson tree in Norridgewock, a few miles from Francis Fenton's orchard; note the vigorous graft

Four: Thompson was planted in 1816 from seed brought to Maine by John Thompson. This is the most plausible, although of course unverifiable, since Thompson had died by the time of Bradford.

Now we look at what Bradford has to say about Somerset of Maine (SOM). Bradford writes that SOM originated in Mercer on the farm of either George Thompson or A. J. Downs, and was first sent to the editor of *The Maine Farmer* by Downs in 1849. This suggests that SOM was not introduced by John Thompson but rather by a man named George Thompson. First names get mixed up now and then, but I'm going to assume that John and George were related but were not the same person. Williams Favorite originated in Roxbury, Massachusetts, in about 1750. Thompson, if it is a distinct apple, originated in about 1815. Here's where I love having access to multiple sources. You get those books all open and piled on top of each other. Flip to this one. Flip to that one. Downing has listings of three Somersets, one of the three being "Somerset, Origin Somerset Co., Me." He describes the apple as "large, roundish, somewhat flattened,...mostly covered with splashes and stripes of bright red." This is Bill's Somerset.

My conclusion was that there are two Mercer apples: one named Thompson, introduced by John Thompson, and a second named Somerset of Maine, introduced thirty years later by George Thompson or maybe A. J. Downs. Somerset of Maine is large, oblate and rusty red. It was dubbed Somerset of Maine to distinguish it from the other unrelated apples of the same name. Thompson is a conical but somewhat chubby, entirely red, Williams Favorite lookalike. Williams Favorite originated in Roxbury about 1750. It was a very popular apple throughout New England and beyond, as far west as Ohio and as far south as Virginia.

Years later when I began to study the USDA fruit watercolor collection, I came across several relevant and revealing paintings. First was of a "Somerset" apple sent in from an orchardist named Hutchins in East New Portland in 1898. East New Portland is in Franklin County, less than twenty miles from New Sharon and Mercer. The East New Portland Somerset is clearly Bill's apple, the one I believe to be Somerset of Maine. The second painting is titled "Thompson or Somerset" from a Massachusetts grower named Hartwell two hundred miles south in Middlesex County, just west of Boston. The confusion made it all the way to Massachusetts. How did the apple get down there? Maybe John Thompson himself brought the scionwood "home" with him. This apple is not Bill's Somerset. It is the apple Francis called Thompson. Lastly there are the seven Williams paintings. Thompson, and Givens, as you will recall, both were said to resemble Williams. Although the Williams paintings are not identical, none resemble Bill's Somerset. Hartwell was growing Thompson (not Somerset) down in Mass, while Hutchins was growing Somerset up here in Maine. While you can't always trust those watercolors, in this case I think they had it right. There are two varieties not one: Thompson and Somerset of Maine.

Somerset of Maine: from the USDA watercolor collection; Reprinted with permission

# Ten

# Briggs Auburn and the Naked Limbed Greening

Now, my dear Watson, is it beyond the limits of human ingenuity to furnish an explanation which would cover both these big facts? If it were one which would also admit of the mysterious note with its very curious phraseology, why, then it would be worth accepting as a temporary hypothesis. If the fresh facts which come to our knowledge all fit themselves into the scheme, then our hypothesis may gradually become a solution.

Sherlock Holmes, *His Last Bow*, p. 875

Briggs Auburn (from Waldo County)

Between 1894 and 1916 about twenty artists painted gorgeous watercolors of fruits and nuts for the USDA. All the paintings are still in good shape and are housed at the National Agricultural Library on Baltimore Avenue in Beltsville, Maryland. There are about 7,500 paintings in all, about 4,000 of which are apples. The collection includes a vast spread of thousands of varieties of apples from many states. The paintings are, as they say on the website, a national treasure.

The entire collection was digitized in 2009 and is now easily accessible online. Any search for USDA and watercolors will bring it up. I refer to it almost every time I work on apple ID's. The paintings are exquisite and detailed, far superior to the plates in Beach and all other images ever produced in the U.S., topped only by some of the classic European pomological paintings.

I first heard of the collection a few years before it became available online. I was deeply into Maine apples by that time and decided to make a trip to see the paintings of my favorite Maine apples firsthand. I caught a ride with a friend down to D.C. in November of 1999. He dropped me off at the library, where I was directed up to the room with the watercolors. An attendant brought out several boxes of the originals, along with a pair of white gloves. I was allowed to flip through them like we used to flip through record albums at the record store in Town and Country Village. They were everything I could have hoped for. Truly magnificent. I pulled out twenty or so varieties that originated in Maine. They color-photocopied them for me. The quality of the photocopies was mediocre by today's standards, but I was thrilled.

I have a link to the watercolor website saved on my computer. Whenever I'm doing IDs, I go the website and bounce around the watercolors, checking out this painting and that. I'm looking at the candidates for each ID I do. At the least, I'm hoping for a little visual reassurance that I'm not barking up the wrong tree.

Briggs Auburn: from the USDA watercolor collection; Reprinted with permission

The website is well designed. You'd never get much done if you had to wade through all 4,000 apple watercolors every time someone gave you an apple. Conveniently written on each painting is the name of the apple, the name of the person who submitted the specimen, the location of the tree, the date of the painting and the name of the artist. If you want to look for the apples submitted to the USDA from New York, for example, you can do so in an instant.

Use the watercolor website with a smidgen of caution, however. Or is the word trepidation? As far as I can tell, no one vetted the identifications. The artists simply used the names given to them when the apples were submitted. Some of the names are what you might euphemistically call "provisional." Some names were apparently invented by the submitter. Some apples were clearly misidentified. You can see some of these goofs simply by looking at all the paintings of a given variety when multiples were done. Other errors are not so easy to pick out. Don't get fooled. It's also true that there can be variations due to where a tree was grown and when the fruit was harvested. Varieties behave differently in different parts of the country. You could have made a correct ID but think you were wrong because the painting doesn't quite match your apple. Ideally, you find a painting of a given variety that was submitted from your region. And you take all of it with a grain or two of salt. Identifying apples, after all, requires accumulating volumes of seemingly unrelated bits of information and then putting them all together.

Several years ago I searched the website for all the apples that were submitted from Maine. There are seventy. My objective was to see if some of the mysteries I'd been working on might be right there in front of me. I was also curious to see the span of varieties that were submitted, as well as where in the state they came from. Although apples were passed around with surprising fluidity a hundred and fifty years ago, for the most part we can assume that you'll find Variety X in a predictable area of a predictable state. Every once in a while we read about some bird that appears hundreds of miles out of its range, like the great black hawk that appeared in Biddeford in August 2018. (It's native range is Mexico and Central America.) Once, I was exploring an old orchard ten miles from Palermo when a green parrot flew right past me through the apple trees and out of sight. Boy, was I surprised. Apples do that too. Even back then, scionwood was sent through the mail or brought back home after a trip to see relatives far away. Where did Ben Davis originate? It's true that you can find varieties way out of their range, but you don't want to count on finding a Baldwin in Virginia or a Red Limbertwig in Massachusetts—or a parrot in central Maine.

Garden Sweet is one of the oldest varieties originating in Maine

Seventy Maine apple watercolors seemed like a mini-gold mine, although I could have wished for seven hundred. That might make my job a lot easier. But despite the fact that there are only seventy, they represent a remarkable cross section of Maine's heirloom apples, and they add another wrench to the ID tool kit. I knew I was on the right track as I examined each of them that day. I had a hunch that I would see some familiar faces in the collection. While it's true that not many apples are as distinctive and visually arresting as Black Oxford, each is distinctive in its own way. The identifier's job is to tease out the recognizable in the subtlety. One of my favorite jazz pianists is Bill Evans. Very subtle. But, once you have listened to his records for a few years, entirely recognizable.

The watercolors have been of particular value in sorting out the apples grown in coastal Waldo and Hancock counties. An orchardist named Charles Atkins submitted ten specimens to the USDA in 1911 and 1912. Evidently Atkins had orchards in inland Kennebec County and coastal Hancock County. While it's not entirely clear from which of these locations the specimens originated, I've been able to come up with a pretty good guess for each of the ten. Several undoubtedly came from the coast, including Wardwell Sweet, Garden Sweet and Prospect Greening. These three are rare. Garden Sweet I'd already found, but the watercolor provided verification. Prospect Greening I'm pretty sure I found after seeing the painting. Wardwell is more complicated; we'll go there later. Suffice it to say, I love the watercolors and, despite their challenges, they are an enormous help to the apple historian and identifier.

JPB and Pete Jenkins at The Fedco Sale, 2017

A history professor from College of the Atlantic, Todd Little-Siebold, was the one who alerted me to Charles Atkins. He and a retired high school teacher named Pete Jenkins encouraged me to ramp up my focus on the many old orchards surrounding the mouth of the Penobscot River, a particularly interesting area for fruit exploration. It was first settled around 1760 with the construction of Fort Pownall during the French and Indian War. Pete has kept up the search mostly in Prospect and inland. Todd's range stretches Downeast along a couple hundred miles of coastline from Prospect to Eastport. Every year they make new discoveries. Sometimes they even provide me with apple names, such as the Northern Greening and Late Duchess, from the old-timers who still remember. Oh, the things that people know. Often they don't think we care. There aren't many of them left who know; when they leave, those nuggets of knowledge will disappear with them if no one grabs the baton. One day when I was out with Pete, he said to me, "I wished I'd listened to my grandfather more."

Todd Little-Siebold hard at work in Lubec, October 2018; Yes, that really is an apple tree!

We found Northern Greening up that way in two locations not far from one another. One of the trees was from the Quimby family in Orrington. The Quimbys gave the name to Todd. They also gave Todd an apple they called Late Duchess and a large, ribbed, unidentified apple I think may be Northern Belle. A second Northern Greening tree stands about forty feet from the oldest existing house in Orland. That apple was brought to me at Fedco's Fall Bulb Sale by a woman named Karen Cote. Northern Greening is a roundish apple that resembles a Mac or a Fameuse in shape and size but is solid green with a brick red blush.

Another time, Todd simply gave me an address on Verona Island and suggested I do a cold call, something I actually enjoy. Verona Island sits between Prospect and Bucksport, not far from Orland. I found the place and knocked on the door. No problem; feel free to check out the fruit. They had about ten trees, half of which were Wolf River. Another was Winter Banana. The rest were a variety I'd never seen before. And what an apple: medium-large, oblate, a swirl of greens and russets with the occasional reddish blush. I took a couple of bags home and began to use them. I was in love again. It was an excellent cooking apple, and also a perfect match for Prospect Greening, one of those Charles Atkins submissions to the USDA. Thank you, Charles.

One of the twenty photocopies I took home with me after my first visit to the Agricultural Library in 1999 was that of a beautiful yellow and green variety called Briggs Auburn. Apparently it originated in about 1820 on the farm of Thomas Record in Minot, part of Auburn back then. The apple was eventually grown and popularized by John and Joseph Briggs, both of whom lived nearby. There is a pretty good description in Bradford that also matches the watercolor, so we can assume the painting's correct. Briggs is medium-large sized, oblate with no ribs, colored yellow and green with some russet, and a slight blush on the sunny side. I was taken with the painting and became determined to locate a tree.

I decided to commandeer the support of Liam Cassidy, a young apple enthusiast from Turner, not far from Auburn and Minot. Liam and a friend appeared at our farm one day in May 1997, looking for leftover scionwood. We had lots. We were on our way out, however, and had just enough time to say hello and goodbye. The two of them spotted the basketball hoop above the double doors on our barn and couldn't resist. We left and they stayed to shoot hoops. When we returned home, they were gone, having taken a lot of scionwood with them. Topworking season was over for us and we were happy to see it go.

That was the beginning of our friendship. Liam and I became fruit exploring buddies. He was probably twenty years my junior, and my first protégé. Turner and Auburn are far from Palermo, half an hour north of Portland. Auburn is one of Maine's larger towns; you could call it a small, post-industrial city. But, despite living a couple of hours apart, we got together pretty regularly. He also wound up working at Fedco Trees for two or three seasons so I saw quite a bit of him there. He liked to call me Thelonius, as in Thelonius Bunk—an honor forever. Eventually, Liam moved to Alaska and I haven't seen him since.

After hearing about my trip to the Agricultural Library, as I hoped, Liam got excited about Briggs Auburn. In the fall of 2001 he did some major snooping around that part of central Maine. He would call me, and I would rush off on one of those spur of the moment trips to look at potential candidates. On October 21, we really thought we'd found it. We were feeling so cocky, we even took along a reporter from the Sun Journal, Cindy Larock.

Cindy wrote a few days later, "And so, last Wednesday, Bunker and Cassidy descended on the promising-looking tree, located on a Durham homestead now owned by Mark Mitchell, who enthusiastically gave his approval…" Sadly, the apple turned out to be just another ancient Yellow Bellflower. Darn.

JPB, Liam Cassidy and Mark Mitchell, October 2001: photo by Russ Dillingham

The Briggs house, Auburn

A couple days later, a fellow named Dan Stearns called me. He explained to me that he and his wife Karen were the current owners of the oldest house in Auburn. It was known as the Briggs house. He had seen the article in the Sun Journal and thought I might be interested. (You never know when one thing's going to lead to another.) He thought that he might have a Briggs Auburn tree. I scribbled down the directions. On November 4, I headed west to Auburn. I had a hard time finding the place because the old country road on the east side of Auburn had been chopped up, dismantled and suburbanized. Sections no longer existed. The modern world had swallowed up another countryside. When I finally found their island of antiquity, Dan welcomed me and took me for a tour of the grounds.

The Briggs house was built by William Briggs in 1797. Briggs was one of the first settlers in the area as part of a land-grant payback for service in the Revolution. The country had no money to speak of, but it did have plenty of property it could dole out in the District of Maine. Briggs and another of the first settlers of Auburn, Samuel Berry, chose lots that stretched from Wilson Pond (now Lake Auburn) to the Androscoggin River on either side of the pond's outlet. Berry had been a scout and guide for Benedict Arnold's excursion up the Kennebec to Quebec in 1775. Both men did well. They each built grist mills on the outlet at what became known as "Berry's Mills," then Minot, and eventually Auburn.

One of William's grandsons was John Briggs (1785-1853), for whom the Briggs Auburn apple was most likely named. John, his wife, Esther Allen, and their family farmed close by in what is now Turner. John didn't discover the apple but had a connection with Thomas Record, another early Minot resident on whose farm the seedling first appeared. Mr. Record passed scionwood of his large green oblate apple to I.T. Waterman who, in turn, passed the baton to Briggs. Briggs brought it to the attention of the Maine Pomological Society in in 1853. The Society came up with the name.

Dan and Karen Stearns and their family had lived in the Briggs house since 1983. The compact white cape and its large attached barn were both still there in all their glory. One look and you knew that this had been the centerpiece for a great farm. But though the house and barn had been restored to their former grandeur, sadly, the farm around it was gone. The house was surrounded by busy roads and strip malls. All that was left of the property was a tiny oasis.

Three old apple trees had also been spared. In the backyard was a russet of some sort; in the side yard was an immense Siberian crabapple tree between the house and the road; and behind the house we found a nearly bark-less old apple tree in a patch of woods and bearing not a single fruit. That was the tree that really interested me.

I returned that winter to collect scionwood from the russet and the bark-less tree. Only one or two sprigs were still alive. I grafted a branch of each onto trees in our orchard. Sometimes when the original tree is close to death or located a long way away, I'll topwork it to a branch at home even if I've never seen the fruit. Not only can I protect the variety that way, I also know I'll eventually see decent fruit even if the original tree dies. I suspected that the old bark-less tree would never fruit again. My only chance to identify it would be to graft it myself. The russet did have fruit although the quality was not great. The branch I grafted at home fruited a couple of years later. I've been calling it Briggs Russet for now—someday I'll do a definite ID. The other bark-less variety took forever to produce. Finally, a couple years ago it did have fruit. What do you know? Northern Spy. A great apple but not a Briggs Auburn.

Earlier that same October, a few weeks before Liam and I went to visit the Durham "Briggs" tree, I had gone fruit exploring by myself on my birthday in a totally different direction. I like to get in the truck and head out for the day as a birthday treat. I prefer to go alone, with no agenda on those celebratory treks. Just looking for the trees and stopping when the spirit moves me. Off I went that day, eventually winding up in the small town of Waldo, just west of Belfast. Waldo is one of Waldo County's twenty-six towns. I stopped by an abandoned orchard I knew well, though I hadn't been back in decades.

The shaker pole, a.k.a. panking pole

The first time I visited the old Waldo orchard was in the fall of 1980, a year after Ira Proctor introduced me to Black Oxford. I was out looking for apples in my VW truck with the flap-down sides, when I spotted the picturesque orchard. I pulled over and parked on the shoulder. From up above the orchard I could see that the trees had a bumper crop. There were no people or houses in sight. Across the road was a small baseball diamond. I knew instantly that this was going to be a great spot. I did a "you-ee" and headed off to find whoever owned the place. I wanted those apples.

I located the right people at a dairy farm a couple of miles away. I found the old man milking in the barn. No, they didn't actually own the orchard. Yes, they did have the use of it and didn't mind me taking all the fruit I wanted. I thanked them and hustled back. Every one of the fifty trees was absolutely loaded. I had my gear with me: shaker pole, two plastic five-gallon pails, a bunch of grain sacks, and some baling twine.

I had made my shaker pole one afternoon out of a thin spruce tree about twelve feet long. I cut the branches close and smoothed it off with a drawknife. Then I took a 60 penny nail, bent it into a u-shape, and attached it to the pole at one end with hose clamps. It was the perfect tool for hooking around a high branch and shaking off the apples. If the apples were ripe, they would rain down, and the tree would be empty in less than a minute. I was happy to collect the apples off the ground. Most of them would get pressed into cider. I would pick some of the nice ones directly off the tree for the root cellar. The rest could be dinged up and bruised. Recently I read somewhere that in England they've been making "shaker poles" for a few hundred years. They call them "panking poles." I didn't know it at the time, but I'd reinvented the panking pole. I still have mine thirty-seven years later.

Topworking #5: insert scion into the slit

I wiggled through the barbed wire fence and headed down into the trees. I did notice that I was sharing the orchard that day with a small herd of young Jerseys. They were grazing off in the distance. I climbed a tree, shook it down, and began to collect the fruit, taking everything except the ones that landed directly into a cow patty. I was in heaven. I was also not alone.

The curious heifers noticed my arrival right away and scampered over to say hello. Before I knew it, they were gobbling up the apples faster than I could scoop them up off the ground. I was racing around grabbing apples. The cows loved the game. This was not going to work. I headed off to another part of the orchard, and of course they came along with me. It was like one of those dreams when it's right there in front of you but you can't have it. The trees were so loaded! Finally, I had to give up. I headed back up the hillside and sat in my truck. So many apples waiting for me, but all of them beyond my reach. Eventually the heifers wandered away, and I tiptoed back. I shook down a tree and this time they didn't see me. Or maybe they'd had enough apples for one day. Before long the truck was full, and I headed home.

These days, grazing cattle in the orchard is a faux pas, especially anytime near harvest. And shaking apples down and collecting them from between the cow patties has become a criminal act. But back in 1980, Escherichia coli O157:H7 hadn't yet been "invented." No one seemed to mind, and none of us ever got sick. I confess that I recall even wiping a bit of cow poop off an apple here and there and then tossing it into the grain sack with the others.

But E coli is dangerous. I no longer lift apples directly from the patties. If you're making fresh cider you'll never have a problem if you pick the apples off the tree. Unless your cows climb trees. Fortunately, for those of us who ferment their juice, all reputable studies show that the fermentation process kills E coli.

Anyway, back to 1980. The picking was great. I returned a week later with friends, and we collected a lot more fruit. The heifers weren't home that day, and we filled two or three trucks. The apples were perfectly ripe. No need for the panking pole. Climb up a tree and position yourself near the trunk, grab hold of two big branches, tilt your head towards the ground and then shake. The apples come thundering down all over you. In twenty seconds the tree is bare. Keep your head down—otherwise you might wind up with a black eye and a fat lip. Then down you go to the ground, filling those plastic five-gallon pails with apples as fast as you can. Three buckets fill a fifty-pound grain sack. Fill the bags, tie them up with a piece of baling twine, and lean them against the tree trunk. Then off to the next tree.

The Green Monster tree has an odd spiraled trunk; it may be a seedling

Those really good picking days were an adrenaline rush. Yellow ones and green ones and every shade of red. I was pressing a lot of cider then. Maybe five hundred gallons a year. I needed a lot of apples. But I was also beginning to get into the fruit itself. I had learned that apples actually ripened in a progression from summer to late fall. What a surprise! I had also made enough bad cider and enough decent cider to come to the conclusion that the best juice came from the apples that ripened late. I was beginning to notice which apples were on what trees and when they ripened. I was wondering about names, but still with no mentors, I was satisfied to welcome the fruit into my life anonymously.

Two apples in that Waldo orchard were of particular interest to me. In an ocean of rusty red fruit, these were the green ones. The first was large and blocky and had an odd taste. So odd that I wasn't sure I liked it, but so peculiar that I couldn't resist it. Why was this tree in the orchard? It was one of a kind. It was definitely in the orchard and definitely in a row, although it was along one edge. Maybe that was significant. Even the tree itself was unique: low and wide-spreading with a stocky spiraled trunk. Despite—or maybe because of—its peculiar flavor, I grafted it onto a tree at home. I tried to identify it but never succeeded. It's big and green so I called it Green Monster. I suppose it might be a seedling despite being placed as though it belongs there in the orchard.

Waldo Yellow, from the Waldo orchard: could it be Briggs Auburn?

Near the Green Monster tree there were two trees with identical greenish yellow apples. Because they both bore identical fruit, I knew they were grafted. The fruit was roundish-oblate and of excellent fresh quality.

Not knowing its name either, I called it Waldo Yellow. I topworked the Waldo Yellow onto one of our trees at home as well.

I returned to the Waldo orchard with the heifers and the Green Monster tree every few years but never again to fill the truck with grain sacks of fruit. I was making less cider. I did collect scionwood from those two green apples now and then. The few times I happened to drive by in the fall, I'd sneak through the fence, run down the hill and grab a handful of fruit. I'd cradle a small stash in my sweatshirt as I hustled back up to the car. The Waldo Yellow was still delicious, and the Green Monster still weird.

Twenty-one years after that first battle with the heifers, I found myself parked by the side of the road once again looking over the bank into the orchard on my birthday fruit exploration drive. I wondered if the Green Monster and the Waldo Yellow were still even alive. Yes, they were. I stuffed my jacket pockets with fruit, said goodbye to the old trees, and headed home.

I decided I would make a big effort to ID the Waldo Yellow that fall. I'd given up on the Green Monster, though I loved the tree. I think it's a one-of-a-kind seedling. But the Waldo Yellow had to be a grafted tree. There were two of them side by side, and the fruit was too good. It was good fresh. It made a good pie and good sauce. I wanted a name for it.

A week or two later I was working at my desk when I found myself staring at that 1900 USDA watercolor of the green and yellow oblate Briggs Auburn. I went downstairs and brought back three or four Waldo Yellows. Then I knew.

The painting was the spitting image of the apple. This had to be it. I had a name with no apple and an apple with no name. Could it be true? It made sense, except for one detail. If the two side-by-side trees in Waldo were Briggs Auburns, then they were a long way from home. Waldo is ninety miles from Auburn. Apples travel, maybe not as far as parrots, but they do travel nonetheless. I was concerned about the distance, but decided in the end that I was correct and that I had found the Briggs Auburn. At least for now.

During that period I had my radar up for another large, oblate, yellow apple in Waldo County called Naked Limbed Greening, which appears in Bradford. Apparently it was commonly grown in Waldo County but rarely anywhere else. Why was that? There was speculation, but no certainty, about its origin. It may have been a synonym for some out-of-stater, or it may have been from Maine. Every so often the thought came to me that the Waldo Yellow was actually Naked Limbed Greening, not Briggs Auburn after all. It fit the description of both.

Another ten years floated by with many other mysteries to solve. Then came the fall of 2011, a banner year for old apple trees. I love those years when the old trees come out of their slumber and produce huge quantities of fruit, giving us one more shot at identifying them before they desert us for orchard heaven. 2011 was one of those years. At the Fair, David and Alice Haines brought me a roundish-oblate green apple. It was from a small orchard behind their house on one of the oldest farms in Minot. I was stunned. This apple could be the real Briggs Auburn, from where it originated.

Next summer I received this email from David and Alice,

> Alice and I gave you a few Greening apples at the fair last fall. You wrote to us in October saying they might be from a Briggs Auburn. We would very much like to stay in touch with you about this tree, including having you visit it any time. Here is what I can tell you about the tree and its location: Our house is historically a part of Minot. Our property is on the Auburn/Minot town line, 1/4 mile north of Minot Corner. The tree is in Minot. Minot Corner was a thriving community in 1810, with 17 mills on the Little Androscoggin. Our property was owned by one of the original settlers, Benjamin Butler. The house was built about 1820.
>
> When I bought the house in 1970 there were numerous old trees scattered about the fields, a few of which have survived. The Greening is one of the healthiest, and seems to be doing well, although probably could use a better pruning than I have given it. The diameter at ground level is 24". It has two branches 30" up: one is 12" diameter and the other is 16" diameter. It is open on one side from losing a branch sometime before 1970. Despite its age, it has been productive. Last year it produced several bushels. It appears to be fruiting well again this year. From what I know about the history of ownership of our property, it is likely that the tree was planted before 1920. Most likely it was planted by the Butlers, the last owners to actively farm the property as a family farm, likely between 1820 and 1900. We would love to have you visit any time to collect scions, to see the tree, and to advise us.
>
> Regards,
> David and Alice Haines

This was my type of email. These people had done their homework. I wanted to solve this mystery.

I was concerned about the Briggs ID I had done in 2001. I felt good about it in many ways, but I still had questions about the location of the tree. Why would Briggs have shown up in Waldo? I knew at some point I would need to locate a Briggs tree in the Auburn area. Although I'd made a big effort ten years earlier with Liam Cassidy, I hadn't spent much time in Androscoggin County since then.

Briggs Auburn from the David & Alice Haines (Benjamin Butler) farm in Minot

But once again the years drifted by. Or maybe they flashed by. Too many apple

Alice & David Haines' home (the very old Benjamin Butler Farm) in Minot

trees to visit in too many places. I didn't hear from David and Alice Haines again until January 2014 when they sent me a box of fruit from their tree. I opened the package and pulled out thirteen apples. The apples were exactly what I was hoping for. It felt as though I was looking at the 1900 USDA watercolor. Round, oblate, rich yellow, fading into patches of iridescent green with an occasional slight blush. Dots were scattered but prominent, stem medium and angled off to the side. Wow! I think we've found the Briggs. This was fantastic. At long last. I whipped off an email to David: "I think you've got it!"

I keyed out the apple, giving careful attention to Munson's 1901 University of Maine description. The process of describing an apple begs for multiple specimens. In this case I had thirteen apples off the same tree. Lots of opportunity to look right through the differences to the essence of Briggs-ness. No need for conclusions yet. Just look.

It's extremely helpful to have thirteen of these Briggs apples on hand—in fact, the more the merrier. It's important to be able to cut up your apple in two directions. Once you slice it from stem to stern, it becomes difficult to slice it around the equator. You also need to absorb the essence of the variety, in this case the Briggs-ness. Individual apples, even of the same variety, never look exactly the same. Describe one Briggs Auburn in detail, and you might hit it perfectly. More likely you'll be close but wrong in some aspect or another. You might even be really wrong. If you look at a few of the descriptions in a good book such as *The Apples of New York* you'll notice that apples of the same variety vary in subtle or even in significant ways. The cavity can be acute or acuminate, the stem can be medium to long or sometimes short, the core lines can be clasping or meeting. All in the same variety. These are apples, not widgets. One Briggs is more or less round while another might be more or less conic. Does this mean that a Briggs is a conic apple or a round apple? It can be maddening.

We're looking for Briggs Auburn-ness. We're looking for Plato's chair-ness but in apples. On its Plato web page, Gonzaga University writes, "This thing that is common to all chairs—that all particular chairs 'participate in'—is called 'the form of the chair,' or 'chair-ness.' " Every Briggs Auburn is a grafted tree. All grafted trees are clones. Every Briggs apple should therefore be identical to every other Briggs apple, and in certain respects they are. But still, every Briggs apple is going to vary slightly in its own unique way, just like all the chairs out there in the world. You know it's a chair but one might be plastic and another, wooden. Tall, short, wide, narrow: still they're all chairs. The leg is bent on this one. There's a ripple or a scratch in the back of that one. They are all the same, but they are all different. Still, you know they're all chairs. Same with Briggs Auburn apples. They're all Briggs, yes, but each is a little bit different from every other Briggs, even on that same tree.

In appearance, a dozen Briggs are more like siblings than twins. This nose is a little longer than that nose. This hair is blonde, but that hair is brown. He's taller than she. And yet, something tells you that they're all related. With more than one Briggs in front of you, you have a much better chance of teasing out the commonality between them. We want to internalize the essence of what it means to be a Briggs. What makes a Briggs a Briggs. We start with the details. From the details we extract the essence. We internalize that essence as a generalization. Then, when someone brings you their apple to ID, you won't need to pull out your micrometer. Even though you've never seen that particular apple before, you have seen a Briggs. You know a Briggs when you see one. You look at this one and you just know.

As Plato would have put it had he been an orchardist, "There is no Briggs Auburn, only that which is common to all Briggs Auburn—that all particular Briggs Auburn apples participate in, is called the form of the Briggs Auburn, of Briggs Auburn-ness." That's what we'll be looking for when we attempt to describe any apple. Cortland-ness, Baldwin-ness, Northern Spy-liness.

A couple of days later I pulled out the box to stare at them once more. What a treat to have thirteen apples in front of me. The match-up with the USDA watercolor is good, although the calyx tube is larger in the painting than in the apples sliced in half on the table in front of me. The fruit is also more oblate than the painting. I also compared the apples to the 1853 Maine Pomological Society written description. That match is good too. It gives the season as September to November, storing later with care. These are still in pretty good shape in January, but I can see some browning in the flesh. That means they are going by—a good sign. I think we've got our apple. Where's Liam Cassidy? In Juneau or Ketchikan? Maybe I need to contact the Lewiston paper. Where's Cindy Larock? I wonder if she still works there. I wonder if the paper still exists. They're dropping like flies these days.

I head to the root cellar to pull out several "Waldo Yellow" Briggs Auburn apples. I'll re-key them and see how the two match up. I'm not sure if I want them to be the same or not. If they are, then I suppose I can feel good about my work. I'd like to think that I was able to identify Briggs Auburn thirteen years ago, even if the fruit was nearly ninety miles from its point of origin. After all, McIntosh originated seven hundred miles from here. If they are different, do I assume that the one from Minot is Briggs Auburn and the one from Waldo is the Naked Limbed Greening?

It may be that neither apple is Briggs or Naked Limbed Greening, but that's not the ending I have in mind. From further examination, it does appear that the two apples are the same. Naked Limbed Greening received its name in Waldo County, presumably from its habit of producing long sections of

Comparing the Briggs Auburn from Waldo (L) and the Briggs Auburn from Minot (R)

fruitless branches. Orchardists call this "blind wood." I have looked at the two trees in Waldo with that in mind. I think that those two trees are too old for me to come to any definitive conclusion. In our orchard we only have a single topworked branch of the apple, and you really can't tell much about the growth habit of a variety from one branch. However, at MOFGA we have a fifteen-year-old tree grafted from the Waldo Briggs Auburn. Emily and I were there pruning and collecting scionwood in 2016. When we got to the Briggs Auburn tree, I checked it out. After fifteen years it's putting on some size now and developing its form. Both Emily and I noticed long sections of limbs with no small fruiting branches. Could these be the naked limbs? Maybe Naked Limbed Greening and Briggs Auburn are the same variety.

From the literature, it appears as though no one knew where the Naked Limbed Greening came from. This might seem odd but was hardly unusual. As we know, these apples were passed around like folk songs. Names were not all that important. Apples picked up names and discarded them as they traveled. Perhaps someone was visiting the Auburn area several generations ago, tasted the apple and loved it, and decided that they wanted it back home in Waldo County. A friend sent a stick or two of scionwood through the mail. Or maybe one of those itinerant grafters showed up one spring with saddlebags full of scions and set them in for a penny a piece. Yes, I'll take a hundred. Please graft me up some good apples.

When the trees fruited a few years later the orchard owner may have remembered the source of the wood. Or maybe not. Even today we have a bunch of those apples here on our farm. ("Yes, that's the Waldo Yellow.") And after that oblate yellow apple had fruited for a few years somewhere in Waldo County, maybe in that very orchard, someone looked at the tree with its long fruitless branches and said, "We'll call it the Naked Limbed Greening."

Royal Sweet tree, with a fifty-foot spread, at Rollins Orchard in Garland: the watercolor specimens sent to the USDA by James Stone in October 1894 probably came from this tree

Royal Sweet from the USDA watercolor collection: Reprinted with permission

# Eleven

# Ben Davis

One forms provisional theories and waits for time or fuller knowledge to explode them. A bad habit, Mr. Ferguson, but human nature is weak.
Sherlock Holmes, *The Adventure of the Sussex Vampire*, p. 1038

Ben Davis

The Butler Hill Road heads north from the southern end of Minot, and follows the Auburn line. It leads you up Butler Hill past the farm of David and Alice Haines and their Briggs Auburn tree, then briefly dips into Auburn before taking a gentle curve to the northwest where it becomes the Center Minot Road, also called the Center Minot Hill Road, which turns into the Shaw Road then back into the Center Minot Road before becoming the Death Valley Road which becomes the Hersey Hill School Road, then the Simeon Road, and finally the Bailey Road before leaving town and crossing into Turner.

These roads are like the old apples that grow along the stone walls that line the edge of the dirt or gravel or pavement. The same road can have one name, then another, then back to the first, then another and so forth. We call it Naked Limbed Greening here in Waldo County and Briggs Auburn there, and who knows what they call it somewhere else. They needed no labels. Everyone knew. These days the roads are mostly identified with signs at the intersections, but not so long ago, this was not the case. When I moved to Palermo in the early seventies there was not a road sign in sight. In fact, I doubt there was more than a handful of road signs in all of rural New England put together. They were like the apples. Who needed to identify them? Everyone knew the Jones Road was the Jones Road that we lived on, but how could you be sure? Everyone knew but us. We weren't entirely sure what road we did live on. It was a secret society of road names. Unless someone clued you in, you had no idea. How did we find anyone? We were given a post office box, not a street address. We had no street address. No one did. When friends were visiting, we told them to turn south off Route 3 at the house with the scaffolding. It would do no good to tell them to turn on the Jones Road. There was no sign. Besides, the scaffolding was up on the house on the corner for over a decade. It was way better than a road sign even if there had been one.

And I bet we could say that everyone knew that the Butler Hill Road was the Butler Hill Road. And that the Butler Hill Road became the Center Minot Road, which became the Shaw Road. It was only with the invasion of outsiders into the back woods that road signs began to dot the landscape. It was only recently that anyone would actually need help in finding the Death Valley Road. Before the signs, everyone knew. Now every driveway has to have a name. Even the fire department doesn't know where you live anymore. And even with all these road signs and smart phones, we still get lost.

Huge old Ben Davis tree; 32′ high with a 28′ spread; Falmouth, 1999

And who doesn't know a Northern Spy from a Baldwin? Who doesn't know a Tolman Sweet from a Pound Sweet from a Pumpkin Sweet? Labels on the trees? That would be like labeling your dogs or your kids or, even worse, the road you live on. In the "preliterate" cultures of the past, we had to remember everything. Every brain was a Google. We didn't have to write it down. Where did you say you lived? Let me check my smart phone.

Ben Davis orchard in Dixmont

The label at the end of our road was the house with the scaffolding. It was a great label and a perfect sign. Fortunately all the apples are labeled too. Wolf River is written all over the huge, oblate, pinkish fruit with the enormous splash of russet surrounding its stem. That's the Wolf River label. Tolman Sweet is written all over the round, yellow-green fruit with the thin, sharp, dark-green line that extends from stem to blossom. You can even feel that line if you touch it lightly with your finger tip. When you're out and about late in the fall, you might see apples on a tree that don't want to drop. Maybe it's even winter, and they still hang on tight. And the tree appears to be one that someone planted long ago. Check it out. Is the fruit about the size of a Red Delicious but a bit more rounded in the corners? Is it a deep rich, orangey-red, quite haunting and beautiful? Does the fruit look tough and strong and determined to fight off anything that might get in its way? If you pull one down, is it heavy in your hand? Do the apples look like you could shovel them onto a wooden cargo ship, like the 800-ton Escort launched from Round Pond in 1854, and sail them off to England, and they'd still be edible when they got to Liverpool? You just found a Ben Davis.

Ben Davis: russet stem splash; hard as a rock; hangs on the tree well into winter

Ben Davis was an apple that many people loved to hate. It was the Red Delicious before there was a Red Delicious. It reliably bore annual crops of late, perfectly shaped, red fruit on perfectly shaped trees. They may have hated it, but Ben Davis was the orchardist's dream-apple-come-true right when commercial orcharding was taking off in the second half of the nineteenth century. Though it may have tasted like cardboard in November and cardboard in June, no one seemed to care. It made lots of money for lots of people. In 1905 Beach called it "the most important variety known in the apple districts of the vast territory which stretches from the Atlantic to the Pacific between parallels 32 and 42." It was that popular. It was iconic. It had at least two dozen synonyms. It was popular throughout nearly all of New York and New England. Thousands of abandoned trees can still be found from Bangor south. We just located another Ben Davis orchard in Dixmont in the fall of 2015. There are hundreds of these orchards melting away into the Maine landscape, grown up with Pine and Maple and Hemlock.

Why is it that so much that's iconic remains cloaked in mystery forever? Who was Shakespeare anyway? Did he really write those plays? Who invented rope? Pottery? Grafting? Who was Ben Davis, and where did one of the most important commercial apples before Red Delicious originate? J. A. Warder in his 1867 *American Pomology* writes, "Long cultivated by Verry Aldrich in Buran County, Illinois, and exhibited as New York Pippin, which name gave an idea of its eastern origin…" Could Ben Davis be from New York? Thirty-five years later, S. A. Beach punts, "The origin of this apple will probably never be definitely known."

The consensus between writers and scholars and other pomologists has been that Ben Davis originated somewhere down South. It may have been named for a Captain Ben Davis of Butler County, Kentucky, they say. (Could it be that the Butlers of Butler Hill, Maine, went south and settled Butler County, Kentucky?) It may have been named for a Ben Davis who lived in Hamblen County, Tennessee. In *Old Southern Apples*, Lee Calhoun writes that "everyone believed it had originated with his neighbor or his granddaddy." It's like the old TV game show *To Tell The Truth*. "Will the real Ben Davis please stand up?" I assumed like everyone else that Ben Davis was from the South, but I also wondered why would a southern apple be so popular in Maine? If it was from so far away, how did it get here? Could Ben have come north as a twig in the backpack of a returning infantryman after the Civil War?

Tim Fanning: photo courtesy of Bill Haynes

In the fall of 2002 I met a fellow named Tim Fanning. Tim was a blacksmith and a surveyor who worked for John A. Belding in Harrison, Maine, in the western part of the state. John Belding and I would later become good friends through our mutual involvement in MOFGA, although at that time we were only vaguely acquainted. Tim did field work with John, but his specialty was title searches. According to John, "he was meticulous." The reason for our conversation, and later our correspondence, was a search Tim was doing for an old property at the corner of the Death Valley Road and the Harris Road in Minot. A client of John's was considering purchasing a chunk of land on a piece of property known as the Ben Davis place. Tim thought I might be interested in what he had learned in the Androscoggin County and Cumberland County Registries of Deeds, as well as at the Androscoggin County Historical Society. It concerned the origin of the Ben Davis apple.

His story sounded a bit far-fetched at first, but also intriguing. I asked him to send me all he could find. He did. In early January, he sent me photocopies of deeds from 1857 and 1859. "Beginning at the northeast corner of the Ben Davis orchards & running north..." He also sent me a copy of his three-page, handwritten diagram of the ownership of the property, dating back into the eighteenth century. At first glance the schematic looked like something you'd see on a physics chalkboard at MIT, with arrows and circles and more arrows and circles. The name Ben Davis was sprinkled all over it. This was great.

Topworking #6:The bigger the stub the more scions you'll need

A month later, I received a second package of photocopies with a note that began, "I was back at the Androscoggin County Courthouse recently researching an old right of way in Minot. I had some time at the end of the day so I took a look through the miscellaneous Minot files in the Androscoggin County Historical Society Library and Museum. Found a few more pieces of the Ben Davis puzzle."

He had sent me a small gold mine, complete with gold, including an article called *Minot in History* from 1971 and another called "Legend Says Ben Davis Apple Originated in Minot" from the December 6, 1941, *Lewiston Journal*. "As you can see from date," Tim added, "events of the next day would concern people more than local history for several years to come." More like sixty years, but here we were at last back in Minot on the Death Valley Road.

In 1778, Zebulon and Tryphosa Davis came up from Ipswich, Massachusetts, and settled in Bakerstown, Maine, later to become Minot. They began to farm and raise kids and apples. Successive generations farmed the piece for 174 years. Zeb and Try named their fifth son Benjamin. Ben Davis grew up and married Sarah Chandler in 1823. They, in turn, named their fourth son Benjamin Franklin Davis. Ben Franklin and Alice Thurlow then named their only son, Benjamin H. Davis. Somewhere along the way someone in one of those three Ben Davis generations discovered a promising wild seedling apple according to *Minot in History*.

Ben Davis with the basin side up (apex)

"The Davis family developed the Ben Davis apple and the popular demand for this variety for shipping to Liverpool, England, made a flourishing and productive business for farmers...This demand for the Ben Davis continued until the heavy freeze of apple trees in 1933-34, when farmers turned to 'macks' and red and golden delicious which we know today...Other varieties of apples were raised in Minot, but needless to say the Ben Davis was the King of the crop for export trade to bring in the sheaves to help pay the taxes."

Ben Davis, a northern apple? One compelling argument for "down south" is that the fruit is generally said be tastier there. That would suggest a Tennessee or Kentucky native son. Flavor, however, could also be an explanation for why southerners decided to grow a northern apple. If it was from the south and it tasted like cardboard up north, why would northern orchardists ever have planted it with so much zeal? In Downing's 1868 edition of *Fruits and Fruit Trees of America*, there is no mention of the apple. In his brother's revised edition just under twenty years later in 1886 the apple has both a write-up and a line-art graphic. The apple made it big sometime during those twenty years, more likely closer to '68 than '86, giving it time to spread around before the revised edition went to press. Here's the first paragraph in its entirety:

> The origin of this apple is unknown. J. S. Downer of Kentucky writes that old trees are there found from which suckers are taken in way of propagating. The tree is very hardy, a free grower, with very dark reddish brown, slightly grayish young wood, forming an erect round head, bearing early and abundantly. In quality it is not first-rate, but from its early productiveness, habit of blooming late in Spring after late frosts, good size, fair even fruit, keeping and carrying well, it is very popular in all the Southwest and West.

The reference to J. S. Downer and its geographical popularity suggests to some a southern origin. But that logic could also suggest that Red Delicious originated in eastern Washington, when we know it came from Iowa. Let's quote Lee Calhoun from *Old Southern Apples* again: "Because of the sudden and overwhelming commercial success of Ben Davis, belated attempts were made to trace its history and origins. By the time this occurred the apple was growing everywhere and everyone believed it had originated with his neighbor or his granddaddy." Maybe J. S. Downer was somebody's grandfather or neighbor. Downing's description could also be used to advance the side of the North. He says it's "very hardy." He calls the fruit "not first-rate." Perhaps most importantly, he says that the origin is "unknown."

So what if we form a provisional theory and then wait "for time and fuller knowledge to explode" it? The first mention of Ben Davis in Maine appears to have been an article in the Maine Farmer in the fall of 1863. Evidently, in October that year, an apple called Kentucky Red was on display at the Hebron & Minot Farmer's Club fruit show. Was this apple Ben Davis?

Let's say it did originate on the farm of a farmer-blacksmith and apple grower named Ben Davis in Minot, Maine, somewhere between 1830 and 1850, about the time that Ben and Sarah Davis were reaching middle age and Minot's population was reaching its peak. Maine was engulfed in the transition from small subsistence farming to commercial agriculture. The seedling orchards of the past were being cut down or topworked to new varieties. Farmers all over the state were on the lookout for promising seedlings to propagate and sell. Ben Davis was one of those farmers. He found a beautiful, large, brightly colored, wild seedling "natural fruit" that hung on late in the fall. You couldn't miss it in the landscape. It was hard as a rock. The family put them in the root cellar, and they kept all winter. One thing led to another, and before long everyone in Minot, and then the county, and then all over southern Maine was growing the "Ben Davis" apple from Minot.

Then from 1840 to 1850, Maine's population collapsed. Minot's did too. It went from 3,500 to 1,700 in ten years. Where did they go? They went to Downing's "Southwest and West" to find better farmland and new opportunities. Could it be that one of those young people from central Maine packed up a small canvas bundle of scionwood from an apple tree on the Death Valley Road in Minot? Did he or she take it to the new southwest in Kentucky or Tennessee or west to Ohio? I suppose we'll never know. Who did write Shakespeare's plays, anyway? Could it have been Edward de Vere, 17th Earl of Oxford?

Were it not for the curiosity of Tim Fanning, this small slice of apple history would likely have remained tucked away in the pages of the Cumberland and Androscoggin County deeds, and in one file folder in the miscellaneous Minot section of the Androscoggin County Historical Society. No one would have ever known or maybe even cared.

John Belding and I served together on the MOFGA Board of Directors for a number of years. John retired from the board and so did I. We remain friends but live far apart even by Maine standards. We still see each other from time to time, usually from across the room at some MOFGA event or across the crowd at the Fair. Recently we did have a long conversation on the phone. He told me that Tim Fanning had died some years ago. I didn't know. John remembered well the Davis place and the title search and

Ben Davis

Tim's excitement about it. He told me there may still be an ancient orchard buried in the woods across the road from the remaining Davis buildings. "I know there were apple trees there." Maybe it was those trees that first sparked Tim's interest in the place. I'll go visit them this fall. I'll be thinking of Tim.

Tolman Sweet is one of the oldest New England varieties. The trees are long-lived and can still be found in Maine. The flavor is unique. (Apex & basin above, base & cavity below: note the typical slightly green suture line from stem to calyx)
photos by Abbey Verrier

# Twelve

# An Old Orchard in Palermo

Much of what I tell you is no doubt quite irrelevant, but still I feel that it is best that I should let you have all the facts and leave you to select for yourself those which will be of most service to you in helping you to your conclusions. (Dr. Watson)

Sherlock Holmes, *The Hound of the Baskervilles*, p. 726

Tolman Sweet

"Father occasionally received apples for identification. One particular one haunted him for several days. When he finally summoned up enough courage to slice it open and have a taste, he decided that the only kind of 'apple' it could be was in his face, a 'pear.' He still thought whoever sent it was sincere in thinking it really was some odd breed of apple."

Sally Dawson, in a letter to the author, November 24, 1996

You've got to know a cavity from a basin, and which end is up and which is down. And you've got to know if it's sweet or tart and if it ripens in August or October. Axile or abaxile? Furrowed or wavy? All these things are key to making that ID.

But before dealing with the details you've got to know if it's an apple or a pear or a carved up piece of cork. It should be possible to accomplish this by looking at your elephant from a few feet away. Forget about the trunk and the tail and the skin for now. Never mind the core-lines clasping. In the beginning an apple tree can look like any other tree. They all blend together. It's got leaves in the summer and none in the winter. It's a tree. Then one day you recognize them. You observe a few hundred of assorted ages, shapes and sizes, and you know. You may not be sure how you know or why you know. One day you won't need to check the trunk. You can tell if it's an apple tree. You can tell if it's an elephant. Of course that's John Prine or John Coltrane or John Lennon or Johnny Cash.

Eyes can't help but scan the landscape. You see them on the roadside. You see them in the fields. You see them on the lawns in town. They speak to you. Sometimes I can hear them as I drive down the road. I get that sense, and I turn my head just in time to see a big old apple tree disappear from view. One summer, after only four months on the farm, our apprentices had a "find the apple seedling contest." They found three hundred in an hour or so, some only a few inches tall. Once you get them in your brain, can you ever get them out? Maybe yes, but probably not. Either way, having apple trees on the brain is a lot more fun than waking up in the middle of the night with the same tune spinning around in your head like a broken record.

Ross Hannon dousing our well with a willow stick

In 1972, a few weeks after we first moved to "The Land," as we called it, one of the selectmen came to visit us. He walked in down the long driveway. He didn't want to risk his car. His name was Ross Hannon. Ross was elderly. He had farmed much of his life next door to our property and now lived a few miles away. "Let's see if you have any water," he said cheerfully. "I'll just cut myself a willow stick." He stepped over to a thicket of small trees, pulled out his jackknife, and cut a Y-shaped branch from a small shrub. "How does he know that's a willow?" I said to myself. Back then, those shrubby little trees all looked the same to me. For the next two hours, Ross doused for water in the yard of our half-built log cabin. He paced along. The willow tip bent and pointed to the earth. We put down our axes and chainsaws. He held my hand while we each held one "arm" of the Y-shaped branch. Again and again the tip bent and pointed to the same spot. "It's forty feet down," he said.

Four years later we drilled a well right there. It was forty feet down. And after a few years I began to see willows when I looked at the small trees outside the cabin. And then I saw the winterberry and the viburnums. These things have names! It was like magic. You look long enough, and they appear. Same with apples. Someone shows you an apple tree, even a small one only a year or two old, and you begin to recognize them any time, any place. You might say that the fruit is the giveaway, but even Morris Towle got fooled one day, and he'd been growing apples all his life. You don't need the fruit. The tree is enough.

Topworking #7: Adjust the number of scions to the size of the stub

In 2013 we purchased a seven-acre piece of land up the driveway and around the corner. We bought the land from our neighbor, who was born with the unlikely name of Harry Potter. People around town call him Dean. Dean and his cousins own most of the land surrounding us. It was all part of a 600-acre dairy farm owned by two brothers, Dean's father, John, and John's brother, Bob. John and Bob ran the farm together for about forty years until the cow buy-out in 1986. At that time there were still two dairies within two miles of us. But that was the end.

Eventually the farm was split up among the next generation, and Harry Dean Potter wound up with a portion of the property next to ours. A few years later we were looking for a small piece of land with some decent soil to create a nursery. We asked Dean if we could lease an acre. He said yes and we hired a neighbor to plow it up. We put up an eight-foot deer fence, and for the past twenty-five years we've grown nursery stock and assorted vegetables, including beans, squash, potatoes and corn. It's a sunny spot with excellent soil. Although we have good gardens down at our place, we don't have a lot of space. Having the use of an acre of Dean's land was a huge help.

The Nursery, as we call it, is on the edge of a four-acre field, bordered on one side by a short, narrow, discontinued dirt road called Finley Lane. Long ago Finley Lane led to two farms, both of which were abandoned by 1900. Over the decades, the field had partly grown up to white pine, hawthorn, sugar maple and red oak. Here and there were old apple trees, now mostly shaded by the huge pines and oaks that grew up around them. The apples were on their last legs but still hanging in there. The land wasn't ours, and I didn't pay a lot of attention to it beyond the Nursery fence. For a few years we pastured our draft horse, Jules, and later our oxen, Red and Rudy, in the field. I put up barbed wire around the perimeter. We called the pasture "Finley Lane."

That was about it. We had plenty of other things going on. In the early 2000's when I was writing about the apples of Palermo, I poked around and noticed that some of the trees were the remnants of an old orchard that lined one side of the field, parallel to the stone wall that bordered Finley Lane. Other than that, I couldn't help but notice the old apple trees now and then as I came and went. Too bad I didn't have time to do something with them.

Then a few years ago, Dean had quite a bit of timber harvested in his wood-lot down beyond Finley Lane. For two or three weeks the skidders dragged trees through the field up to the Jones Road, right past and in between the old apple trees. Almost as an afterthought, they cut the dozen or so huge gnarly pasture pines in the field, leaving the old apple trees untouched. In Maine, loggers generally don't cut apple trees. They tiptoe around them even with their enormous machines. Maybe that's because there's no good wood in apple trees, or because the loggers like to leave the fruit for the deer, or because the apple trees are sacred. They just don't cut them. But they cut nearly everything else along Finley Lane, and when they finished, the overgrown field was transformed into a ragged orchard. Old apple trees were revealed for the first time in generations. The apple trees looked like a bunch of Rip Van Winkles, just woken up after eighty years in the shade. Old trees, all covered with crooked dead branches like Rip Van Winkle's shaggy uncut hair, yawning and blinking and looking around. What just happened? Sun? What's that?

Sometimes that burst of sun can actually kill an old apple tree. But not down Finley Lane. For the most part, they survived the shock. The field was looking like a field again. I was getting inspired. There were a couple dozen old apple trees. As a good neighborly gesture, we began to bush-hog Finley Lane once a year. We pruned a few of the ancient trees. Some of them had spent too much time in the woods and had gone to orchard heaven. Others looked pretty good. They seemed to enjoy seeing the sun again for the first time since Friday nights meant dances at the Branch Mills Grange, and TV was yet to be invented. A few of the trees looked very good. We thought we might even want to own Finley Lane someday. It made sense. We'd been using the Nursery for twenty-five years.

In the fall of 2012 I sent a note to Dean and his wife, Debby. I told them that if they ever thought about selling the Finley Lane piece, we'd be interested. A few months later we received a call. "Yes, we'd like to sell." In the spring of 2013 we made the purchase, and then Cammy said, "Let's plant an orchard this spring!"

Although it might have been smarter to take a deep breath and a leisurely year to create a plan and prepare the soil, we decided to go for it. Make hay when the sun shines? Plant your apple trees when you find yourself owning a new piece of property? Spring was approaching and we didn't want to lose a year. Time to make a plan, quick. How many trees do we plant, and where do they go?

I had made a basic map of Finley Lane when I was doing my Palermo orchard book ten years earlier. That map was hardly complete, but I could use it as a starting point. My goal was to determine how we would interface our new orchard with the existing trees. Even though the trees were old and tired, we weren't going to cut them down. We wanted them to be part of our new orchard. We would prune them during the next few years. Eventually we'd do a complete map of all the trees and then go about identifying them.

A brown paper bag and a Sharpie are all you need to make a long lasting map

I've made dozens of orchard maps over the years. I do one for nearly every site I visit. It could be three or four trees, and it could be three or four dozen. There's no point in spending all that time identifying an apple and then losing track of the tree. The map will guide you back should you ever return. You might want scionwood.

Stand and look. Pace each way. There are new handheld devices for doing this work. My preferred handheld device is a brown paper bag and a Sharpie or a pencil. If it's a single tree by the side of the road, I'll note the phone pole number or the name of the nearest side road or a nearby business. If I remember, I keep a roll of flagging tape in the glove compartment. Abbey and Angus have an apple they call Triple Pink. Pink flesh? Pink skin? Pink cider? Pink sauce? Actually, it's none of the above. They labeled the roadside tree with three pink ribbons so they could find it when they returned for scionwood the following winter.

Maps can often tell you who planted your tree. Was it a non-human, or was it one of us? This is essential information when doing an ID. Far more important than the cavity and basin and the core-lines clasping. Although there are few absolutes in the world of apples, we know that every apple tree that grew up from a seed is new and unique. These we call seedlings or chance seedlings or wild apples or natural fruit. They are all apple trees from seeds. These seedlings are always unidentifiable. They have no name until you or someone else gives them one. If the tree is a seedling, you cannot do an ID because there is no ID to do. While occasionally people plant nameless seedling apple trees, it is safe to assume that when planted by people, the tree is almost certainly grafted. Grafted trees have names, and named trees are grafted. Planted by a deer or a rabbit or a squirrel or a bear? It's a seedling and has no name. Planted by a human? It's a grafted tree and has a name. The most important question the identifier must ask is, "Do you have a seedling or a grafted tree?" It all begins there.

Although my initial goal was to create a map for planting a hundred and fifty trees that spring in our new field down Finley Lane, I would have been doing all the same stuff if I was identifying the trees in your back yard. Out I go on a warm afternoon in March with some paper, a couple of pencils, and a clipboard, and I start to measure and draw. I pace out everything. A typical pace is three feet. Nature plants apple trees in complex patterns that we may never understand. The old timers typically planted their trees in grids, and they did so with predictable distances between the trees. Very convenient.

The piece of land we bought is roughly four hundred feet wide and six hundred feet long. To the north is the Jones Road, to the east is Finley Lane, and to the south are stone walls. Most of the property is open thanks to the logging operation. There's brushy growth along the stone walls, mostly young hardwoods, including cherry, birch, maple, red oak, hawthorn and white pine. To the west is a chunk of somewhat wetter ground, grown up more thickly into woods. This area is about two hundred feet wide and four hundred feet long. We've got a rim of apples around the perimeter of the field. They love those edges. There are a few apple trees out in the center of the field and a dozen or so buried in the woods to the west. I stand here and there and look at the trees and hope that patterns emerge. I sketch.

Initially, I'm not concerned with the fruit. I'm more interested in the trees themselves. I can learn a lot from them. Until they get pruned and see some sun again, their fruit is likely to be skimpy in quantity and unrepresentative in quality. It's true that wild, unpruned apple trees will bear on their own with no human intervention. 2015 was a great example of that, when every tree in much of New York and New England was loaded. But when left to their own devices for many decades, fruit production will often slow down to a crawl interspersed with the occasional boom year. Their objective is to drop a few seeds now and then for the next generation of trees, not to feed us. Fruit buds form on relatively young branches. Pruning is the best invigorator of new growth and fruit bud production. No pruning means less fruit. The cleanest pruning is done with a saw and a pair of clippers, but bears, porcupines, snowstorms and heavy winds all accomplish the same thing. You often get a great crop of fruit a year or two after a big branch breaks off in a storm or a couple of porcupines have a feast on a summer night.

Unpruned for sixty years; where do you begin?

Having lived not far from Finley Lane for over forty years, I'm pretty sure that I'm the only one who's done any pruning on any of those trees in a very long time. I pruned up there once. I was in my twenties. I had just learned a few basics and asked Dean's uncle if I could prune some trees. He said yes. I didn't know much about what I was doing, though I did have a fun afternoon. Not only that, the old tree I pruned is still there. I didn't kill it! You also might recall that it was down Finley Lane that I did some of my first grafting many years earlier. Alas, I never found a single take.

In the new warmth of that March afternoon, I make a reasonably accurate map, including distances between the trees, location of the stone walls, roads, water features, cellar holes, lilacs, and all the other details—whether they seemed relevant or not.

Initially I won't be concerned about which are grafted and which are from seed. I will do that later. Eventually I'll identify any that were grafted. The map will be very useful as we get to know the trees. We'll number and tag each of them like cows in a feed lot. Not very glamorous, but it will allow us to refer to them and know which one is which. The map will reveal patterns of planting. Those patterns will be useful in reconstructing the layout of the original orchard if there was one. Not only will that help in sorting out which trees are grafted, it may also help to determine when they were grafted. All these bits of information will begin to reduce the pool of possibilities, thus making eventual IDs possible.

So there I am with my clipboard and a couple of pencils and a small notebook. No rulers or tape measures or iPhones. Let the map reveal. It shows a rim of apple trees randomly placed along the stone walls. Any of these trees under sixty years old I'll call seedlings. Because the trees haven't been pruned or cared for in at least six decades, it's also highly unlikely that any have been planted or grafted since then either. Therefore, any trees under sixty go directly into the seedling bucket. Not grafted. Not planted by humans. No names. That was easy. A dozen or so others along the stone walls are older than sixty. Those could be grafted. Take note and get back to them.

A quick word about notebooks. When new apprentices arrive on the farm, we give them a short pencil and a three-by five-inch spiral notebook. They're cheap. I keep one in my pocket all the time. I even have an assortment of dedicated notebooks. I now have one for the Finley Lane Orchard. They can also be a catch-all for whatever you find yourself compelled to write down. There's a guy at work who makes his lists and notes on squares he cuts out of cereal boxes. It's that nice thin cardboard that holds up well. You can even stack and sort the cards. Pencils break, but they're easy to sharpen. I have friends who have special little pouches for their short pencils. This brings up the most important tool that every fruit explorer should have: the jackknife. Years ago my mother-in-law, GG, gave me a small orange Leatherman. It fits in my pocket without being too weird or uncomfortable. It has a decent blade and a nice pair of pliers with a wire cutter. Knife, pliers and wire cutter. What fruit exploring orchardist could ask for more?

There are eight apple trees in the open field. Again I pace these off and add them to the map. Although I still haven't examined the trees very closely, other than to take note of their placement, the map reveals a lot. The trees in the western woods do not appear to have been planted in any recognizable pattern. For now, I'll say that they're wild apples, possibly topworked but most likely not so. The stone wall apples are just that: stone wall apples. The half dozen oldest of these could be grafted. The rest must be seedlings since they age from thirty or forty years ago.

Now to the eight in the center. I sight the trees. Six of them are in two rows, three in one row and three in the other. An orchard! The trees are thirty feet apart, or in thirty foot increments. The rows themselves are thirty feet apart. Thirty by thirty was the most typical distance between trees in central Maine orchards before 1950. Those six must be part of a small planted orchard. Now I pace off and count the "missing"

Renegade One (a.k.a Half Moon) is a typical two-trunked seedling producing small low-acid fruit

trees from that old orchard. From the northern-most tree to the southern-most is roughly three hundred and sixty feet, enough room for thirteen trees. Times two equals twenty-six. Twenty-six would assume that neither row extended beyond the current end trees. In 2004 when I was working on my first book, I guessed then that there had been thirty-six trees in the orchard. We'll leave it at twenty-six for the moment.

During the late thirties the U.S. Department of Agriculture took aerial photos of every town in Maine. Palermo's were done in 1939. Most of these photos still exist somewhere. Ours are at the USDA office in Belfast, our county seat. Your town's aerials probably exist somewhere as well, unless they were thrown out. Waldo County's were nearly trashed forty years ago. These photos are a priceless treasure. They have been a huge help in locating the orchards in Palermo. The detail is excellent. You can often count the apple trees in the orchards. Using the photos, I've been able to reconstruct partially remaining orchards and discover trees I didn't know about. It's like google maps before Mr. Google was even born.

I had counted eight trees in the middle of the field. It's evident that six of them were part of this old orchard. There are two others that don't fit into our thirty-by-thirty grid. What about them? Why aren't they part of the pattern? Is there any chance these renegades were part of the planted orchard?

Renegade Number One is gigantic. The trunk is not massive but the branches spread out more than fifty feet. The height is over thirty feet. This is a really big tree. For the moment I avoid looking for grafting clues. I'll just stick with the placement of the tree. Renegade One is forty feet over from our two-row orchard. That means that if it was once part of the orchard, its row would have been spaced forty feet from the next row. That makes no sense. There's also no other indication that any additional trees were in that "row." It's unlikely that we'd have two rows thirty feet apart followed by a third row, forty feet away. While this is not impossible, it would be so rare that I'm going to call it impossible. Eliminate the impossible! The form of this tree is also entirely different from the six orchard trees. It's wide, arching, spreading and expansive. From a distance, it resembles a red oak, not an apple tree. I write "seedling" on my map.

Abbey in the hollow Renegade Two

Renegade Number Two is about eighteen feet off the pattern. I can't help noticing that this tree is really old. Our six orchard trees are all about the same diameter. The trunk of Renegade Two is twice the diameter of any of the others and entirely hollow. We took several photos of Abbey inside the tree. We lifted her up, and she slid down in. Only a few tufts of branches are left on the entire tree. One look and anyone would say "old." I decide that Renegade Two is not part of our orchard either. It's got to be older than all the other trees on Finley Lane. Most likely it dates back to when the land was first settled. It's one of the oldest trees I've ever seen.

When I'm back at the house, I go to the basement and pull out the gigantic photo montage that I taped together years ago, using photocopies of all the 1939 Palermo aerials. I take off my glasses and stare at the dots in the Finley Lane field. I pull out a metric ruler, which I've been teaching myself to use in recent years. Metrics? They can be useful, I suppose. I stare and measure. Years ago I had thought that the orchard was thirty-six trees. Today I'm thinking twenty-six. Squinting at the tiny dots in the photos, I can see ten trees in two rows. That means four have died since 1939. It appears as though the rows were longer, extending beyond the current "ends." How about two rows of twenty-five trees each? This makes sense. Our farmer decided to plant a small commercial orchard. Not too big, not too small. Something to create a bit of additional income for the farm. How about fifty trees? Perfect.

The very old "Renegade Two" on Finley Lane, 2015 (without Abbey)

As I lay on the floor and stare at the little black dots, I notice that the images of our Finley Lane apple trees are smaller and less focused than those in the other orchards on the Jones Road. All the others are larger and darker than ours. Why is that? Perhaps the camera lens caught the light poorly that day as the pilot circled overhead. That's certainly possible, but here's another thought. The orchards up the road feature varieties common in the late nineteenth century. I knew that. I've found Ben Davis, Stark, Baldwin and Northern Spy up there. Those trees would have been mature at the time of the photos. Maybe ours were not. Maybe this was a young orchard in 1939.

The trunks of the trees in our orchard are large, though not huge. They are hollow but still substantial. I'll make a guesstimate that the orchard was planted in about 1920. That would make the trees only twenty years old at the time of the photos. A sixty- to a hundred-year-old tree on a standard rootstock would be considerably larger than a twenty-year-old. Older trees equal darker dots? A twenty-year-old tree would be a fainter dot? This could be another useful clue.

So where are we? It appears as though we have the remnants of a fifty-tree orchard that was about twenty years old in 1939. Forty of those trees were gone by the date of the photo. We know that 1933-34 was the last great "test winter" during which millions of apple trees died in New York and New England. Over three hundred thousand apple trees died in Maine alone. Another three hundred thousand were rendered useless. These terrible winters come along every so often and inflict tremendous damage. Other recent test winters in Maine were 1856-7, 1904-5, 1906-7 and 1917-8. They're sort of like earthquakes; we're due for another one anytime now.

The 33-34 test winter was only six years before the aerial photos were taken. It's likely that forty of the trees died that winter. They also could have been killed by apple tree borers or been eaten by deer. Of the remaining ten, four more have died since the photo was taken. Six now remain.

Now let's go a little deeper. For some unknown reason after the die-off, no trees were replanted. Otherwise, we'd almost certainly have a few of those re-plants. We have none. The six that remain today correspond to dots on the photo. Why were none replanted? Perhaps because our farmer was middle-aged or older. Maybe he or she died when the trees were young. Maybe the orchardist was a man who was lost during the Second World War. This is unlikely. The trees were gone by the time of the photos in 1939. They probably died in '34. He would have had at least two years to replant before 1941 when we entered the war. More likely, he would have had seven years. Why were they not replanted?

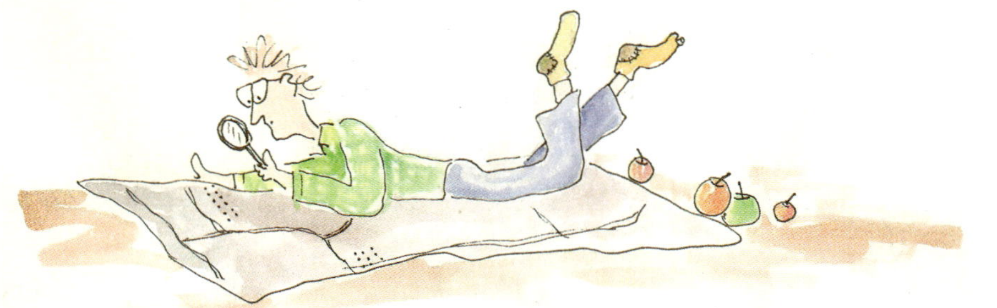

The orchard map is like a magic carpet: get on and ride

# Thirteen

# On Finley Lane

Well, we can adopt it as a working hypothesis and then see how far our difficulties disappear.
Sherlock Holmes, *The Valley of Fear*, p. 790

Let's take a shot at a history of the Finley Lane orchard. A few additional resources are worth mentioning first. These include local histories, old maps, the Maine agricultural census, and the registry of deeds. Our Finley Lane history will be a work in progress. We hope it will be useful in identifying the apple varieties we have. Kinks will emerge in our story. We'll adjust. Some of the difficulties won't disappear. That's okay. We will have begun the process of sorting out what we have. We reserve the right to adopt a new "working hypothesis" at any time. If we get the history right, doing the ID's should be easier.

During the past seventy years Palermo has had two local historians who've written short but detailed and useful accounts of early life in town. You probably have some equivalent wherever you live. Milton Dowe published two pamphlets, *History, Town of Palermo* (1954) and *Palermo, Maine: Things That I Remember in 1996.* Millard Howard published *An Introduction to the History of Palermo, Maine* (1976). From these accounts we know that the first European settlers arrived in 1779. Stephen Belden and his family built a log cabin and began to clear land less than a mile from Finley Lane. Many more settlers arrived after the Revolution. We can assume that the land down Finley Lane was cleared at about that time, and then pastured and mowed throughout the nineteenth century. You don't have to look hard to find the stone walls. And all the while, apple trees were planted from seed.

Topworking #8: a 4 inch stub takes 3-4 scions

Years ago I was able to borrow and photocopy a huge wall-size 1856 map of Waldo County showing every town with all the residents and businesses labeled. Although some names are not easy to read, most are. The map shows that there were two farms down Finley Lane at that time, one on either side of the narrow dirt road. On the east side was "J. Bowler." Our land on the west side was owned by "O. Jones." The name Jones makes good sense since Finley Lane is off the Jones Road and the Jones family owned land in the immediate area for many generations. We don't know what the name of the road would have been at that time. It didn't become Finley Lane until later.

In 1850, '60, '70, and '80, agricultural inventories were taken of every farm in the state. The Maine Agricultural Census is available on microfilm at the Maine State Archives in Augusta. Although they don't mention specific apple varieties, the town by town spreadsheets do include the dollar value of apples sold from each farm. They also include a vast amount of additional information. You can see all the census charts on a viewer at the archive, and print copies right there. Maybe now they're also online.

Years ago I photocopied all four years for Palermo. I pull out a file folder and scan through the 1850 census. There's no sign of an "O. Jones" although there is a Joel Bowler. Perhaps he's our J. Bowler. Then I turn to the 1860 census. The entries were all handwritten on June 22. We can assume that the census takers simply walked or rode on horseback from farm to farm. We want to find a Bowler and a Jones on successive lines. Score! On page 5, entries 6 and 7, we have Oliver Jones and Joseph Bowler. This has got to be them. Jones owned forty acres and Bowler, seventy-five. Jones is listed as having twenty dollars in sales of orchard products. Bowler none. That doesn't mean that Bowler had no apple trees, only that he and his family derived no income from whatever trees they did have that year.

One ancient apple tree still remains on the Jones side of the Lane today. It's the one Abbey climbed into. Most likely that tree seeded itself next to an outbuilding or a fence or some other protected location where it avoided the axe. Seedling apple trees are a bit like cats. They sneak into nooks and crannies and other out-of-the-way places and then they hope that you don't notice them. Unlike cats, however, the trees don't ever move. They just grow. That tree is now large in diameter, completely hollow and mostly broken down. It's the one we call "Renegade Two" for now. It's considerably older than any other tree on the property and probably dates from before the Civil War. Maybe some of Oliver's twenty dollars came from Renegade Two. If Oliver or his wife planted it, the tree would be about a hundred and sixty years old. Wouldn't that be fun?

Since neither Jones nor Bowler appeared in the 1850 census, they both probably moved onto the land between 1850 and 1856, when the map was made. Perhaps they were the first owners, which would mean the land was cleared about mid-century. I then pull out the 1870 census. There's no side-by-side entries Jones and Bowler. There is a Joseph Bowler and there's a Jones or two, but no Oliver. Maybe Oliver died sometime between 1860 and 1870. Maybe he fought down south and never returned. We check "the Adjutant General's list of men who died 1861-1865 in service or in consequence thereof." No mention of Oliver, but Nelson Jones, age 18, is the first entry: "killed in battle." Perhaps Nelson was Oliver's son. So much to learn.

19th century traveling tree "agents" sales book

Both families almost certainly planted small orchards during that time. Except for Renegade Two, those trees are all gone. Over the years additional apple trees randomly sprouted and grew from seed along the stone walls. Some of these may have been topworked. A few still remain, all dating from thirty or forty years after Jones and Bowler.

One day I find myself in Belfast with some time to spare, so I go to the registry of deeds and poke around. Maybe somewhere between the lines the deeds will tell us more. Lots of difficult handwriting to wade through. I dig into Book 206 and find that C.A. Erskine appears to have purchased the land in 1883. I'm somewhat lost but willing to say that Erskine purchased both the Jones and the Bowler farms. "...at a stake and stones and birch tree by the wall..." Erskine then sold both properties in 1889 to John Finley for whom the lane was presumably named.

Finley's daughters, Christina Campbell and Ida Moody, sold the "Erskine Homestead" and the "Bowler Place" to Ada Northrup in 1897. Ada Northrup bequeathed the land to his two daughters, Lila Leavitt and Christine Northrup, sometime between 1897 and 1946. There's no record of a specific deed transfer to the daughters, only that it happened. This, I learn, was not unusual during The Depression. Like many other rural areas, central Maine was hit hard. Official records from that period are sketchy. People disappeared, deserting their land and everything on it.

Meanwhile, I can see my brain adopting a working hypothesis: Fifty apples trees were planted sometime between 1915 and 1929. By that time, Ada Northrup would have owned the land for roughly twenty years. The war to end all wars was over. The economy was doing well enough to trickle down to Palermo. The Northrups were feeling pretty good and the family had a little extra cash in the cookie jar. In about 1927 they purchased and planted fifty trees in two rows of twenty-five each, as a business investment.

Who cares if they're not from Maine, I'll take ten of each! (1. York Imperial; 2. McIntosh Red; 3. Yellow Transparent; 4.Fameuse or Snow); photo courtesy of Jim Arsenault

The varieties would have been the modern commercial apples of the day, bought from one of the "tree agents" traveling across Maine showing off their magnificent trade cards. For over fifty years, beginning about the time of the Civil War, a fruit illustration industry grew and thrived in the Rochester, NY, area in parallel with the emerging Rochester nursery industry. Purchasing fruit trees grown several states away—often of unfamiliar varieties—was a new concept. Until then it was nearly all local. Traveling salesmen, called agents, began to go farm to farm throughout the country showing off their colorful "nurserymen's plates," in hopes of enticing skeptical customers into purchasing their nursery stock.

The planting of the Finley Lane orchard dates from the time when the age of the agents with their books of bright illustrations was giving way to the nursery catalogs familiar to us today. One might imagine some agent appearing in his wagon a hundred years ago on the Jones Road, with a pocketful of some of these very plates in tow. With colors like these, who could possibly resist? Put me down for fifty trees. I'll take some of those McIntosh, Fameuse and Gravenstein. Where's the check book? So a new orchard went in. Then Wall Street crashed. The economy collapsed. The terrible winter of 1933-34 arrived and killed forty of the fifty trees. Northup had neither time nor money to replant. He moved away or died, and his daughters assumed possession of the land. They also never replanted. Maybe they weren't even aware of the orchard their father had put in. They sold everything to Dean Potter's paternal grandparents in 1946 who used the property exclusively for pasture while living nearly half a mile away. Finley Lane and the remaining trees were abandoned. The remnants of the original farm buildings collapsed and disappeared, and no one lived there ever again. Four additional apple trees died over the years, and the remaining six lived to become large and unruly. Those six are all that's left of the 1920's orchard. The large ancient "Renegade Two" and "Renegade One" also survived, as did an assortment of seedling apples along the stone walls and out in the field itself.

Wow, check these out! Even Shakespeare would want these in his orchard!

Until we came along, the only visitors were the deer, the hunters, and the occasional cow. You can still find stretches of crumbling barbed wire. The edges of the field grew in over time. Pasture pines sprouted and began to take over. The apple trees received no care of any kind. Being far from the barns and beyond a wet, woodsy area, even the cows rarely came down this far, although, according to Dean's cousin Bruce Potter, for some years the area was used to pasture the farm's heifers. The apple trees were mostly spared browsing except by the deer. In the 1980's the cows were sold, and the barbed wire rusted and fell apart. The pines and other trees migrated in from the edges towards the center of the field, shading out successive apple trees. Apples don't do well without sun. Some of the old Finley Lane apple trees died, but others fended off the darkness to see another day when a small army of loggers arrived in 2010 to liberate them from their jungle prison camp. They were free at last, back in the sun, just in the nick of time.

In April 2013 we were prepared to replant Finley Lane. We had determined which trees were part of the old orchard. We laid out our new rows using those six trees as a guide. We also used the center "Renegade One" tree as part of one of the rows despite the fact that it's forty feet over from row two, not thirty, as it should be. We'll have a nice big alley between rows two and three. The three remaining trees in row one were still somewhat buried in the woods, so we'll wait for now and plant that row in a year or two. This year we'll plant five rows, numbers two through six. We decided to plant the trees fifteen feet apart within the rows, despite the fact that most of them are on standard rootstock and standards like more space. A tighter placement will allow us to get twice as many trees into the ground. Some might not make it anyway, and if they all thrive, fifteen feet will be plenty of space for the next twenty-five years or so. If the orchard is too crowded at that time, someone can cut out every other tree. I'll be delighted if I'm still around to face that challenge. On April 14 a bunch of friends came over and we planted a hundred and fifty trees in a few hours.

With the planting complete, I could now give some attention to creating an inventory of the rest of the apples scattered around Finley Lane. Time to complete a new map with distances between all the trees. We label each tree with a tag made from a scrap of vinyl siding, hung from a convenient branch with a piece of wire. "Vinyl is final," as they say, and pencil won't fade like Sharpie. We could call them "By the

Gate," "Next to the Large Pine," "Mr. Leaning Over," and "Broken Limb," but we'll resist the temptation and wait until we get to know them. Then we'll come up with catchy names. Instead, for now, we'll just number them. We score the vinyl siding with a utility knife, and then snap it. Then we pierce a hole in each with a jackknife. We write a number on each tag with a #2 pencil. #910 is not very glamorous but it will do for now. Now we can keep tabs on each tree as we observe and learn. Create a map and label the trees. You can do this stuff any time of year. It's OK that we still haven't seen the fruit.

The next big question is: which trees are grafted, and which are seedlings? We're assuming that numbers 1 through 6 are all grafted. They will therefore have names that we should be able to determine. That being said, there are exceptions to every rule, especially in orchards. Maybe the graft failed on one or more of our six orchard trees. Maybe it's a seedling impostor disguised as a McIntosh or a Spy. Recently a fellow named Chris Beach brought me a bitter apple from an old orchard in Dixfield, Maine. He calls it "Allen's Cider Apple," and he hoped I could identify it. He assumed it must be grafted because it was part of an orchard. Although his logic was good, I was skeptical. It didn't taste or look like any variety I'd ever seen. Eventually I was able to look at photos of the tree itself and I was convinced: despite being positioned perfectly in the old orchard, the tree was a seedling.

Why would you find a seedling set in an orchard of grafted trees? There are reasons. Young grafted "whips" are usually set out in the nursery to grow for one or two seasons before they graduate into the orchard. Sometimes the graft never takes on one tree or another. Usually the propagator catches the error and re-grafts or pulls out the offender. Other times, the botched graft gets overlooked. In that case the rootstock grows and thrives ungrafted, looking rather innocently no different than all the rest. After a couple of years, the impostor gets transplanted out in the orchard as though it were a Northern Spy or Baldwin. No one notices until one year it fruits and, lo and behold, the apples are not what they were supposed to be.

You can find these now and then in every commercial orchard. At the Apple Farm in Fairfield there's a excellent example of this. A few years ago Abbey and I were picking fruit one afternoon late in the season. Above us was one of those beautiful mackerel skies. Around us all the Macs had been picked clean—except for one. The apples on that tree were about the size of a McIntosh, but it was no Mac. We tasted it. It was definitely not a Mac or any other common commercial apple. The tree itself was about the size of all the surrounding semi-dwarfs. Abbey named it "Mackerel Sky" on the spot. We picked it clean and used it in our cider that year. I asked Steve about the tree next time I saw him. "It was a botched graft," he said.

Before the days of dwarf trees and large modern commercial operations, orchardists resisted purchasing grafted trees or even rootstocks from away. They preferred to do it themselves using the seed they had. They believed that the best rootstock came from their own land. It was the "terroir" version of apple rootstocks. Homegrown rootstocks from wild seedlings from one's own land were considered to be the hardiest, healthiest stock. The trees grown elsewhere were unsuitable, as far as most orchardists were concerned. They spread their cider pomace in their nursery beds and let it germinate, then they selected the most vigorous seedlings and set them out in the orchard. When these seedling trees were about one to two inches in diameter at chest height, they were sawed off, and two scions were inserted with a cleft graft. If both grafts took, the orchardist would typically snip off the less vigorous one and allow the stronger of the two to take over. Sometimes both grafts were left to grow. When you see two trunks emerging from one base at about chest height, think graft. Even after a hundred and fifty years, those double-grafted trees are easy to pick out once you know what to look for.

Sometimes after the old tree dies, the rootstock sprouts below the graft and forms new trunks

In the case of the occasional seedling in perfect position, perhaps the grafter got distracted or went home for the day and never got around to grafting that particular tree. Or maybe the graft failed. These mistakes are not always that easy to spot. A rootstock bud can pop out right below the graft and grow undetected. It resembles a grafted tree, but it's not. In other cases the graft takes, thrives for a while and then dies for some unknown reason. Maybe a crow lands on the young grafted sprout, and dislodges it. The tree grows merrily away, free to be its own self, new and unique and unnamed forever. It becomes a case of mistaken identity.

Sometimes, after a grafted tree gets old and dies of natural causes, the rootstock makes a seemingly miraculous comeback, sprouts up from the ground and forms a brand new—now ungrafted—tree. Usually when this happens, the "new" tree puts out several sprouts close to the base of the dead trunk. The resulting tree will be multi-stemmed, looking more like a clump of birch or maples than an apple tree. Sometimes the original tree will still be standing, though dead. Other times, the old dead trunk in the center will completely decay and disappear. All that remains is a circular clump of apple sprouts that may each be a foot or two in diameter.

What clues suggest that you have a seedling tree? First is the age of the tree. I always want to know who owned the place at the time the tree was planted. Could they have planted that tree? Sometimes I visit an old farm with apple trees that are not large. It's pretty clear that they date from well after the last farming took place. Could anyone have planted that tree? Often that's all the information you need. If no one was here to plant them, then no one planted them. They are seedlings!

In our case on Finley Lane, we've decided that any tree under sixty years old, or about 12" in diameter, is an ungrafted seedling. Since Ada Northrup never replanted the trees that died in 1934, I'm willing to say

The astonishing Tolman Sweet tree in New Sharon; it's likely over 200 years old

that no grafting was done since that time. Combine that with the fact that maples, pines and oaks were allowed to take over much of the space. It's highly unlikely that any orcharding has been done in the past eighty years. Even being conservative, it's pretty clear that nothing has been pruned or grafted in sixty years. We can safely say that any tree younger than that can't be a grafted tree.

How do you correlate a trunk diameter with age? How do you know that a 12" apple tree is sixty years old? You don't. It's actually impossible to determine the age of apple trees simply by measuring the diameter. They grow at different rates. Some grow fast; some slow. Some never get large, and some get huge. You guess. You see a lot of apple trees; occasionally you learn the age of a particular tree or a whole orchard, and you begin to gain that sense. I am always on the prowl for trees of a known age. Francis Fenton's father planted the original Sandy River Orchard in 1906. Francis knew the date. I've stared at those trees a hundred times or more. I try to embed the look of those trees in my brain. Not necessarily the diameter, which I could measure. It's the look. This is what a one-hundred-and-ten-year-old apple tree looks like.

Now and then someone else shows me a tree of a known age. That's great. Photograph it. Memorize it. Visit it now and then. Sometimes they know who planted it but not the date. If their grandmother planted the trees, do the math. Sometimes they'll have an old photo of the farm with the tree coincidently in the background. Find out when the photo was taken. Other times they know the age of the building next to the tree. If the original dwelling was built in 1840, maybe the trees are that old as well. They're probably not older. They could be younger. All details are relevant.

What about doing a core sample of the tree? Nice idea, but all old apple trees are hollow. Without a core, you can't do a core sample so give up on that one.

A hundred-year-old apple tree sounds like an old tree. It's actually not. A hundred years is pushing it for you and me, but not for some apple trees. Why some? Why not all? Traditionally, named apples were grafted onto seedling rootstocks. These trees are known in the trade as "standards." There's nothing uniform about standards because every seedling is unique. But this method of apple tree propagation was considered the standard, so standards were the standard even if they weren't really standardized. The apples grafted onto standard apple trees will routinely grow to be a hundred. Many will grow to be one-fifty, or even two hundred or more. The Black Oxford tree at the Vaughan estate in Hallowell is two hundred and twenty years old. It looks like an Italian olive tree planted by St. Francis.

Since about 1950, apples have been typically grafted onto "size controlling" rootstocks. These "dwarfs" and "semi-dwarfs" have names like M7, M9, M111, Bud 9, Bud 118, Geneva 11, and Geneva 16. There are dozens of different rootstocks kicking around these days. They are clonally reproduced by the millions, which means they are genetically identical. So when you graft your trees onto a Bud 9, for example, every tree will be roughly the same size.

These rootstocks have many advantages for today's commercial orchardist. Their smaller stature means no ladders, and therefore easier picking, pruning and spraying. They are also precocious, bearing fruit at a young age. But dwarfs and semi-dwarfs have their drawbacks. Being dwarfed, they are not well-rooted. They are often planted on wires like grapes in a vineyard. Every few years it snows in October in New England when the leaves are still on the apple trees and the ground is not yet frozen. Perhaps you've seen photos of thousands of dwarf trees all laying on the ground. When a few start to lean, they take their neighbors with them, like dominoes.

You can still see the graft about two feet above the back of the bench on this ancient Porter tree on Route One in southern Maine

With their wimpy root systems, dwarfs are also more susceptible to drought and soil deficiencies. Most notably, they are not long-lived. This is not a huge issue for commercial orchards; they are used to tearing out blocks of unproductive trees and planting new trees by the hundreds or even thousands. Modern orchardists treat their trees like semi-perennials. The trees thrive for twenty or thirty years, then get pulled out, and new ones are planted. Friends of ours planted a one-acre hillside orchard near Palermo in the mid-1980's on M9, the smallest dwarfing rootstock of the day. This was about the time I planted my first standard trees here on the farm. Their trees grew and bore lots of fruit long before I got my first apple. I was so jealous—why didn't I plant trees on dwarfs? Then, about the time that my trees were beginning to kick in, their dwarfs began to lose their vim and vigor. They looked bad. They lost productivity. They were done. So our friends pulled them all out, and now what was once a beautiful orchard is a field again. No more trees. Meanwhile our tortoise trees marched along, bearing a few more apples each year, until now they are giving us excellent crops. We climb in them. We dance around them. We collect piles of scionwood from them. They won't reach their peak for another eighty years or so. In a hundred years, they'll still be cranking out Starkeys, Black Oxfords, Briggs Auburns and Somersets of Maine.

What relevance does this have to identifying fruit? When you find dwarf trees, you've almost certainly found grafted trees, not seedlings. It is also highly unlikely that the trees are more than thirty years old. Otherwise, they'd be toast.

If our ancestors had grafted their apples on dwarfs, most of the old varieties would be long gone. A standard tree can sit by the sidelines neglected or even totally abandoned for decades until one day someone comes along and decides to prune it up, cut a few sticks of scionwood, and save the variety for a new generation. During long periods of neglect, any dwarfs would have quietly passed away, taking with them their genetics and their stories forever.

Porter arrived in Maine about 1850; yellow, conic-ovate and ripens in fall

Knowing the age of the young trees in your life is just as helpful as knowing the age of the old ones. We planted our first dozen trees in the early '80s. They're now pushing forty. We plant a few more every year. When you see someone else's tree, you'll know its age because you've been watching your own trees. The same goes for kids—when you see someone else's, and you know instantly, "She's about eight months." How does that work? Something about observation.

The trees I began to pick in 1972 looked really old then. They had to be considerably older back then than our forty-year-olds are now. If they were forty then, they're now eighty. If they were eighty then, that makes them 120 now. Every time you learn the age of another tree, absorb it.

Not only can the tree's age help with determining if it's grafted or from seed, it can also help you narrow the pool of possible candidates when you do find a grafted tree. Different varieties got planted during different time periods. Exceptions are inevitable but rare. Learn which varieties were grown when. If it's in Maine and it's really old and it's yellow and conic and ripens in late summer, it could be Porter. If it's yellow and conic and ripens in late fall, think Yellow Bellflower, or as our friend Logan calls it, Belle-Fleur. If it's really old and green and round, think Tolman Sweet.

Anyway, maybe you established the age of the tree, and it took a lot of research to do so. Maybe it only took you a second or two. You looked and you knew. You knew it had to be a seedling because there was no one around to plant it. Or maybe you're baffled. What next? Let's look at the physical configuration of the tree itself. Its form will tell us a lot, maybe even more than its age.

# Fourteen

# Do You Have a Seedling or a Grafted Tree?

It is just these very simple things which are extremely liable to be overlooked.
Sherlock Holmes, *The Sign of Four*, p. 136

Fameuse a.k.a. Snow

When an apple tree springs forth from a seed, it takes on its own unique persona not only in fruit but also in the shape of the tree itself. This is the tree form. We should be able to recognize a seedling apple just by looking at the tree form. Every one of them develops their own unique personality. They may resemble one or both of their parents, but the key is that the shape of the seedling tree will rarely conform to the norm of the cultivated orchard. It will look uncultivated, noticeably out of place in an orchard world where conformity is the priority.

Nan Cobbey's huge multi-trunked seedling apple tree; it might look like it's in Africa, but it's actually growing in Belfast, Maine

A seedling apple is like a musical improvisation. In the world of music, one parent would be the original composition and the second would the musician. When the musician sets aside the printed page and plays the tune extemporaneously, the result is something new. The improvised creation will bear the mark of the musician as well as that of the original piece of music. You may or may not recognize the tune. It may be pleasant to the ear, or it may be discordant. Musical improvisations may be endlessly fresh and inventive like those of Bill Evans, Django Reinhardt or John Gilmore. They may also be stale, cliche and uninteresting. However, whether or not you like them, each will be new and unique.

Same with apples. Some seedlings will be stale and uninteresting, while others will be quite wonderful. Each is an improvisation. It may look a bit like a Mac and a bit like a Spy, but something tells you it's something else.

In the sixties, the Grateful Dead improvised their shows. Each rendition of a particular tune was similar to the original composition, and you knew it was the Grateful Dead, but it was also new and unique. A friend of mine confessed that he stormed out of a Grateful Dead concert in 1970 because they didn't play it "like the album." They never did.

The recorded version of a song is like the apple variety. Like the improv and the seedling it began as a chance. But once recorded, it becomes fixed in time. It's a snapshot. It can now be exactly reproduced on vinyl or tape or CD. Same with the grafted tree. The first McIntosh was just another chance, one of the millions of apple improvs, like a Sun Ra performance of *Space is the Place* or the Grateful Dead's *Dark Star*. But once that Mac was grafted, it entered into a new world, the world of the replicable grafted variety.

Seedlings love to lean out into the sun

Catch the seedling, graft it, plant it in the orchard, and tame it. It then becomes a recording. It becomes a variety. Because it will be primped and trimmed and pruned it will begin to resemble those three-minute recordings that take months to make. Every toot and tap and thump and hoot and shout has been placed and replaced until they are perfect. I love many of those iconic three-minute pop hits, but improvisational they are not. They are nothing like Coltrane screaming away on his tenor sax in 1963 while Elvin Jones, Jimmy Garrison and McCoy Tyner hang on for dear life. Like Coltrane, the wild seedling branches out according to its own plan. It leans and spreads and twists and turns. Like Henry David Thoreau, the seedling marches to a different drummer. Sometimes you won't see a branch for twenty-five feet. Other times, it grows into a tangled bush. The branches themselves tend to be spiny and unforgiving, as though they are saying, "Leave me alone. I do my own thing."

The seedling is an escapee. It knows no boundaries. It chooses to grow where it wants to grow. Seedlings don't like straight rows. They mostly avoid fields unless there's some protection and a bit of shade. They definitely don't like the deep forest; not enough light for them there. They usually don't care much for lawns either. Seedling apples like edges. Like improvs, they're edgy. They love the side of the road, the stone wall, the stream edge, the woods' edge. You can even find them between the boulders by the sea. They like almost any spot where they can have their feet in the forest and their head in the sky. They don't mind leaning. You can see they love to lean out of the forest. They love to check out the action out there in the sun. They can hold amazing postures for decades. They are the ultimate yoga practitioner. When you see a tree leaning over with branches on one side and none on the other, you know you've found a seedling.

Seedlings often have several trunks

Seedling apple trees are frequently multi-trunked. They can disguise themselves as a clump of white birch or red maple. Sometimes you'll find a multi-trunked apple tree with one trunk grafted. Be on the lookout for that variation. It'll be on someone's front lawn. When you come across an ornamental crab in town with pink flowers and white flowers on the same tree, look down. I'll bet it has more than one trunk. Trunk #1 is the grafted crab. Trunk #2 is a renegade white-flowering rootstock sprouting from below the graft.

Grafted trees, on the other hand, have a dignified and civilized look. Many of the most famous varieties were originally selected in part because of their desirable form. They were born conformists. Born of privilege. Selected. Many received their early training in a nursery. Clipped and tended and fertilized. Primped and poofed. Then set out in straight rows, thirty feet each way, and then pruned to fit the mold. Don't step out of line! No branch out of place! Individuality not recommended. They're shaped by a pruning saw. Someone long ago wanted to give them some form in hopes of one day picking the fruit. Even after long periods of neglect, that form should be still visible, even if the branches are gone. Often you must use your imagination to reconstruct the original shape of the tree back in its day. Does it look as though it's ever been in partnership with people? This is about seeing the past in the present. Does the tree look as though someone took care of it? Or does it resemble a maple or the birch?

Typical shape for old grafted trees; tree on right shows "water sprouts" that can grow to be very large

Grafted trees will often have a single trunk with a lower tier of horizontal branches or the remnants thereof. That first tier may be as low as three feet or as high as six feet. Once upon a time each one had a regimented form. Some of those branches may still be there. Sometimes large verticals called "water sprouts" have emerged and risen from the lower branches, distorting the shape of the tree. Conversely, the tree may now have no branches at all. Look for holes in the trunk indicating lower branches have rotted off. This is an old grafted tree.

Old trees will sometimes look like umbrellas or palm trees with no limbs for the first twenty feet, topped off with a few branches and a scattering of fruit only a very tall ladder or a long pole picker can reach. Look at that beautiful fruit that is totally inaccessible. Is this a seedling or a grafted tree? You can have two apple trees that both look like an umbrella yet one was clearly grafted and the other is clearly a seedling. From a distance your tree could be either, but up close it should be possible to figure out. If it has evidence of lower branches, such as large holes a few feet off the ground or rotted broken stubs of branches, then it could easily have been grafted. Having not been pruned for many years, the old grafted tree grows taller and taller. The top growth shades out the lower growth which then atrophies and drops to the ground. You're left with an umbrella of new growth hanging over a trunk that used to support branches but is now stripped bare.

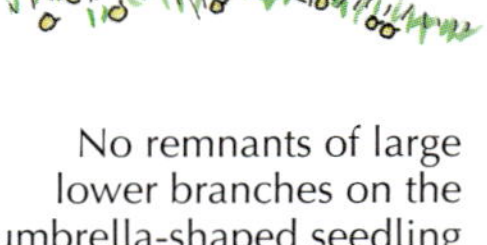

No remnants of large lower branches on the umbrella-shaped seedling

Holes and stubs of large old branches still visible on the umbrella-shaped grafted tree

If, on the other hand, your umbrella has no indication of any substantial branches down low, you probably have a seedling. In this case the tree grew tall rapidly like a pine tree in the woods. The pine tree growing in the center of a field is like an orchard apple. It produces substantial horizontal branches. These "pasture pines" are branchy, picturesque creatures. On the other hand, a pine tree growing in the woods appears to be a totally different species. Shaded by the surrounding trees, it grows straight up towards the light and sheds its branches before they're more than an inch or two in diameter. Forest pines often have no substantial branches for thirty or forty feet. Same with wild apples when they grow in the shade of their own upper branches, or the shade of other trees. In these light-challenged conditions they too shed their branches early on. Branch remnants are typically twiggy and small in diameter. Wild seedling trees develop into an apple version of that clear limbless pine in the forest. Even if the woods are later cleared away, you can still recognize this seedling-umbrella-palm-tree apple because there will be no indication of any old lower branches. If you want fruit, you'll have to do some serious climbing or get out the panking pole.

Grafted trees always have a graft line. Sometimes it's obscure and nearly impossible to see, but it can be visible even after a hundred years. Sometimes the rootstock is more vigorous than the scion. In that case, the tree rises out of the earth, and then dips in dramatically at the graft line. Other times, the scion is more vigorous than the rootstock. In that case, the scion bulges out at the graft line. Sometimes the dip is so pronounced that the graft is like a bench. Other times the bulges and the dips are subtle, just a ripple around the trunk. Sometimes the wood rises like a mountain range where the scion meets the rootstock. Sometimes the bark color varies right at the graft line. Look for the graft.

Old grafted tree: two scions still visible from an old cleft graft

When the rootstock is more vigorous than the scion

How high is the graft above the ground? I have an etching of a French grafter inserting a scion into a cleft graft. It's an old copy of the famous 1865 painting *Man Grafting a Tree* by Jean Francois Millet. He's setting the graft at waist height, probably about three and a half feet off the ground. Three to four feet is probably average, especially for the period before 1900 when farmers were growing rootstock in place and then cleft-grafting them on the spot. Why three or four feet? It's easy to work at waist height, be it a work bench or a kitchen counter or an apple tree grafting project. A cut-off tree of that height will also be visible to you and everyone else on the farm. All the better to see you with, my dear. Less likely to step on, drive over, or mow with your scythe, cutter bar or Lawn-Boy. It will also grow above deer browsing range more quickly if it's grafted high. But waist height was by no means universal. You will find trees grafted up by your shoulders, or down by your feet. So look high, look low, and look in between. Once you find one graft in the orchard, chances are good the rest are going to be about the same height.

Variations in soil type don't seem to matter to the seedlings. In a good year, you can go exploring for wild apples in every town in the state. Aroostook County is a particularly fun place to find them. Aroostook seedlings routinely cover acres of old fields. You can sample hundreds of different apples in a single afternoon. Taste and spit. Taste and spit. Better not swallow if you don't want a bellyache. Down our way, you can find long tangled hedgerow colonies of seedlings along the roadsides, occasionally stretching for a half mile or more. On the coast, they'll grow a few feet from the high-tide line. Apples love freedom.

When the scion is more vigorous than the rootstock

When I come across seemingly random trees that do not fit an orchard pattern, I ask myself if this might have been a place for a cultivated tree. If it's on a stone wall, the answer is "maybe," especially if the tree is old. Stone wall seedlings were sometimes topworked to known varieties. Along a roadside, again the answer is "maybe," especially if the tree is old. Along streams or along the ocean, the answer is more likely no. I have never seen grafted trees in either location. These are your true escapees.

Deeper in the woods, I look for orchard patterns overgrown by the forest. I look for cellar holes and stone walls and other indications that someone lived here long ago. Even in the woods with many of the original trees dead and rotted away, it is usually possible to see the pattern if one ever existed. Once I locate the

Four-trunked seedling in Cape Elizabeth predates this housing development by a hundred years

Fruit from the Cape Elizabeth tree; you won't find it in your local Whole Foods anytime soon

grid, I can then pick out the seedlings that have interspersed themselves over the years among the grafted trees.

I take a few hours and look at the remaining apple trees in our Finley Lane orchard. I stare at each one. Could this be a tended tree? Is there any chance that someone would have grafted this? Can I detect a graft line? Does it look as though anyone ever pruned this tree? I decide they're all from seed, with the exception of the remaining six from Northrup's fifty trees. Next, the fruit.

The Hyk house in Belfast, Maine, built in 1835-36 by Robert White, Jr. The photo is dated 1916. Look to the extreme left and you'll see the top of the apple tree in the background, already as high as the peak of the three-story barn. In 1916, the tree must have been *at least* seventy years old to be that tall, dating its planting to about the time the house was completed: photo courtesy of Christopher and Diana Hyk

The Hyk tree in 2016 from the back yard; in 1999 the circumference was 104", the height was 53' and the spread was 39'. The tree appears to be a seedling (not grafted); it probably popped up behind the barn in 1837 or so. That would make it 182 years old this spring!

# Fifteen

# Gravenstein, McIntosh and a Basque Snowflake

No, no: I never guess. It is a shocking habit—destructive to the logical facility.
Sherlock Holmes, *The Sign of Four*, p. 93

Six trees from one orchard in one town in one state, all planted around 1925. Just knowing the state in which the trees are located dramatically reduces the potential pool of varieties. Knowing the town and the approximate planting date reduces the pool even more. Without examining a single apple, we're able to reduce the pool of choices from more than 20,000 down to less than a hundred. There is always the slight possibility, of course, that whoever planted the Finley Lane orchard was an eccentric collector of odd varieties and that everything we come up with will be wrong. But for now we'll say we're on track. We've eliminated many impossibilities.

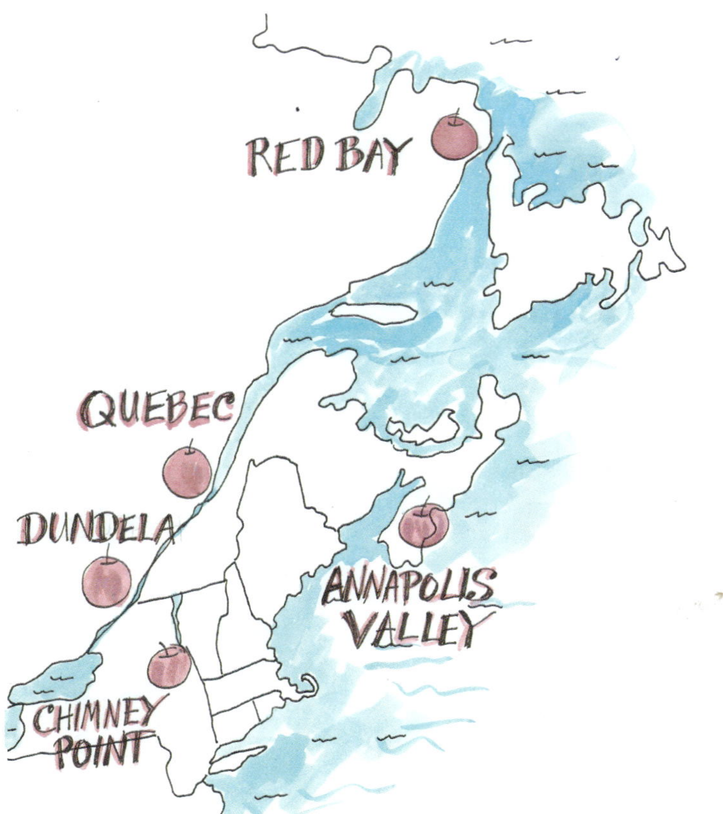


If you hear a tune on the radio and you know it came out in 1960, you know it's not the Beatles. Their first U.S. release wasn't until 1963. You also can assume it's not Fats Waller or King Oliver. Both were dead by 1941. We're guessing that our orchard was planted in 1925. We're also guessing that, by then, our orchardist is thinking commercial rather than nostalgia. What were the popular Maine varieties in 1925? Cortland wasn't being planted in Maine until at least 1920. Our orchardists would have had to be on the cutting edge if they planted Cortland. Macoun was released in 1923. If we have Macoun, the orchard would have been less than ten when the big freeze hit in 1933. Empire and Gala both came well after WWII. Honeycrisp began to be planted around 2000. We can be sure that we have none of these.

We can also assume that we don't have any of the ancient Maine varieties. Bell's Early, Black Pippin, Warren Russet and a whole bunch of others aren't mentioned after the early nineteenth century. It would be fun if we had them, but I'll bet we can discount them as well. They probably weren't seen anywhere in Maine after 1830, a century before our six trees were planted. Even popular old-timers such as Yellow Bellflower and Maiden's Blush are unlikely. Let's eliminate from contention all varieties no longer commonly being grown after 1880 and all varieties still unreleased in 1920. Then let's assume that this was intended to be an income producer for the farm and that our orchardist planted the popular varieties of the day. Let's narrow that window even further and say 1890-1920. That gives us a window of thirty years of varieties to consider.

Topworking #9: We use PVC grafting tape

Apple varieties come in predictable combinations. Whether it's six trees or six hundred, apples were typically planted with the other popular apples of that time and place. Orchards are like anthologies of pop music. They are apple anthologies carved out of pomological history. Most of the time they follow patterns that make sense if you know the history. These were planted by regular folk who had access to the

Chenango Strawberry: a delicate dessert apple "recommended as an excellent variety for the home orchard." S.A.Beach, 1905

regular stuff of their time. Wolf River and Ben Davis are often found together. Northern Spys were planted with Stark and Baldwin. When you find a Mac, Cortland may be a few trees away while Baldwin is nowhere in sight.

One of our neighbors has a tree next to his old farmhouse grafted with two varieties. The tree looks to be about a hundred years old. Why is the tree not in the orchard? Why is it next to the house? Why are there two varieties on the tree? Why did they choose those two varieties and not any of a thousand others? The current owners weren't around of course, when the tree was planted and grafted. In fact, they were not familiar with either variety, though they do appreciate the tree.

The two apples are Chenango Strawberry and Sweet Bough. Both ripen in the summer. Chenango is conic, light-pink, and best for dessert and sauce. Sweet Bough is a round, green, cooking apple. Summer apples do not keep. Use 'em or lose 'em. You snooze you lose when it comes to summer apples. The tree wasn't planted up in the orchard because it would be out of sight and out of mind. You get distracted picking tomatoes or maybe a screen door needs to be repaired; the next thing you know you haven't been out to the orchard for a few days, and all your Chenangos have dropped and are now rotting on the ground. You want your summer apples next to the house so you can keep an eye on them. You can pop out of the kitchen to grab a few and not burn up dinner in the process. Two varieties are on the tree because not many families can use up a tree-full of any summer apple. It's too much. So you topwork two varieties onto the same tree. Why these two in particular? Each has a different use, and each scion was undoubtedly available at that time. It's a perfect setup. You have a tree with two summer apples handy to the house.

Sweet Bough a.k.a. Large Yellow Bough: "Valuable chiefly because so early. When fully ripe quality very good." Maine Ag Yearbook 1874

Would you find a winter keeper like a Ben Davis or a Baldwin or a Northern Spy right next to the house? Occasionally, yes. Keep an eye out for the Baldwin next to the front door, but don't bet on it. You're way more likely to find a Chenango Strawberry, Sweet Bough, Red Astrachan or Yellow Transparent. This is not rocket science. You also won't find ten summer varieties in an old orchard or even in the yard by the kitchen door. No one could deal with all those apples. You will, however, often find ten or even fifty winter keepers in an "orchard away at the end of the farm," as Robert Frost put it. The summer apples are for home use. The keepers are mostly for sale.

Typical seedling leaning towards the sun on Finley Lane; we pruned the top of the lefthand trunk to get more light to the lower branches

Some farms come with diaries or journals or receipts or maps, occasionally with dates. These are all tools for reducing that pool of possibility. There is an old commercial orchard about fifteen miles from home that I visited for the first time thirty years ago. I was out fruit exploring that day and met the caretaker by chance, and we got to talking. "Come with me," he said. We hopped into his truck, and five minutes later we were there. The huge barn and farmhouse were still standing, in excellent shape. Unfortunately, the orchard was buried in the woods. We wandered together through thick pines where once over a hundred apple trees had sat perched above China Lake. Now most of the apples were dead, victims of the darkness of the forest that had grown up around them. Then he took me into the barn. There on the white barn wall was a large map of the orchard. Fantastic. I copied it all down. It was the like liner notes on the back of the LP. Here's what upscale orchardists were planting in China, Maine, in 1900. If you can identify one or two, and you can guess the year, you know what else to look for.

I've kept all the Fedco Trees catalogs, and most of the St. Lawrence Nursery catalogs. And I have an assortment of South Meadow, Stark Bros, Millers and other nursery catalogs of the past forty years. These are also useful. When your client knows the year they bought the tree, you can sometimes pull an ID out of the catalog like a rabbit out of a hat. The receipt is even better. Even if they bought ten trees. Eliminate the impossible! "I bought this tree from St. Lawrence in 1995." Let me consult my crystal ball.

Back to Finley Lane. None of our trees have been pruned in decades. The trees are overgrown and, until recently, filled with dead wood. Most are still surrounded by small maples and oaks and conifers. Some are totally buried in the woods, and most are bearing no fruit. Over the past several years we've pruned them. We cut out all the dead wood. You can do that any time of year. In many cases that's really all you need to do. That leaves you with a rather open sparse tree ready to soak up the rays and put on some new growth. Dead apple wood makes top quality firewood. We cut out the competing trees, but we don't do it all at once. Apple trees crave light but get used to decades of shade. You can drown them in light. So we release them from their competitors over the course of a few years. In some cases we also remove some live wood from the apple tree itself, such as large overhanging branches and secondary trunks that have grown up from water sprouts and now shade the growth below. Old apple trees tend to grow tall and then spread out in the light. Remember those tree top umbrellas that shade the lower branches that in turn atrophy and die? You wind up with one of those apple trees that look more like a palm tree than a fruit tree.

Should you cut down your old trees? No! Never euthanize an old apple tree. When they are ready to retire to orchard heaven, they will. They don't need you to do it for them. Every now and then I get an email or phone call from someone who has been told to cut down their old tree. "They harbor disease." You and I harbor disease. Your cat and your dog harbor disease. Germs are everywhere. Fungi lives. That two-hundred-year-old Black Oxford in Hallowell is still producing a few bushels every year or two.

Prune that old, neglected, disease-ridden tree. Pruning stimulates growth. On most of the old Finley Lane trees, growth had slowed to a crawl. In the first growing season following our pruning, we got lots of new growth. The trees are coming back to life. Remember Woody Allen in *Sleeper* when he goes into the cave and finds a VW covered with dust? He turns the key and it starts right up. Turn the key. The trees respond to your pruning by regaining vigor. New growth translates into new fruit buds in the second or third year. The fruit buds translate into fruit. The wheels are turning; the engine is picking up speed. More coal in the fire. Full speed ahead. The ol' gal ain't dead yet. A couple of years after the first pruning, we get the first fruit in decades. Now you're ready to do an ID.

The old Fameuse tree on Finley Lane began to produce excellent fruit again two years after a thorough pruning: photo by Abbey Verrier

It's been six years since we took over Finley Lane. I've looked pretty closely at most of the fruit. The fruit on the suspected seedlings appears to be correct: wild seedling fruit. The apples are small, in the two inch diameter range, round, and have almost no basin. Seedlings often have no basin. The coloring varies from red to yellow, but mostly in the yellower range. Although a few of the trees produce low-acid "bittersweet" fruit, the rest are more acidic. This is about what you'd expect: mostly acidic, sometimes a little bit bitter, and every so often low in acidity. Nothing you'd want to eat fresh, and nothing that resembles any cooking apples I know. There are one or two that could be old grafted trees. The jury is still out. I will keep an eye on them in particular. They will respond to our pruning. The fruit will gain some size and achieve its potential. As it does, I will reassess.

Keep in mind that the acidity and bitterness of many seedlings translates into a flavorful apple sauce. In addition, acidic apples tend to cook up quickly. Try making single-variety sauce and try blending them. You may be pleasantly surprised. Eat the sauce with your oatmeal. Eat it with your pancakes. Eat it with your burger, lamb or pork. Don't forget to eat it with your ice cream. You could also eat it alone, but why bother when it goes so well with practically everything.

All these acidic seedlings should make pretty good cider too, although cider usually benefits from the addition of some sweeter, low-acid apples. I'm especially interested in the one bittersweet I've found so far. Bittersweets are what some cider makers are looking for these days. They are not that common. Of the twenty plus seedlings down Finley Lane we appear to have two or three bittersweets. Some of the trees haven't produced a decent crop yet. Maybe I'll find one or two more, but one in ten might be the ratio you'd find in nature. It's no coincidence that Dabinett, Yarlington Mill, Harry Masters Jersey and other European bittersweets are so famous in cider circles. Low-acid apples with bitterness and astringency are not common. Many cider-makers are scouring the roadsides and stone walls for interesting seedlings that will be the foundation of our future cider industry. Stay tuned.

Fameuse, a.k.a. Snow: photo by Abbey Verrier

Our six orchard trees stand ready to be ID'd. They've been singled out and pruned. Mowed around. Competing growth removed. Time to take a look. Apple #1 is Fameuse, a.k.a Snow. I knew it as soon as I saw the fruit.

The overall look of Fameuse is small, round and ruby red. One of its synonyms is Sanguineus [sic], presumably because its skin looks like blood or maybe because its flesh is sometimes red-stained. Technically Fameuse is small-medium in size, usually about 2 to 2½" in diameter, somewhat smaller than a Mac. The shape can be conic or oblate but mostly it's round. The skin can be striped darker red and can include patches of green ground color, but the stripes generally disappear into the blood-red blush that covers most of the surface of the apple. When you cut it open, the flesh itself is snow white. In the fall of 2017 I gave a talk at a Lion's Club dinner in Clinton, near Fedco. As I was gathering up my stuff to leave, a fellow came up and told me about his grandfather, Eugene Salsbury, another one of the old-time grafters. One of his favorite apples was Snowflake. "That white flesh is *really* white." Most growers assume that's how it obtained its alternate name, Snow. In French, it's called Pomme de Neige. Neige may actually refer to an early location—it was grown in Québec, which may even be its point of origin. Fameuse is delicious fresh, but also makes excellent sauce. It is not a pie apple. It keeps well into January in the root cellar.

Fameuse is one of the most ancient of all North American varieties. Its history is mostly lost in the fog of yesteryear. We do have decent documentation of its whereabouts after about 1700, by which time it was already "famous." Pomologists have tried and failed to determine its origin. Hogg says that "any attempt to identify it must be mere conjecture." Some assumed it was a French variety brought to Canada during the seventeenth century. They hoped they would find early records of it growing in France. They never did. That effort was also complicated by a couple of factors. North American apples were routinely brought back to Europe and added to European collections. Fameuse was one, which makes it difficult to determine if it originated in Canada and was taken to Europe or vice versa. It also turns out there were other apples in Europe called Pomme de Neige, all of them with white *skin*, none resembling the apple we call Snow.

The story became that it grew from seed in the seventeenth century somewhere in eastern Canada. This is where Saint-Ferreol-les-Neiges might fit in, not far from Québec City. Maybe the seed came over in the pockets of Samuel de Champlain and Pierre du Gua de Monts. We can imagine Sam and Pete randomly tossing cores overboard as they mapped the coast of Acadia and New France in the spring of 1604. It only takes a few seeds washing up on the rocks, and next thing you know, a little apple tree. Beach, in *The Apples of New York*, quotes a lengthy history of the apple written by a Mr. Chauncey Goodrich of Burlington, Vermont, in 1851. That account is worth reading in its entirety. It is detailed and interesting but I was unable to find any smoking gun. Goodrich simply declares that it originated in Canada from French seed.

In November 2017 I went to Spain for the first time. On day one, I heard a great deal about the Basques and their whaling, cod fishing and barrels of apples off the coast of Newfoundland a hundred years before Champlain. Some believe the Basques sailed across the Atlantic as early as 1400. During that first evening, my thoughts drifted towards Fameuse and then my brain clicked into overdrive. If no one can agree on the origin of Fameuse, and if it didn't come from France, maybe it came from Spain with the Basques. It would have had plenty of time to establish itself by the time the French arrived. Being the wonderful apple that it is, it would have been waiting for the French to spread it all over eastern Canada, northern New England and the Great Lakes. It would have had several decades to get famous.

Pepe Madiedo pouring us a glass, 2017

Two weeks later I met a nurseryman, cidermaker and fruit explorer named Jose "Pepe" Madiedo in Villaviciosa, Asturias, not far from the Basque country. Pepe took us on a tour of his nurseries and kiwi plantings. Before we left he gave me a bag of his favorite apples, a medium-sized roundish red fruit from an ancient seedling tree he had discovered in the hills outside the city. He called the apple Madiedo, and it was gaining a following as a cider apple in the area. I can attest that Madiedo made an excellent single varietal *sidra*. He served us his last two bottles.

When he laid out eleven Madiedo apples on the table, it was *déjà vu* all over again. I thought I was looking at Fameuse. I took the apples back to the hotel, photographed them and keyed them out. When I returned home a few weeks later I compared the results with Fameuse. Aside from the fact that Madiedo's flesh is brilliant yellow-green and astringent, the apples are nearly identical. Could the seeds that spawned the original Fameuse have come over in a Basque whaling ship? And if Fameuse came from Spain, what about St. Lawrence, the other great mystery apple in Canada's past? Maybe it's Spanish too.

Tastes like the perfect cider apple but looks a lot like Fameuse

In French, *fameuse* is the feminine form of "famous," "celebrated," "excellent'" "first-rate," "precious." A likely explanation is that it received its name from its ability to produce good fruit, relatively true to type, from seed. This would have been a bonus for any early orchardists who couldn't graft. And most could not. Spread the apple seed on the ground in the fall and wait for them to sprout next spring. Everyone was doing it back then. An apple that produced decent fruit from seed would be incredibly useful and would certainly become famous. Some Fameuse seedlings received their own names and became recognized varieties in their own right. These are thought to include Bloom, Brilliant, Canada Baldwin, Fameuse Green, Fameuse Noire, Fameuse Sucre, La Victoire, Louise, Hilaire, Shiawassee and...McIntosh. In Maine there is one called "Winter Fameuse," which keeps in the root cellar until January.

After returning from Spain, I dreamed up a "working hypothesis." Here it is: The famous apple Pomme de Neige, a.k.a Snow, a.k.a Fameuse originated somewhere around Red Bay on the coast of Newfoundland from Basque seed, brought over by whalers in the early sixteenth century, about the time that Ignatius from the Basque city of Loyola was founding the Jesuits. Call it 1540. The high quality seedling apple possessed the enviable characteristic of coming relatively true-to-type from seed itself. By the time the French arrived in Canada *en masse*, it was already established. Wherever the Basques camped out and rendered whale oil, they planted the seeds of this blood red, white fleshed fruit. It became famous.

The French mapped the coast, found their way up the St. Lawrence River, and attempted to contact the native populations. When those efforts failed, they solicited the Jesuits to assist them. In 1625 Jean de Brébeuf, Ennemond Massé and Charles Lalemant arrived. By then Basque and French Jesuits may have been coming across the Atlantic for over fifty years. The three of them, and other Jesuits to follow, put

Fameuse seeds in their pockets, headed off into the wilderness to live with the First Nation peoples of Canada and spread the famous Basque apple throughout Québec, out to the Great Lakes and down into Vermont. Fameuse seeds or grafted trees were planted by 1700 on Chimney Point at the southern end of Lake Champlain, a stone's throw from New York State and not far from the eventual sites of Fort Saint-Frédéric at Crown Point and Fort Ticonderoga at the portage between Lakes George and Champlain. The apple then picked up another synonym, Chimney apple. From Chimney Point, it was a hop, skip and a jump to New Hampshire and Maine and everywhere else in New York and New England.

Fameuse: maybe it originally came from Spain?

What did Fameuse teach me about the Finley Lane orchard? I've come across Fameuse elsewhere in Palermo and around central Maine. It was a common variety here in the nineteenth and early twentieth centuries. It was also common in most, possibly all, northern New York/New England districts. While Fameuse is one of the oldest varieties grown in New England, its presence doesn't necessarily indicate a very old orchard. As recently as 1910, Maine growers were reporting the apple to be one of their most popular. Add another ten or fifteen years for the heck of it, and it makes sense that our Finley Lane orchardist would have planted Fameuse in 1925. The tree itself is pretty beat up, core-less and split open. At some point it leaned over and one limb stabbed itself into the ground. When we pruned it, we left that limb, which resembles a flying buttress holding up Notre Dame Cathedral. There was hardly an inch of scionwood on the entire tree, its growth having slowed to near zero. Now it's growing vigorously again, producing lots of new shoots and we're harvesting good fruit.

The next tree on the inside row is larger than the Fameuse. It featured a huge second trunk that grew from a water sprout that Ada Northrup should have trimmed off in 1939. We cut it off a couple of years ago. We probably scared it into submission because it's had multiple impressive crops since then. The fruit is larger than Fameuse, slightly more oblate, and is distinctly striped red over a yellow background. I assume it's a common red and yellow striped commercial variety of the period, such as Gravenstein. I take several fruit home. I open up *The Apples of New York* and read. Turns out I'm correct—the second variety is Gravenstein.

Gravenstein is the epitome of the red dotted and red-striped, medium-sized, roundish apple. It's a long-time Maine favorite. Be careful not to mix it up with Duchess and Wealthy, two other late summer/early fall apples. All three varieties are medium-sized. All three are roundish and red-striped. All three make excellent pies. All three have redder skinned sports (mutations). Fortunately, all three also have their differences.

Duchess is most common in northern districts. In Aroostook County, you'll find them in nearly every old farm. They are less common in central Maine and rare in southern New England. In the mid-Atlantic states, growers disapprove of them. Duchess is a northern apple. The fruit shape is roundish, somewhat oblate and regular. It's only the slightest bit lumpy or irregular, if at all. The ground color is light green. The stripes can be narrow, wide or obscured by blush. The core is axile. Commercial Duchess plantings are exceedingly rare.

Wealthy is also hardy and can be found in northern districts. It is more common than Duchess in central locations and, like Duchess, less common in southern New England. The fruit is smooth and roundish and somewhat oblong. The shape, size and color are classic. The ground color is light green covered with thin stripes that fan out evenly and densely from the cavity like dozens of spokes in a wheel. The stripes are not short and choppy; they are long and thin. Even in "Red" Wealthys, the stripes radiate out from the cavity evenly and distinctively. There is a small russet patch around the stem and numerous tiny white spots scattered across the surface. The core is axile. Sometimes the flesh is red-stained. Commercial plantings of Wealthy are not uncommon.

Classic Maine Gravenstein from an old tree in Blue Hill

Our Gravenstein is the most oblate of the three and the most irregular. It is often lumpy with a furrowed cavity and a furrowed basin. It has short, choppy, wide stripes overlaid on top of a blush composed of thousands of red dots, and a rich, greenish-yellow ground color. The contrast of red and green is pronounced. The basin is furrowed, not regular. The cavity is large. Its core is abaxile. Old Gravenstein trees can be found in central, southern and coastal New England districts. Commercial plantings were common in some areas. Think lumpy. Our tree is not Duchess or Wealthy. It's Gravenstein.

Gravenstein may be much older than Fameuse. The name is probably only one of many it's been called over the centuries, picked up in northern Europe long after it originated. It was first cultivated at least five hundred years ago in Italy, Germany, Sweden, Denmark or Russia. No one knows. It was a migrant. Its similarity to many Russian varieties has prompted some to suspect that it came from Russia, although conservative pomologists lean towards Germany, Denmark or Italy.

Regardless, Gravenstein was grown throughout Europe. It became popular across Russia, probably during the reign of Alexander I of Alexander-the-apple fame. Eventually it was brought by the Russians to Fort Ross, in what became Jenner, California, sometime after 1812. It loved Jenner and all of Sonoma County, still the Gravenstein capitol of the world despite the fact that the orchards have now been largely cut down and replaced with Chardonnay and Pinot Noir. A few old plantings remain along the coast of northern California and up into Oregon and Washington. A highlight of a trip to California in the summer of 2017 was an excursion into Gravenstein country, where we visited ancient Gravenstein trees with local fruit explorer and author Darlene Hayes.

A great place to go in August: reprinted with permission

About the same time as it landed in California, Gravenstein also arrived in New England and the Maritimes. Oddly enough, it also loved the East Coast, especially the Annapolis Valley in western Nova Scotia, where it's still commercially popular. Old trees can also be found all along the coast of Maine down into

Chad Frick with one of his ancient Gravenstein trees; note the grass already dead and it's only June: Sebastopol, California,

Red Gravenstein is always less striped and more blushed

southern New England. It became popular in Maine by the mid-nineteenth century and was still commonly grown commercially all over New York and New England well into the twentieth. It's almost certainly the only historic apple to have been introduced to North America from both the East and the West.

All apple identifiers must make themselves aware of the multiple strains of Gravenstein. Often they are simply referred to as "Red Gravenstein" and "Green Gravenstein" kind of like "chile" in New Mexico. "Red or Green," is what they ask you in every New Mexican restaurant no matter what you order for breakfast, lunch or dinner. Green Gravensteins would tend to be the older strains, closer to the original Gravenstein of five hundred years ago. One of the best of these seems to be the super-stripy Rosebrook Gravenstein of California. Of the reds, the first in modern times may be Banks, a red-blushed mutation discovered in Berwick, Nova Scotia, in 1876. Crimson Gravenstein which followed in 1912, was also from the Annapolis Valley. Red Gravensteins are shaped like the original but may have sacrificed fruit quality to get that red blush. Still they produce an excellent pie. It looks like a Gravenstein, it tastes like a Gravenstein, but it doesn't have the stripes. Old Red Gravenstein trees do exist in Maine and other New England states. They are most common in modern Gravenstein plantings where red seems to win out over green. If your number one priority is flavor, you should look for the old-fashioned green Gravenstein, although you probably won't be disappointed with either.

Unlike Fameuse, Gravenstein is an excellent pie apple. Some would say that it's the best pie apple in the world. It is also well known for its sauce. I first got to know the apple through Dick and Connie Sweetser of Sweetser's Orchard in Cumberland, Maine. Dick and Connie grew Red Gravensteins. I wanted to offer them in the Fedco catalog. We got to talking, and they told me about Mary Blenk. She had won the Maine State Pie Contest using their Red Gravensteins, so, naturally, I decided to contact Mary. About that time I also met Bob Shafto, a hospital administrator and the husband of a former college professor of mine, India Broyles. Bob and India live on a beautiful restored farm in Falmouth, not far from Sweetsers. The Sweetser Gravenstein trees are not old, but Bob's are. He has two or three hundred-year-old Red Gravenstein trees that he prunes and cares for. They are beautiful. He's provided us with scionwood now for many years.

Mary Blenk is a chef who loves to make pies. I don't recall exactly how she learned about Gravenstein. My guess is that she'd made enough pies to figure out that the right apple will make the best pie. So she went to the Sweetsers, and they introduced her to Gravenstein. She made a pie and entered it in the Cumberland County Fair, a mile or two up the road from Sweetser's Orchard. She won. That bought her a ticket to the Maine State pie competition. Problem was the state competition wasn't until winter. She knew how to make a good pie, and she knew that the Gravensteins were a key part of the equation. Talent and ingredients are a nice combination. So she crossed her fingers and saved a bag of Sweetser's Red Gravensteins until January. She won again.

What makes a good pie apple? It must have great flavor and great texture after being in the oven for sixty minutes. Why sixty? More or less and the crust will either be underdone or burned. The apples must conform to the demands of the crust. The apples must be cooked all the way through, hold their shape, and not turn into sauce. You don't want your crust to collapse. All the best pie apples fit within a fairly broad window of texture and flavor variability as long as they hold up the crust and taste great. Some might be a little mushier, some thicker, some softer. Some lean towards the sweet side like Black Oxford, and some lean towards tart like Duchess. Some land right in the middle which, in the case of the perfect pie and the occasional politician, is not a bad place to be. Like Gravenstein.

The old Gravenstein tree on Finley Lane gave us a nice crop three years after renovation pruning: photo by Emily Skrobis

Can you taste an apple raw and know if it will be good in a pie? No, but you can get an inkling. If the apple is devoid of acidity, don't bother; otherwise, it's worth a try. Best to start with a single variety pie. If you like the single variety version, try it in combination with other apples. As far as I know, no laboratories are breeding or selecting new pie apples. We could use a few more. Are there other good pie apples? Try Duchess, Wealthy, Spice Sweet, Twenty Ounce, Redfield, Northern Spy or Black Oxford. Any of those will make you happy. Mary Blenk and Connie Sweetser will tell you Gravenstein is the best. Will you ever find good pie apples at the grocery store? Probably not. Another reason to keep an eye out for the heirlooms.

So here we are in this remnant of an orchard in a small town in central Maine, and we are two for two in the iconic apple department. Seeing Gravenstein triggered thoughts about the "New England Seven." Maybe what we have is an orchard planted in response to the declarations of the New England Cooperative Extension agents. They were given the job in the 1920's of whittling the New England apple pool down from several hundred to a select group of varieties for the burgeoning commercial orchard industry of the new century. These were dubbed the "New England Seven" in 1927.

The list of the chosen seven consisted of Baldwin, Delicious (later Red Delicious), Gravenstein, McIntosh, Northern Spy, Rhode Island Greening, and Wealthy. We assume it was a list created in good faith. It included some good apples which had stood the test of time and still do. As far as I know the specific criteria by which the seven were selected will forever remain a mystery. Disease resistance was apparently not one of them. The list includes a pretty good spread as far as season and use, and it shows some diversity in its genetics. That being said, given that hundreds of varieties were known in New England at that time, the focus on seven apples represented a gigantic implosion of diversity. Ninety-nine percent of available apples were swept aside.

It became an unrelenting slaughter of old varieties; even the New England Seven were not safe. Gravenstein held on for a while but eventually succumbed. One of its most enviable characteristics contributed to its demise. In Maine it ripens over several weeks beginning in September. This was an advantage for the small farm of the past because it meant you didn't have to pick them all at once. You harvest your Gravensteins weekly and then sell them or use them right away. It sounds like lettuce or string beans or broccoli. Pick weekly, and enjoy a long season. A little income two or three times a month adds up.

The McIntosh strain from Francis Fenton's orchard dates from 1906

Sadly, modern day orchardists have been forced by economics to give up on Gravensteins and other long-ripening varieties. When you have a paid crew out there picking and the clock is ticking, it's painful to send them back to the same tree over and over again. You want them to pick the entire crop all at once. Even if Gravenstein is the best pie apple in the world, it was tossed into the dustbin of pomological history.

Let's return to the Finley Lane orchard and our attempt at identification. Four trees to go. One of them is a second Gravenstein. That was easy. If I'm correct in my thinking so far, then we'll also find at least one McIntosh or two. This was a small commercial planting in 1925, and McIntosh was the Honeycrisp of the day. Everyone wanted it. Of course, I'm hoping for some super rare one-of-a-kind variety not known in Maine for two hundred years. Tree #3, next to the Gravenstein? McIntosh!

Not that I don't like a good Mac. They earned their reputation in part because of their excellent flavor. They also make a great sauce. I just never imagined I'd be growing Macs. Yes, three of the remaining four trees are Macs.

McIntosh is both an heirloom and a modern variety at the same time. It's way younger than Fameuse and Gravenstein. In 1900 it was considered brand-new. There are twenty-one Mac submissions in the USDA watercolor collection. The earliest is 1895. The average date is 1921. The paintings came from all over the eastern U.S. The earliest submission from Maine is dated 1911. While this doesn't tell us for sure when our orchard was planted, we can guess that it was not before 1910.

It's generally accepted that the wild seedling apple tree that would eventually be named McIntosh was discovered in about 1800 by John McIntosh in Dundela, Ontario, fifty miles southeast of Ottawa near Williamsburg, not far from the St. Lawrence River and only twenty miles north of Potsdam, New York. The apple became a local favorite within a decade or two, but it wasn't until about 1870 that it began to gain any widespread popularity. The McIntosh family recognized that they had a great apple, but they were not aware of the importance of grafting. Consequently the first Mac trees were propagated from seed, and therefore not true-to-type. John McIntosh's son eventually learned about grafting and began to disseminate grafted trees.

The apple is not listed in Downing's first edition of his comprehensive *Fruits and Fruit Trees of America* (1848). Twenty-eight years later, the Second Revised Edition, edited by Downing's brother Charles, does list McIntosh Red; it says that the apple "is not widely known." By 1905, "Red" had been dropped from the name. Beach says of the apple, "It has not been sufficiently tested to demonstrate fully its value for commercial purposes but it is regarded by many as one of the most promising varieties of its class for general cultivation in New York." Seventeen years after Beach, McIntosh is listed in Hedrick's *Cyclopedia of Hardy Fruits*. What a transformation. "If one were compelled to choose the apple of apples as the season's varieties pass by, choosing in respect to the qualities which, united, gratify the greatest number of senses, few would hesitate in naming McIntosh sovereign of all." Within ten or fifteen years, every orchard in the Northeast was clamoring for trees.

One night twenty years ago, I was visiting my neighbors Donna and Steve Haskell. Donna's family had lived on the road for generations and had several orchards dating from before 1900. (She was a Jones.) I think she was born in 1933. I asked them about the old varieties. Yes, they had "Baldwin, Northern Spy, Banana, Wolf River, Snow, Pound Sweetens and Ben Davis." Steve then said, "and McIntosh?" Donna replied with a scowl, "We didn't grow McIntosh!"

If your apple tree dates from before 1870 in most of the Northeast, it is almost certainly not a Mac. If it dates from about 1900, there's a chance it's a Mac. If it's younger than 1920, it could very well be a Mac. Dates matter when doing apple ID's. Knowing a tree's age can be a huge help.

Who were Mac's parents? With most old varieties there is only speculation. Even many modern varieties are "orphans." For a while the University of Minnesota thought they knew the parentage of their star introduction, Honeycrisp. All the early published descriptions called it a cross between Macoun and

Honeygold. Like McIntosh, Macoun is a Canadian native, beloved throughout the Northeast. Honeygold is an earlier University of Minnesota introduction with high flavor and even higher scab susceptibility. After some genetic testing however, they determined that they had it wrong. Eventually they settled on Keepsake, another Minnesota introduction, as one parent. The second parent remains a mystery.

Evidently it's difficult enough to keep track of apples bred by technicians and then set out in carefully labeled university test plots. You'd think they'd know what they had. Imagine what it was like in the wild world of apple seedlings generations ago. The thousands of varieties we now call heirlooms were discovered only after they began to fruit. "Hey, check out this one!" "Wow, that tastes pretty good. Let's call it Black Oxford!" By then the newly found trees might have been ten or twenty years old. The parents could have been long gone. Maybe the seed was deposited by some bear from the other side of the mountain. Who knows where Mom and Dad might be? Most of those early apple parents were, on top of all that, unnamed seedlings themselves. Seedlings of seedlings of seedlings. When we describe the parentage of most heirloom apples, we just say, "chance seedling" or "natural fruit" or "wild apple." All euphemisms for "parents unknown."

Blue Pearmain and two of its most famous children

For those who care to speculate about the parents, look around and make a guess. Seedlings sometimes resemble one of the parents. That can be helpful. Duchess seedlings often resemble Duchess, sometimes closely. Wolf River and Northwestern Greening both more of less resemble Alexander, one of their probable parents. That would make them siblings, or more likely half-sibs. And yet Northwestern Greening and Wolf River do not resemble each other. This suggests that they have two different fathers. Other times the children bear no resemblance to the one parent we do know. Although recently brought into doubt, for many years they said that Wealthy (medium-sized, round and red-striped) is the child of Cherry Crab (small and yellow). Rolfe and Nodhead (both medium-sized and red) are said to be the children of Blue Pearmain (large, purple and russeted).

Topworking #10" Keep the tape tight as you wind

We can safely guess that our ancestors didn't have much interest in knowing the parentage back in the days before intensive breeding programs and high-stakes marketing campaigns. Farmers were more interested in the fruit qualities than the parents. Does the apple make a decent pie? Still, sometimes you do find information about the parents of old varieties. With Macs, there has been the occasional guess. Beach says that the apple "belongs in the Fameuse group," suggesting Fameuse as a possible parent. Others have proposed St. Lawrence or even Alexander. The first two are reasonable guesses. Both Fameuse and St. Lawrence were popular in southeast Ontario in 1800. Both would have been growing on or nearby the McIntosh farm. Both vaguely resemble McIntosh. Both Macs and Fameuse are quite susceptible to scab. Maybe McIntosh picked up its scab susceptibility from Fameuse.

In his *North American Apples: Varieties, Rootstocks, Outlook*, W. H. Upshall writes that "Alexander is a possibility because of its similarity in foliage characteristics to McIntosh." Still, it's unlikely that Alexander would be a parent. Alexander was not imported into the U.S. until sometime between 1817 and 1830. This would mean that it was still in Europe at the time of the supposed birth of Mac. We can nix Alexander. St. Lawrence was likely at the right place at the right time but our best bet would be Fameuse. We'll pencil it in.

Rogers Red McIntosh: the most famous red McIntosh sport

Like Gravenstein, McIntosh is a variety with many mutational strains. If I could identify which one we have, I might be able to pinpoint the date of the orchard. These mutations are different from the multiple seedlings of Fameuse. They are not from seed. A branch or a whole tree of a given variety will sometimes exhibit some variation or another. It will mutate. If you graft from that branch, you will replicate the mutation. If by chance you find a Mac that's redder, earlier ripening, heavier bearing, larger, or rounder, such mutation is called a "sport." The more trees out there of variety X or Y, the greater the chance that there's going to be sports along the way. It's the law of probability. These occasional sports have been much sought-after by orchardists over the centuries. You graft from that branch and you've got a whole new variety. Those Red Gravensteins are sports.

By 1920, different Mac strains were available. Already there were solid-red Macs and striped Macs. The most famous of all Mac sports is "Rogers Red," an entirely red-blushed bud mutation discovered by Isaac Rogers in Dansville, NewYork, just west of the Finger Lakes. It was introduced in 1932 just in time for the catastrophic winter of 1933-34. Which Mac we have will have to wait for another year. If it's a Rogers Red Mac, that would place the orchard at 1936 or so, destroying my working hypothesis. I don't have any decent specimens right now because it's springtime. Even if I did, some of these strains are not easy to tell apart. For now I can say that we have some strain of McIntosh, likely planted after 1910.

McIntosh tree, Kent's Hill c 1998; thought to have been grafted with scionwood taken from the original McIntosh tree in Canada

# Sixteen

# Red St. Lawrence

You see but you do not observe.
Sherlock Holmes, *A Scandal in Bohemia,* p.162

You can observe a lot by just watching.
Yogi Berra

I am the observer who is observing.
Van Morrison, 1997

Red St. Lawrence on the left and the original St. Lawrence on the right

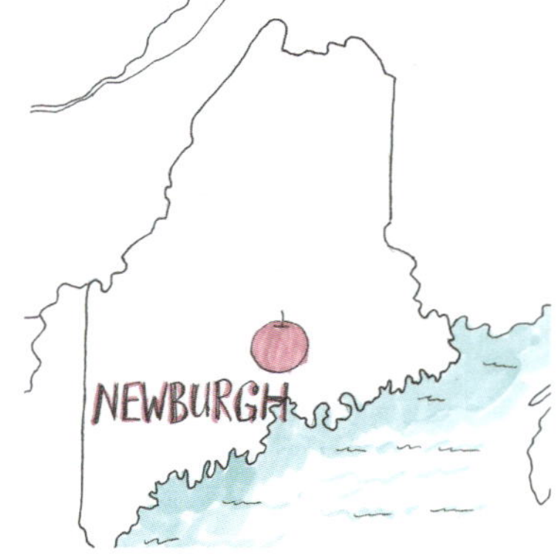


When a plant person talks about sports in the orchard, it's usually not about croquet or badminton or snuggling under an apple tree. A sport is about unexpected variations, a spontaneous plant mutation that usually emanates from a single bud. A bud sprouts, and the resulting branch acts differently from the rest of the tree. No one seems to know where the word comes from. Because of its association with a single bud, they're sometimes called bud sports. If you propagate from that twig or branch, you retain the variation.

Keen observers in the ornamental plant world have been searching for sports for centuries. What they particularly desire are unusual flower colors, multi-petalled "double-flowering" blooms and variegated foliage. Those clumps of witches' brooms you occasionally see in conifer trees are also often the result of bud sports. They are highly coveted by conifer geeks who sometimes actually shoot the witches' broom branches out of the tree tops. For the orchardist it's all about the fruit. Spot a branch with apples unlike those on the rest of the tree; take a scion from that branch and graft it and the variation will be passed on. It's like starting a new variety but not from a seed.

Long ago I was visiting an old homestead in southern Maine with an old Baldwin tree. High up in the top, in the sunlight, something caught my eye. It was a branch of russets. I had discovered my first sport! I was so excited. I went home that night and pulled out Beach and read that "Cases have been reported where Baldwin has sported and developed fruit with russet skin. Since these apples appear to show no advantage over the smooth-skinned Baldwins, they are seldom propagated." So much for observation.

In 2013 Abbey and I were walking through rows of a yellow variety we weren't familiar with at the Meyerhans' orchards in Fairfield when we came across a tree with two different apples on it. This was not a case of two varieties grafted onto one rootstock. On most of the tree the fruit was all yellowish-green, while on one particular branch all the apples were slightly smaller and entirely russeted. We'd found a sport. The law of averages suggests you'll find sports among the most popular varieties. If you plant many thousands of Golden Delicious, Gala, Fuji or Red Delicious trees, and if you keep a close lookout when you walk the rows, you've got a decent chance of finding one of those mutations.

Sports in the orchard; the sport is on the left

Virtually all large wholesale nurseries are thinking sports when they sell trees today. Commercial orchardists want them. Most sports have been selected for their solid red color. A few of the most famous have also been selected for their productive dense "spur" type fruiting habit. Mostly, however, growers want full color earlier in the season. Better color sooner allows the fruit to be harvested before it's ripe. Fully ripe fruit has already begun the process of deterioration. That may be okay when the fruit is in your backyard, but not a good plan when you're about to ship it to a supermarket. You want the fruit to be picked before it's too ripe and begins to rot, but you also want it to look good. You want it fully colored. But color and ripeness go hand in hand, so how do you accomplish this? The answer is to find a sport that colors up early. Voilà! Now you can ship the beautiful, unripe apples to a big box grocery store two thousand miles away, or seal them up in a warehouse for several months before dragging them out and loading them onto a string of eastbound boxcars. And they still look good.

Looking for the original Delicious, Gala or Fuji? At this point you can assume they'll be difficult to locate. You might find them in some collection of rare apples somewhere, but otherwise the apples sold commercially under the names Gala, Fuji and Delicious are all sports. Although each sport is technically its own variety with its own name, most of the names are unknown, even by the orchardists who grow them. In the world of Gala, Royal Gala is a name that many people know. But what about Gale Gala, Waliser Gala, Obrogala, Dalitoga, Galaval, Cherry Gala, Banning Gala, Smith Gala and twenty others? The list goes on. In the grocery store they're all simply Gala.

Cortland's introduction ushered in the modern period of apple breeding at Agricultural Experiment Stations, interrupting the long tradition of farmers selecting new varieties from wild seedlings

The much maligned apple we now call Red Delicious was originally discovered growing as a seedling in a pasture in Iowa in 1872 and called Hawkeye. Hawkeye wasn't even a red apple; it was red-striped and good tasting. Sadly for the commercial people, it didn't color up until too late in the season. Poor color translated into lackluster sales. While changing its name to Delicious generated some enthusiasm, the discovery of the first red sports were the kicker. Now you could pick them "green" and they'd be red! During its subsequent rise to astonishing fame and glory growers selected over fifty sports. Many were sports of sports, or sports of sports of sports. There are also close to thirty Fuji sports, each one a little bit redder than the last. We all want red. But at what price? What about flavor?

And what do sports have to do with identifying apples anyway? A lot, actually. The first two apples you've got to know if you're from the Northeast are McIntosh and Cortland. If you're going to devote a large portion of your limited hours to identifying apples, you don't want to be spending your time struggling with a Mac or a Cortland. You probably want to reserve that time for the tough ones. Macs and Cortlands are a dime a dozen, but beware, they can trick you. Both come in multiple sports. There are different striped Macs and different solid red Macs. Same goes with Cortland. Don't let the variations bog you down. Learn to recognize them, whether they're entirely red-blushed, mostly green with a few red stripes, or some gradation in between. There are at least three red Cortland sports, the most famous of which is called Redcort. (Don't you love those modern apple "names?") Red Cortlands look nearly identical to the old original Cortland in shape and size; they're just more red. And when it comes to Macs, they all taste like Macs, so if in doubt, take a bite.

Mark Fulford teaching his annual grafting class at the Scion Exchange

Don't forget, you also need to be able to tell a Red Gravenstein from a green one. The Gravensteins are trickier. Something tells you they're both Gravensteins once you learn them, but the red sports can look fundamentally different from the greens ones. Are the green ones sports too? With Gravenstein, maybe so. The variety is so old that none of us may have ever seen the original. Embrace them all and you'll learn to tell them apart.

Not all sports have been found in commercial mega-orchards. After all, I did find that russet branch on a Baldwin years ago, and Abbey and I found that small russet sport in Fairfield. And not all sports come at the expense of the good things that many apples have to offer. Be on the lookout. You never know when you might find a sport like the Red St. Lawrence.

Mark Fulford is not a big guy. He looks a little like an elf, and he chuckles like one too. He had a ponytail for a long time, longer than most of our contemporaries. He has a close-cropped beard. Mark has spent his life eking out a modest living doing a combination of farming, agricultural research, teaching and consulting around the world. Sometimes in the winter I see his wife Paula working over at Fedco Seeds.

Mark and I and a few others—Jack Kertesz, Will Bonsall, Tom Vigue—caught the fruit exploring bug at about the same time and began to get together a few times a year. Jack started publishing a newsletter under the name, "Maine Tree Crop Alliance," à la J. Russell Smith. Smith was our tree guru. We all bought and read his *Tree Crops*. If you were around back then, you probably did too.

When I needed a quote to go with the T-shirt I designed for Fedco Trees way back then, I closed my eyes, opened up *Tree Crops* to a random page and pointed. You could do that. It all sounded great. "An idea, a small yard and one tree may be made to produce astounding results." See, pretty good: I just opened up the book and stabbed.

When I started Fedco Trees in 1984, Jack Kertesz was a huge help. He and I drove to upstate New York to pick up most of the trees that first year. We had ninety orders. All the trees we sold fit into the back of a big pickup truck that Jack borrowed from his neighbor. Mark Fulford and his brother, Jonathan, started a tree nursery about that same year. It was good timing. For the next several catalogs Mark and Jonathan grafted and grew out the bulk of the apple trees we offered. For a few years we processed the orders in the root cellar beneath their barn in Monroe until we outgrew the space and purchased an abandoned chicken house with a huge basement in Clinton.

When I began to do apple identifications, Mark would occasionally join me. In late September 1997 we spent an entire evening together in his kitchen identifying apples. For a moment in time, we were the "Click and Clack" of apple identification. We became the NPR apple guys, his wife Paula our audience. We plowed through the apples one by one while Paula laughed and cringed. We did pretty well as I recall. Then we waited for the call from Public Radio. Darn.

Roger Luce out in his garden in May 1997 with the lilacs in bloom: photo courtesy of Paul Tukey

Like the rest of us, Mark had gone out and found his own mentors. MOFGA was just getting rolling. The apprentice program was new. There was no easy way to sign up for a mentor. (Has there ever been?) You had to go find them on your own. Mark found Roger Luce, a retired school teacher who had returned to his family's farm in Newburgh in southern Penobscot County not far from Mark's place. Mark began to bring Roger to our fruit growing events, including the Scion Exchange and the Fall Fruit Swap. Roger was a friendly guy, and we all found him incredibly endearing. He was also extremely knowledgeable about plants.

Decades earlier Roger had left the farm and gone to Connecticut where he taught school for many years. By the time Mark found him, Roger had retired and moved back home where he had transformed the family farm into an arboretum. Winding through the old orchard behind the farmhouse were paths just wide enough for a small Pony tractor to squeak through. The paths were lined with shrubs and trees, many of them one-of-a-kind. Some places were open just enough to see the sun. Others were so overgrown that the sun almost never touched the grass.

Van Eseltine has a narrow upright vase shape

Roger had never been married and had no children. He had been collecting plants since his was a young boy. Although he traveled widely and made several plant collecting trips to Korea and China, he and Kublai, his trusty malamute, always seemed to be home working outside in his gardens and nurseries. He was happy to welcome visitors, although he asked to be called first. We'd arrive, say hello, and immediately head for the back door. The maze of gardens completely covered the hillside that sloped away from the connected house, sheds and huge old barn.

Roger's gardens were sprinkled with plant gifts from collectors, breeders and explorers. He had a Ginkgo tree right outside his kitchen door that someone brought him from China. His specialties were crabapples, magnolias, lilacs and primroses. He had dozens of each. Some were common, and others were rare. Some were of his own creation, either by design or by chance.

One of Roger's horticultural contemporaries and close friends was Lyle Littlefield, the long-time, much beloved University of Maine professor of ornamental horticulture. Someone—probably Roger himself—told me to go visit Littlefield's four-acre trial gardens that had been established on the University of Maine campus in Orono in the sixties. I swung by those gardens at least once a year. Although Littlefield had died by then, the trial gardens had also been transformed into a small arboretum. I studied it all, particularly the crabapples. By then I knew Roger's trees pretty well too, and it was clear that the two men collaborated. It was fun to see where their collections overlapped. Like Roger, Lyle Littlefield was clearly into crabs, lilacs and magnolias.

The classic weeper Red Jade spreads and hangs to the ground

Roger's crabapple pruning style was intentionally laissez faire. He allowed his trees to express their individuality. Not only does each variety have its own unique flower and fruit, each also has its own unique form. Each has personalty as a tree. When you minimize your pruning, you give the tree permission to do what it wants to do, akin to the wild apples down Finley Lane. Roger had specimens of crabapples with all possible personalities. With my

Roger's Semi-weeping Red with its undulating form

pencil and pocket notebook in hand, I would scramble along behind him as he rattled off strings of comments. His Pink Spires stood straight up like a pear tree. "So fragrant it sounds like a bee hive… birds like the fruit. They clean it." His double pink Van Eseltine spread its limbs in the perfect vase shape. His Pink Cascades was aptly named: "It doesn't weep, it cascades." He had one unknown crab he called "Semi-weeping Red." He purchased it at a garden center down on the coast somewhere. The branches didn't weep; they crawled and undulated. What a great tree. Of the double flowering Brandywine: "Darn nice crab—I wish it didn't have fruit that was so large." (Cider-makes have discovered the large bitter Brandywine fruit and love it.) He probably had the most beautiful Red Jade tree that's ever been grown, even more perfectly shaped than the one at the Brooklyn Botanic Garden where it originated. It was the quintessential weeping crab. The branches rose straight up to a height of about twelve feet, arched and then hung to within a few inches of the ground. Underneath he had placed two chairs and a small table.

Once you get to know your apple varieties, you shouldn't need to see the fruit or the flowers to do an ID. Once you've seen a few Red Jade trees, you'll know it, even in the dead of winter. Or a Pink Spires or a Van Eseltine. And it's not just the crabapples that have distinctive forms. All apple trees have their own tree shape. They may not weep or be stiffly upright or vase-shaped. But if you can tell brothers apart or sisters apart or cousins apart, you should be able to tell a Mac from a Cortland without the fruit. You may think they all look the same, but look again. Commercial orchardists can tell a Mac from a Red Delicious from a Cortland in an instant just by glancing at the trees. They see them over and over again; they can recognize them from a hundred yards away.

There's nothing like Brandywine in bloom; the fragrance is intoxicating; the large fruit is exceedingly bitter and becoming popular with cider makers

For the true diehard, there's *Bulletin No. 403* from the Massachusetts Agriculture Experiment station, written in 1943 by Professor J. K. Shaw. It's called *Descriptions of Apple Varieties*. The report describes in detail the tree form, twigs, bark, and leaves of ninety-one common apples of the early twentieth century. It also pinpoints specific vegetative differences between similar varieties. No mention of the fruit is made at all. Shaw was doing all his IDs from the trees, twigs and leaves. His goal was to assist orchardists with identifying young trees before they fruit. Incredible.

Unless you went through Roger's kitchen like we usually did, you entered his gardens around the side of the barn down a dark tunnel through a massive cedar hedge that lined the road. As you emerged back into the light, immediately on your right was the deep-red flowering Lemoine. In front of you was the pink-flowering Redfield, the first of his crabs to bloom every year. A cross between one of Maine's most popular apples, Wolf River, and the red-fleshed *Malus niedzwetzkyana*, Redfield was to introduced in 1938 as a commercial sauce apple, something to add color to your applesauce. But, with the invention of cheap red dyes, Redfield became irrelevant. It did attract some interest as an ornamental crabapple, but growers didn't like the messy McIntosh-sized, pink-fleshed fruit. It also had no allure as a dessert apple. Too tart and weird. But then someone discovered it really does make a fabulous pie and excellent sauce. Simultaneously it become renowned for its fermented cider. Now everyone seems to want it. It is the signature apple for West County Cider in Colrain, Massachusetts. As one of our apprentices once said, "That's the worst tasting apple and the best tasting pie I ever ate." (She says I said it.) As a diehard sauce maker, I love the apple. It's also a wonderful tree.

Redfield is one of the first to bloom every year; excellent pies, sauce and cider: photo by Laura Sieger

Part of Roger's backyard was loosely sectioned off as a nursery. The nursery consisted of half a dozen beds, each about a hundred feet long, packed with a little bit of everything, including rootstock that never got grafted, liners that never got planted out, and seedlings of this and that. Although Roger was still starting new plants when I got to know him, most of the nursery was filled with the ones that got away. You might have described it as a jungle. Some of the nursery trees were thirty feet tall.

During one tour of his gardens we passed a spindly crabapple that had seeded itself in the nursery bed some years earlier and then weaved its way through the chaos of other plants out into the light like Boney Maronie. The tree was engaged in the cleverest of apple tree strategies—worming its way to the sun by any route possible. The tree was also dotted with a few round, glowing, red-purple fruit, each one about three quarters of an inch in diameter, a little bigger than a chocolate egg. Roger plucked off a fruit and sliced it in half with his jackknife. The flesh was electric red. We picked enough to make one small jar of jelly. When we returned some weeks later, we brought it to Roger, and he loved it. He gave me scionwood, and I grafted it at home. A few years later with his permission, we offered it in the Fedco catalog. We needed a name. Roger asked us to call it Henry Luce, after his father. Henry Luce has beautiful red flowers, bronze-red foliage, red-fleshed fruit and makes a wonderful red jelly.

There is no other variety with stripes like the original St. Lawrence

Because of the hundreds of ornamentals that Roger planted over the decades, it was difficult to see that Roger's backyard was actually the remnants of the old family orchard. You'd be walking along through scores of lilacs or down a long nursery bed, and all of a sudden you'd be face to face with an ancient apple tree. They seemed to be randomly placed, but they were not. If you'd plotted them out, you'd get the pattern. The orchard had just blended into Roger's arboretum. Near the nursery area was one of those old trees. The tree was Red St. Lawrence, a rare sport of one of the most important of all historic northern apple varieties.

St. Lawrence originated in Canada, first documented in the early nineteenth century, but quite possibly much older than that. It was likely discovered somewhere near Montreal on the banks of the St. Lawrence. Some say it's a seedling of Fameuse. Others say it's one of the parents of McIntosh. It's extremely hardy and became well-known throughout much of eastern Canada, northern New York and northern New England. You can still find the apple in a few of Maine's commercial orchards. It ripens a couple of weeks before McIntosh, is a decent dessert fruit and good for sauce and other cooking. It makes a very tasty pie about mid-September.

In most regards, the original "green" St. Lawrence fruit is rather commonplace. It's medium-sized and roundish oblate, like a few hundred other varieties. The stem is medium in length and on the thin side. The coloring, however, is unforgettable. The apple has a very light-green ground color punctuated with evenly spaced red stripes. Unlike most red and green apples, there is no blush on St. Lawrence. The green ground and the brick red stripes alternate in stark contrast to one another. It's the stripiest of the stripiest. It's one of the best "first apples" to learn to identify. Once you've seen a few, you'll never forget it. That's St. Lawrence with its incredible stripes.

Red St. Lawrence is identical to the original in all ways but one. Instead of red stripes on a green ground, the red stripes are set on top of a light brick-red ground. While the contrast is not as bold as the original,

Red St. Lawrence is still quite beautiful. Maybe even more beautiful than the original. Unlike some notorious sports, like the dreaded Red Delicious, Red St. Lawrence didn't sacrifice its taste to obtain its beautiful red skin. Red St. Lawrence was one of the contestants in our 2015 Camp Merryweather single-variety pie "taste-off." It won.

As far as I know, the tree in Roger's orchard was grafted from the original sport which was itself discovered on the farm many years earlier. Primarily due to the efforts of Mark Fulford, Red St. Lawrence can now be found scattered about the state. Mark was the first to introduce the apple to me, and he was also the first to disseminate trees through his own Teltane Farm catalog and through Fedco. Now it has spread to many locations. I topworked two trees at our place years ago and more recently put it in the Maine Heritage Orchard. It's also Penobscot County's representative on the 2009 MOFGA T-shirt.

Roger died in 2002. His gardens slipped into a state of wildness. While this is not necessarily a bad thing, gardens like Roger's do best with some human companionship and collaboration. Apparently they've now found some. I've heard in recent years some young people have taken over the old place and are renewing that partnership with the plants for another generation. Meanwhile, many of Roger's favorites live on in hundreds of gardens and orchards through the magic of plant propagation. If you've ever come across a St. Lawrence apple that looks redder than it should be, maybe you've found a Red St. Lawrence. If you have, there's a good chance that it came from Roger's farm and that Mark was the one who grafted it.

The original St. Lawrence surrounded by seven Red St. Lawrence

# Seventeen

# The Boy With no Name

"Then what clue could you have as to his identity?"
"Only as much as we can deduce."
Sherlock Holmes, *The Adventure of the Blue Carbuncle*, p. 246

Garden Royal

Although some apples, like Red St. Lawrence, Red Gravenstein and Rogers Red McIntosh are mutations, the vast majority of varieties have two parents, just like you and me. Most of us know both of our parents. Some of us know only one. Many orphans know neither. The same is true in the apple world, though the numbers may be flipped. Most apples are orphans.

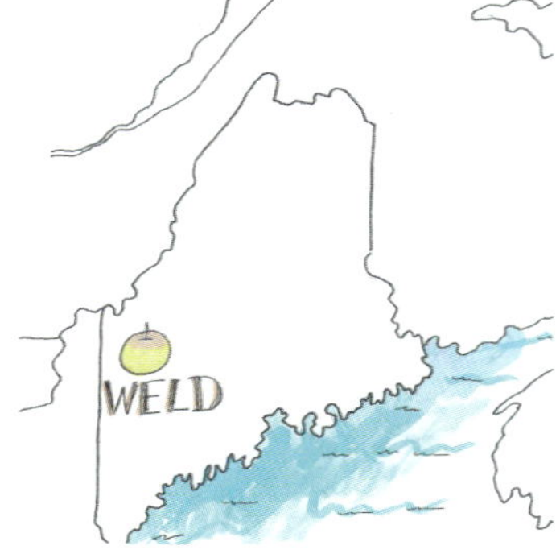


Just like all apples, each of us remained nameless until someone—usually our parents—gave us a name. For some of us, our name awaits us at birth. "Junior!" For others it comes quickly, and for still others it takes a few days or longer. Same with the apple tree, although many of them live out their long lives rather contently with no name at all.

Bethel originated in Vermont but can be found in Maine

My one desire on my forty-ninth birthday was to go to see some old apple trees. I had wanted to meet David Hutchinson up in Franklin County for a number of years. This seemed like a good day to make the trip. David had written to me over the years about that old Sudbury, Massachusetts, apple, Garden Royal. Garden Royal was one of the varieties grown by Earland Goodhue, one that resembled Starkey but ripened early in the fall. Earland had found it near his farm in Sidney. In a good year it's one of the best dessert varieties in the world.

The Dots tree in Belgrade; Note the two rippling grafts by Janetha's shoulder; it was cleft grafted about 1860

One of Garden Royal's defining characteristics is its dots. They are very prominent. In apple lingo, dots are said to be "prominent," "very prominent," or "not prominent." Although the dots are not particularly prominent for most varieties, some apples can be practically identified by their dots alone. The conspicuous dots that cover the purplish-red surface of the large Vermont apple, Bethel, is that apple's most distinguishing feature. It wouldn't be a Bethel without the dots!

While I was researching the apples of my childhood stomping grounds around the Belgrade Lakes in central Maine, I spent a lovely afternoon with Russ DeJong and Janetha Benson. The two of them gave me a tour of their old farm that sits between Great Pond and Messalonskee Lake. One of their favorite apples is a Baldwin-type from an ancient tree in their side yard with a head-high graft that looks like a ripple of sagging flesh wrapped around the trunk. Not knowing the variety, they simply call it "Dots." When

"Dots"is a grafted variety but still unidentified for now

you see it, you'll know why. I suspect that it may be one of Joseph Taylor's introductions, most of which never made it beyond Belgrade or maybe Rome. Despite the fact that I may never identify the deep red, white-speckled apple, we decided to include it in the Maine Heritage Orchard collection. For now, it's Dots. More recently I was asked to ID an old tree in Lincolnville with dots so prominent, the owners, Rose and David Stebbins, affectionately call it, "Freckles." It's a great apple. I think it may be the old Massachusetts variety Beefsteak. Freckles has also found a home in the Maine Heritage Orchard. Now count the dots. Just kidding. But as you get to know an apple, you do want to note how numerous they appear to be. They could be prominent and numerous or prominent and scarce. Dots come in every degree of prominence and quantity, from zero to many.

I always wondered why Garden Royal was not a lot more popular. It should be in every orchard. Over the years I put together a possible explanation for its failure. Garden Royal is an early (mid-September) apple. That's a bit too early for many people. The fruit size is too small. That makes picking more expensive. It's not a regular cropper. You might get a modest harvest every two or three years. Lastly, Garden Royal is a skimpy grower. While many varieties are ready to dig after a single year in the nursery, Garden Royal can take as long as three or four years to be ready to dig. That makes the trees three or four times as expensive to produce.

But despite all that, I loved this delicious apple. Most likely its drawbacks only became an issue when apple growing became a commodity enterprise. Commercial growers needed large volumes of annual fruit. The idiosyncratic apples lost out in the process no matter how good they were. Hard to live off the income from an apple tree with no apples on it. As a result, almost no one knows about Garden Royal.

Beefsteak, a.k.a. 'Freckles' from the old tree in Lincolnville

Now count the dots!

When David Hutchinson wrote to me, it was a pleasant surprise. Despite having never met, here was a kindred spirit who appreciates a truly great apple. He was looking for Garden Royal, and I was able to send him trees. He told me of another variety that was grown in Franklin County, a personal favorite of his called "Golden Ball." I wanted to see the apple myself and taste the fruit. And so I made the two-hour drive to the Maxwell Road in Weld on a magnificent October day in 1999.

Although A. J. Downing calls Golden Ball "a favourite in the state of Maine," its place of origin will probably always remain in question. (Saturn?) Downing believed it originated in Connecticut. Elsewhere, however, he wrote that it was born up here. Cole doesn't say where it came from, and Beach doesn't mention it at all. In the 1852 American Pomological Society report, the Maine apple report attempts to straighten out the confusion while providing a firsthand glance into the folk process, "Golden Ball —... often supposed to be a native of Maine, but is not. Some fifty years ago [1800] the scions were brought from Connecticut without name, and for thirty years or more [it was] known only as the Connecticut apple."

There is at least one other Golden Ball apple, this one across the Atlantic. When I was on an apple trip in England some years ago with Todd Little-Siebold's college-level apple history class, we had dinner with a West Milton cider maker named Nick Poole. Nick told me of an old Dorset cider apple called Golden Ball of Netherbury. Golden Ball of Netherbury had been rediscovered and identified in the 1920's by the once famous, but now closed, Long Ashton Research Center near Bristol. Nick described the apple as looking rather unimpressive until it was at peak ripeness. At that moment it was perfect to press; its juice being the color of gold.

Dave Hutchinson met me out by the driveway of his old family farm. We didn't bother going inside. He showed me the Golden Ball trees he had grafted. The farm's three huge, ancient specimens were gone. These were younger, but nevertheless mature. He showed me his Black Oxford and his Red Astrachan, and as we stood in the shady front yard, I found myself wondering about the Garden Royal and the Golden Ball and the names of apples. Who is the keeper and who is the dispenser of the name?

Dave Hutchinson, with Golden Ball apples, outside the family farmhouse in Weld, 1999

When Earland first introduced me to Garden Royal, I thought it was another cute Honeycrisp-type name. Then I learned about the history of the Garden Royal and about the apple itself. Then the name made perfect sense. It was an apple for the garden, not the orchard. It was a small, imperfect tree, frail when young, but give it some TLC and it will reward you. . . . When that small tree in your garden fruits next year or the year after, the fruit will be fit for a king and a queen.

Although Garden Royal is one of the best dessert apples ever, you'll never find the small round fruit in the grocery store

And what about the Golden Ball? I didn't know the apple yet, but as I stood there with one in each hand, I knew the name must be correct. There could be other Golden Balls out there lurking somewhere in Connecticut or southwest England, but if there was ever a well-named apple, this was it.

Dave Hutchinson's Franklin County Golden Ball is medium-large in size, more oblate than round, colored a rich, deep school-bus yellow and sprinkled with a conservative number of prominent russet dots. There are no stripes or blush. The apple has a wide, deep cavity with a small splash of russet that surrounds the stem and sometimes extends up over the sides. The calyx is majorly open. It ripens in October. It is tart.

I wasn't in a hurry that morning. I had a lunch date with friends who lived a couple of towns away and another orchard to visit later on in the afternoon, but I didn't need to rush. I wanted to hear more about this family and the apples they raised. Dave didn't seem to be in much of a hurry either. We stood under those ancient maple trees that lined the driveway, and he told me a story about names. This one was about his grandfather, the fellow who had grafted those old family trees. His grandfather was "the boy with no name."

Golden Ball usually has a splash of russet around the stem

He was born in 1850 in West Weld or maybe it was Carthage. He weighed two pounds. "He could fit in a quart canning jar," Dave explained. His mother could "slip her wedding ring onto his arm and up to his shoulder." No one thought he had a chance to live so they didn't bother to give him a name. They just called him "Bub."

Time went by, and the Civil War came and went. Meanwhile the baby with no name became a young boy with no name, and the young boy with no name became a twelve year old with no name. Still they didn't give him a name. By the time he was twelve, the boy was driving a team of oxen on the farm. Still he had no name. He was like one of those seedling apple trees growing along the stone wall. "Wow, this fruit is great. I wonder why nobody ever noticed it before. Maybe we should give it a name?" It was about that time that the boy decided that he needed a name. And so he named himself.

Topworking #11: Tape it like an Ace bandage

Throughout history, people have changed their names from time to time. You've got your Mark Twains and your George Orwells. Then, of course, we have Prince and Madonna and Lady Gaga and Marilyn Monroe and Bob Hope and my brother-in-law's uncle, Garry Moore. For some it's about being a celebrity. For some the reasons are devious. For some it's a matter of self-preservation. Many of the French up in northern Maine anglicized their names to avoid deportation or worse by the Brits.

Few of us, however, get to name ourselves the first time around. It's probably quite rare, to put it mildly. But "Bub" was faced with that prospect, and he rose to the occasion. He named himself Elisha Turner Hutchinson. No one knows where he found the name. Who was Elisha Turner? Was he a real person? Was the name an imaginative fabrication? If anyone ever knew, it's all been lost. There were Turners in Carthage at that time. It's possible that some old farmer named Elisha Turner could have been a mentor for the boy with no name. Dave has spent some time in recent years searching for a record of anyone by that name but without success.

Elisha Turner Hutchinson grew up and then left home and went west, along with his new name and thousands of other Mainers who deserted the state during and after the Civil War. Maine was running out of wood to cut, and it was running out of farmland to farm. The upheaval and displacement of the war itself had triggered an early incarnation of the mobile society. The draw for many, including Elisha and his brother, was the logging boom in the upper mid-west.

Although he made it as far as the Dakotas where he worked as a carpenter, it was in Minnesota that he met a young lady school teacher. "I'm from Maine," he told her. "So am I," she replied. "I'm from Weld." "So am I."

These days, many of us don't even know our next door neighbors' names, let alone the names of folks a few miles away. We might assume that years ago everyone knew everyone else in town. You'd have thought that the boy with no name would have known everyone's name in Weld. They all must have heard about

The Golden Ball (L) and Pound Sweet (R) trees in Harpswell are all that's left of an old orchard; now a driveway runs though it

him. But maybe having no name made Bub invisible to all but the keenest observer. Maybe he was like that unnamed seedling apple leaning out of the forest's edge.

Back then, you might also never have a reason to travel from one side of Weld to the other. As it turned out, she was from North Weld. How many times a week do you really want to hitch up the team, just to go to North Weld? By our standards, the roads were lousy. They were hardly designed for commuting and shopping. They were, at best, two parallel paths set about three feet apart, one for each wheel. Probably the closest thing we have to them these days are the driveways we call "camp roads" that still radiate out from a thousand Maine lakes. But even those camp roads are probably highways compared to what they had. It was a long way from West Weld to North Weld. Most people had no need to make the trip. Everything they needed was on the farm. It was Biodynamics when Rudolf Steiner was still playing in a sandbox somewhere in Croatia.

Elisha married the North Weld school teacher, and they farmed for fifteen years in Granite Falls, Minnesota, on a hundred and sixty acres—two plots, one ninety and the other seventy. Dave's uncles were born there. In 1901 they returned to Franklin County, Maine, first to Phillips and then to Weld. His mother-in-law was not well and needed help on the farm, so they came back home. Much later, Elisha liked to say, "If I'd stayed out west, I would have had dollars. Now I'm back east, and I have no sense."

Elisha Turner Hutchinson mowing hay at age 94 in 1944: courtesy of the family

After some years in Phillips, they traded farms with folks in Weld and wound up on the farm where Dave and I stood that afternoon. The farm on the Maxwell Road had a small orchard as every farm did back then. The orchard included Blue Pearmain, Golden Russet, Yellow Transparent, Red Astrachan, Fameuse and two or three Black Oxfords. It also included three huge old Golden Balls.

Elisha continued to farm with horses into his nineties. "He always had at least four horses." He was meticulous. "He never left a blade of hay in the field." Elisha was eighty-three when Dave was born. Before long, he too was raising horses. "I was twelve when I inherited a drag rake," he told me. Elisha died in the spring of 1949 at the age of ninety-nine. As he lay in his bed, he observed, "This must be the full of the moon. Better get your garden in."

Some years later I returned to visit Dave and his wife, Jeanette. Cammy came along with me. We sat together in their new house next door to the farm. Dave picked up where he'd left off as though I'd never left. He'd learned to graft and worked in the paper mills in Jay and later Hinckley. Like his grandfather, he loved horses. He raised Morgans, including a registered stallion named Pine Tree Playboy. "We called him Red." Red lived to be thirty-three. Dave pointed to a package wrapped in cloth. "That's one of my violins." He had taken up violin making at age seventy, and this was his latest. "My grandfather was a quiet man. I wish I'd asked him more."

Dave and Jeanette showed us a photograph of an elderly Elisha Turner Hutchinson. He is standing in his orchard with his scythe. The story goes that a young fellow saw him there and asked him his age. "I'm 94." "What are you doing with a scythe under an apple tree?" the young guy asked. "What should I be doing?" Elisha Hutchinson replied.

# Eighteen

# Naming Apples

It is a capital mistake to theorize before one has data. Insensibly one begins to twist facts to suit theories, instead of theories to suit facts.
Sherlock Holmes, *A Scandal in Bohemia*, p. 163

Cora's Grand Greening

Because there are so many spontaneous apple trees out there in the landscape, mostly planted by one animal or another or by passing motorists who toss their cores out the window, and because most people haven't yet grasped the fact that apples do not come true to type from seed, it's inevitable that many unnamed seedling apples are brought to me in the fall with the innocent expectation that I will know the name. Alas, these are the orphans. These are the apples with no names. These are the Bubs. Even if I knew the names of all the apples in the entire world, I can't help when the apple grew up from a seed. Sorry, your apple has no name until you give it one.

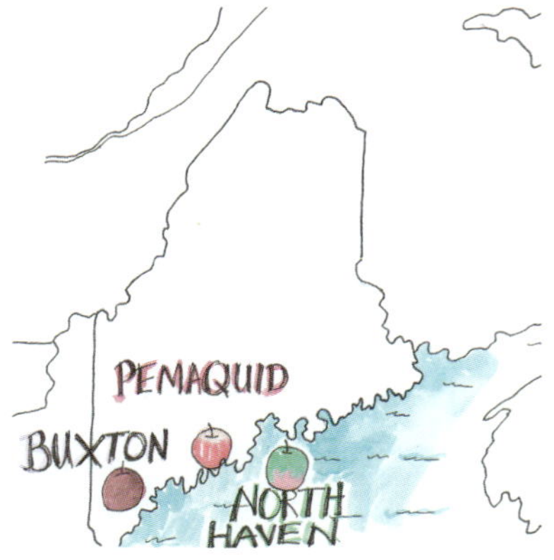


To some of these apple owners this news is not good news. First there is confusion. You mean, my apple has no name? I explain about the deer and the apples thrown from the car. But this may not help. How can it be that an apple can have no name? It's a cruel reality that can seem like a fate worse than death. Some even get angry. If I bring it back next year, could you try again? Please? And so they trudge away. Let down. Despondent. Betrayed.

You'd think the cider glass was less than half empty when it's really half full. In fact, it's way more than half full. It is full. You've got a tree. You've got a tree with fruit. You've got a tree with fruit that presumably you like enough to collect and stick in a bag and bring to me. Meanwhile, in backyards all over America there are thousands of poor souls struggling to get an apple tree started in the middle of their lawn. They can't get it to grow. It's riddled with borers and scab and getting eaten by voles and deer. But at least it's got a name. It's got a name! They bought it cheap on sale; the tag says Winesap, and they're so happy they found a heirloom at the big box store for only eight dollars. What a great deal. They pamper the little guy for ten years, and when it finally fruits, it's yellow. They just spent ten years thinking they had a Winesap, and it turns out to be another Golden Delicious!

Meanwhile, you've got these beautiful seedling apples growing in your woods or by the side of your driveway or along your stone walls. They're mature, they're fruiting, and they're healthy. But for some reason they don't seem worthwhile because they don't have a name. Little Bub made it twelve years without a name. Your apple tree can too.

A seedling apple is a naming opportunity, and it's an easy one. Fret not. It's not nearly as complicated as those expensive thoroughbreds. There are no rules. There are no regulations. There is no International Federation of Apple Authorities. Unless you plan to go head to head with Honeycrisp or SweeTango, there's really nothing you have to do at all. Chances are pretty good that no one will complain or sue you. Chances are that no one will ever know or even care. You get to name your seedling apple tree whatever you like.

In his 1862 *Atlantic* magazine article, Henry David Thoreau provides names for the various seedling apples he finds during his fruit exploring adventures around Concord. "There is…the Apple that grows by the old Cellar-Hole (*Malus cellaris*); …the Truant's Apple (*cessatoris*), which no boy will ever go by without knocking off some, however late it may be; the Saunterer's Apple—you must lose yourself before you can find the way to that; The Beauty of the Air (*decus aeris*); December-Eating; the Frozen-Thawed (*gelato-soluta*)…the Slug-Apple (*limacea*); the Railroad-Apple, which perhaps came from a core thrown out of the cars…"

As I searched the Waldo County orchards in those early fruit exploring days, I quickly began to amass a collection of favorites. Needing a way to refer to my discoveries, I named them. Some of them turned out to be real varieties. That was okay. Others were seedlings. I just needed a point of reference. I was beginning to be able to pick out old grafted trees, but I had no idea about any of the names. Dimmock's was just up the road, Foster's Red was a couple miles away, Sunrise Motel was down in Lincolnville. Not as creative as Thoreau, but they worked. Eventually I learned the true names of some of them. Dimmock's turned out to be Northern Spy. Foster's Red became Stark. Sunrise Motel turned into Yellow Bellflower. I still remember those provisional names thirty-five years later.

Harmon apples, a.k.a. Davis Purple

Some of those faux names stuck longer. Davis Purple was one. It's a good apple. I grafted a whole tree here on the farm. The fruit was brought to me twenty-five years ago by a fellow named Don Essman of Standish down in southern Cumberland County. The fruit is very dark purple and good for fresh eating. It's not Black Oxford although they do resemble one another. Every so often I'd pull out the old books and rummage around for a couple of hours contemplating Davis Purple. You'd think that a dark purple, top quality dessert apple would be easy to ID. It wasn't. Though I never saw the tree, I doubt it's a seedling. I did see Don again a few years ago, and he said the old tree died in 1999. Fortunately we have the wood so the variety won't be lost.

So it remained Davis Purple until a few years ago when I was having one of those contemplation sessions. If Don found the tree in Standish, as we can assume he did, maybe the variety originated in one of those nearby York or Cumberland County towns. I'd been looking for two Buxton apples for a long time: Harmon and Narragansett. Both were selected and named by J. H. Harmon in the later half of the nineteenth century. Buxton is in the northeast corner of York County, not far from Portland and bordering Standish to the northwest. I don't know the actual location of the old tree Don discovered, but it can't have been more than a few miles from Buxton.

Is Davis Purple really Harmon? Could it be Narragansett? I have good descriptions of both apples. Of the two, the more likely candidate is Harmon. Narragansett is too large and the stem too short. The mystery lives on. It will until someone finds some journal reference or an old-timer who just happens to know. For now I say, Davis Purple/Harmon. And if I never find out the original name, Davis Purple will do. Of course I can't help but wonder, who was Davis? Perhaps Don Essman found the tree on the old Davis farm. There must be a story there too.

Cora's Grand Greening with its bright pink blush

Some of those provisional names will never get sorted out. Instead I'll join the long line of re-namers throughout the history of apples. The large, old tree between the house and the barn on Bill and Becky Bartovic's Cider Hill Farm on North Haven Island could be one of those. Becky sent me a box of the fruit in the mail in 2006. I was wowed. The huge, deeply ribbed, olive-green apples look like gigantic green peppers. She wrote to me, "They make wonderful applesauce, cider—tart! But no mashing is required to turn them into sauce—not good for pie!"

A year later Becky introduced me to the tree when I was out on the island visiting orchards and giving a talk. Just across the road from the Atlantic Ocean, the windblown tree has been leaning southerly for a hundred and fifty years. Two beefy, y-shaped cedar logs are doing an excellent job holding it up. It was an apple I'd never seen before, and one I doubt I'll ever see again. It needed a name. Becky and I talked about the naming process, and she suggested Cora's Grand Greening. It was Cora Ames who farmed the place generations ago. She was a midwife who never married. Her father built the house in 1867. It might have been Cora herself who planted the tree. Cora Ames' Beach is just across the road. According to Becky, there's a photo of Cora riding her horse by the tree when she's in her twenties. Cora's Greening would have worked for a name, as would have Cora Ames, but Becky wanted everyone to know more about the apple itself. It is grand. So we settled on Cora's Grand Greening. It could be a seedling, or it could be grafted. Either way, it's old, and it's large, and we've saved it.

Cora has a deeply ribbed basin and an open calyx

These days when I'm out visiting old orchards, I keep a lookout for seedlings. I want to find them. Seedlings are the future generations of the apple. Each seedling is a new mix of genes. Each represents the potential for new blood in the orchard. It's fun to name the seedlings we find. It's a good way to keep track of them. Especially with so many fruit explorers out there searching for new cider apples, there's no reason not to enter into the name game.

Some of our favorite new discoveries are destined, no doubt, to be international hits. You can bet that Bitter Pew will be one of them. Bitter Pew is a rootstock sprout from an old, dead apple tree growing near one of those simple white-clapboarded Maine churches. The original tree was part of a small, grafted home orchard. That tree is rotted

The Cora's Grand Greening tree has been leaning downwind for a hundred years

away and gone, but a sprout from underground has grown up and become a new tree itself. The fruit is very bitter. Shavel Sharp comes from the old Sharp farm in southern Maine, now owned by the Shavels. The tree is only thirty or forty years old and is growing right in the middle of their lawn. Maybe it's also a rootstock; I doubt it could be any named variety. The fruit is like the name: sharp. Howlin' Wolf is colored like a Wolf River and makes you howl when you take a bite. Fuel Service shows promise as a cider apple. It's a soft, yellow bittersweet. The tree is located in the yard of an oil delivery place. Fuel up! Avon Calling is a tree that's not much more than a large shrub growing by the side of Route 4 in Franklin County not far from the Avon town line. My apprentice Laura said we should stop; we wound up picking the entire tree. Vermeaux is a green-skinned apple from up past the Town Office in "Paler-meaux." It's sharp, and it's prolific. Melody Maker is just that. It's a good-sized russet that will really jazz up your cider and put a song in your heart. There are countless more seedling apples waiting for names by the roadsides, along stone walls, and on the lawns of ten thousand Maine cottages and camps. This is our big chance.

Among the dozens of apples I was asked to identify in the fall of 2010 came one from Hedy and Ann Ertman of Pemaquid. I learned of their request through a series of emails. Hedy wrote to MOFGA. MOFGA forwarded his email to CJ Walke, the MOFGA orchard guy, and CJ forwarded the email to me. The Ertmans were summer people who had hoped to bring an apple to the Great Maine Apple Day in late October, but unfortunately they were going to be out of town. Could someone help us identify our apple? One thing led to another, and I wound up with a bag of their apples a few weeks later. I had a look but was stumped. I got back to them later in the fall and told them so. Perhaps it was an old, coastal variety. Perhaps it was a seedling with no name.

The following August I stopped by their place when I was down that way. It's a beautiful spot. Pemaquid is one of the five villages in the town of Bristol at the southern end of the peninsula below Damariscotta. That area of Maine was one of the first to be settled by Europeans, about 1625. Being located close to the mouth of the Kennebec, the inhabitants found themselves in a tug of war between New France and New England. We know how it all turned out, although not without a lot of unpleasantness that included various forts being built and burned along the way. The beautiful strip of land that flares out into the blue Atlantic Ocean also proved to be a temptation for pirates, including the infamous Dixie Bull. The buried treasure he swiped from the villagers still awaits the lucky digger somewhere out on one of those islands in Casco Bay.

Hedy and Ann gave me a tour of their modest place which was a newer version of the original cottage. For decades they'd been summering there near the tip of the peninsula, just a stone's throw from Fort William Henry. The cottage had been in Ann's family for nearly a hundred years. Before then, when the vast majority of the local inhabitants were sheep, it belonged to a shepherd. The tree was just about the same vintage. It was located less than thirty feet from the large, open cottage porch that looked out beyond the road to the ocean.

Topworking #12: Slip the tape inside itself to tie it off

It was clear the Ertmans loved their place, their tree and its fruit. In August they collected the drops everyday and set them on a small table by the road with a sign for the neighbors and any strangers who might go by. "Free Apples— Good Pies, Crisps, Apple Sauce." The tree had done remarkably well with almost no care. All it really needed was a name.

After seeing the tree and looking at the fruit again a second year, I decided that it was a seedling—an orphan with no name. The fruit is about the size of a Mac, roundish-conic in shape and somewhat ribbed. The basin is deeply furrowed. The stem is not long, medium at best. The skin is mostly red-blushed, overlaid with short, darker-red stripes. Some specimens have a pronounced russet splash that fills the cavity and spills out over the rim. While not a dessert apple, it cooks up quickly into a nice creamy sauce. The pie we made a week or two later was delicious. It had a peachy texture and a tart subtle flavor. This was a good apple. It did deserve a name.

Ann and Hedy at the 2012 Common Ground Fair with their apple in need of a name

Hedy and Ann are retired. They spend their winters in the southern California desert, but in the age of email it was easy to keep in touch. I contacted them that fall after they headed west. I hoped they wouldn't be too disappointed. It turned out they were

Everybody celebrated while the Robinson Pemaquid tree looked on with delight

far from it. Their enthusiasm was immediate and infectious. They snapped into action. This was going to be great. They would name their apple and then host a neighborhood naming celebration.

In December Hedy wrote to me, "Our first reaction is that "Pemaquid" is too broad a term since it is such a wide area and this is only one tree. We're inclined to the name "Robinson," since they were the visionaries who bought the earlier cottage many years ago. Do apple varieties ever have more than a one word name? e.g. the "Robinson/Pemaquid" apple might give it both location and history. What is the formal process of naming the tree? Does MOFGA do it, or do you, as apple czar, do it? … or …?"

By the next summer when they returned to Maine, the plans were already mostly in place. On September 1, 2012, I headed back to Pemaquid. It was a beautiful, clear blue, late summer day. About forty friends, neighbors and relatives, most of them attired in their festive best, sat under the shade of the old apple tree in front of the Ertman cottage porch. Hedy was the emcee, decked out in a snazzy patchwork coat of many colors. The porch was decorated like Yankee Stadium on the Fourth of July. A federal appellate judge from Kentucky read a proclamation. A string quartet played a medley of apple songs. A local poet read a poem written for the occasion. There was food. There was cider. I contributed the five-minute version of the history of apples in Maine; I even wore a tie. It was a lot of fun. To top it off, an engraved plaque was hung around the old tree's mid-section.

Robinson Pemaquid

Here longer than a century,
grown tall enough to watch the sea,
I've never had a name.

I've shaded families over time
and spread out limbs so kids could climb —
they never called my name.

I've nodded at all passers-by
whether or not they noticed my
salute or asked my name.

Happily deemed "not in the way
of nuthin' " I could safely stay
deep rooted, though unnamed,

through hurricanes, nor'east destruction,
"cottage" de- and re-construction —
surviving, still unnamed

and still unbowed. And though I lean
a bit these days as you have seen,
this prop is not a name.

My blossoms showered fragrant, white,
each spring to end each winter's blight
but not my nameless plight.

Fruit, of course, I've mass-produced
to make great sauce and pies and juice
that never had a name

except for apple this or that,
generic, but with no eclat,
no scientific name

until now…when I welcome fame
a state pomologists proclaim
that ROBINSON PEMAQUID IS MY NAME

by Anne Johnson Mullin 8/26/2012

# Nineteen

# Deane or Nine Ounce

He picked it up and gazed at it in the peculiar introspective fashion which was characteristic of him. "It is perhaps less suggestive that it might have been," he remarked, "and yet there are a few inferences which are very distinct, and a few others which represent at least a strong balance of probability."

Sherlock Holmes, *The Adventure of the Blue Carbuncle*, p. 246

Deane, a.k.a. Nine Ounce

Is it fair that some apples have no names and others have two? Is it fair that some apples have lots of names? In the world of heirloom apples, synonyms were a dime a dozen. Consistency in names was never a high priority in those early days of North American orcharding. Even when a single name was agreed upon, and it rarely was, spellings varied.

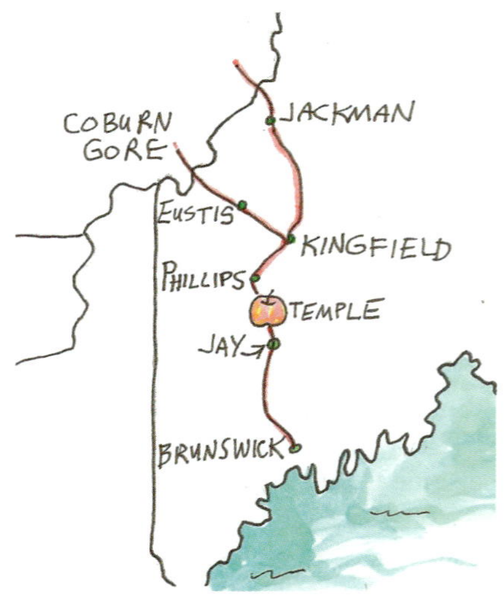


The synonym in the world of apples is like the alias in the world of crime. To avoid being identified, start by changing your name. Bad guys do it on purpose. Apples never had to. They picked up new names wherever they went. Some of those synonyms were recorded, others not. The names changed routinely from state to state or county to county or even town to town. Some apples are still thought of as two or even three different varieties when they're actually just the same apple with multiple names. The name was simply a convenient way to refer to what you had in your orchard. Most varieties were only for sale locally if at all. Having a universally agreed upon name was irrelevant. It only mattered that when you told your neighbors about your apple, they knew what you meant.

Early Harvest is a yellow summer apple that probably originated on Long Island in about 1800, and then spread around throughout much of New York and New England. Before it ripened in mid-summer, it was the preferred choice for "Green Apple Pie." The fruit is roundish-oblate. Sometimes it's small. Sometimes it's large. Sometimes it's angular; sometimes it's not. One source lists twenty-seven synonyms for the apple. Whenever someone couldn't identify it, they just made up a new name. Why not?

Some synonyms are simply spelling variations from a time when communication was largely verbal and spelling was largely casual. It was a time of spelling fluidity, long before the days of autocorrect. One of the most classic New England heirlooms is what we now call Tolman Sweet. I've also seen it as Tallman Sweet, Talman Sweet, Talman's Sweeting, Tollman Sweeting, Tolman Sweeting, Taulman Sweet, Tomey Sweet, and Tallman's Winter Sweet.

Some synonyms bear no resemblance to one another. It could be like playing the telephone game we used to play as kids. I lean over to my neighbor and whisper, "Woodpecker." A few whispers later it becomes "Pecker." By the time is gets to you it's "Calville Butter" or "Steele's Red Winter" or maybe even "Felch." Eventually it becomes "Baldwin." All the same apple. Names morphed over time. McIntosh was originally known as Granny's Apple, then Gem, then McIntosh Red. As its popularity swelled, the Red got dropped and it became McIntosh or simply Mac.

"Quite tart unless fully ripe." Maine Agricultural Yearbook 1874

In the late nineteenth century there was an effort by the American Pomological Society and others to limit apple names to one word apiece. Up until then no one seemed to mind growing apples called Westfield-Seek-No-Further, Hubbardston Nonesuch or Duchess of Oldenburg. But with the publication of William Henry Ragan's *Nomenclature of the Apple* in 1905, the age of the streamlined name was upon us. Westfield, Hubbardston and Oldenburg became the new official names. I like the old long names better. The one-name-campaign coincided with a separate effort to coordinate names across state lines. Generally, it was the Maine varieties that took the hit on that one. Don't want two apples with the same name, even if we have to make one of them longer! The Somerset apple in Mercer became Somerset of Maine. Collins became Cherryfield. Northstar became Dudley Winter. Locally, however, they still call it Northstar, and they still call it Somerset and rarely bother with the "of Maine." One of Joseph Taylor's nineteenth century introductions from Belgrade, Maine, originated in the nearby town of Rome. With a wink to Ragan, T.T. Lyons and the American Pomological crowd, Taylor dubbed his apple Rome of Maine, just in case you might confuse it with one of the most iconic of all apples anywhere. I have no decent description of Rome of Maine, but I love the name, and I'm determined to find it. Probably in Rome.

The Woodpecker apple eventually became known as Baldwin, America's most important apple for 150 years

The Baldwin monument in Wilmington, Mass. Long ago known as Butters, Woodpecker and Pecker...

Not knowing the synonyms can make the life of an apple identifier difficult. Knowing them can be useful. Not long ago a good friend, Liz Lauer, brought me an apple she called Rubicon. It came from an old orchard north of Bangor in southern Penobscot County. I knew that I'd heard the name before, but couldn't picture the apple until I learned that one of its synonyms is Paw Paw. Paw Paw is an apple I'd seen in the Tower Hill Botanic Garden collection down in Massachusetts. Another time, friends brought me an apple they called Martha Stripe. Again I didn't know the apple until I looked it up and saw that it's also known as Lyscom.

Some synonyms have been sorted out while others may never be. Two of my favorite apples, Canadian Strawberry and Sweet Red, are likely Maine synonyms for out-of-staters. But which ones? I've never found either in print. Will I ever learn their real names? The apple identifier can take comfort when attempting to ID either of them. Maybe no one else can identify them either. We call them

Canadian Strawberry and Sweet Red for now. Those were the names given to us. Someday we'll learn the "real" names. Or not.

What is a real name anyway? Is the real name the first name given to an apple? In that case, New England's most famous apple would be called Woodpecker, not Baldwin. Is it the name that became generally accepted over time? But what if two names have become generally accepted over time like Newtown Pippin and Albemarle Pippin. Northerners call it Newtown and the Southerners call it Albemarle. Which one is right? Maybe it's the name that's been used the longest. Those in Solon, Maine, have been calling it Canadian Strawberry for generations. And what about the twenty-seven different names for Early Harvest? Are they all correct? Are they all wrong except one? Which one? A rose by any other name?

Topworking #13: Seal the graft with TreeKote

Plurals can be challenging too. Most apples are straightforward in their plurals. Black Oxfords, Baldwins, Garden Royals, and Golden Balls are all pretty easy. Should Wealthys and Starkeys be Wealthies and Starkies? I think not, and I hope that you'd agree. Northern Spy, however, is problematic. I prefer Northern Spys over Northern Spies. The later reads like someone named Northern is engaged in sneaky activities.

What about varieties that sound like they're plural even though they aren't? Collins is one of them. Is one Collins apple a Collin? Are three Collins apples, Collinses? On one particular trip up to Aroostook County, I kept hearing about an apple called "Dutch." All of a sudden I realized that if a single apple was a Dutch, then the plural would be Dutches—or as it's usually spelled, Duchess. What a revelation. All these people who had probably never seen Duchess in print were assuming that Duchess was the plural of Dutch. Of course, it should be. If we talk about Spys, and Macs and Cortlands, and we mean plural, logic would tell us that Duchess is the plural for Dutch. Most people who know and use these old heirloom varieties rarely see them in print. Why would they? They're using the apples, not writing about them. So the single of Macs is Mac, the singular of Spys is Spy and the singular of Duchess is Duchess? No. The single of Duchess should be Dutch!

Sometimes an apple has two names and becomes equally well known by both, such as Newtown Pippin and Albemarle Pippin. On the other hand, some, like Gano and Black Ben Davis, are considered by some people to be separate varieties, and by others to be one and the same. In other instances an apple, such as Somerset of Maine, kept its original name and picked up a geographical tag line, à la Leonardo da Vinci, to distinguish it from all the other Somersets (or Leonardos) in the world. As many names were shortened, there was the occasional, and inevitable, confusing consequence. Officially Duchess of Oldenburg became "Oldenburg" but sometimes it became Duchess. Before long, even many experts assumed Duchess and Oldenburg were two different apples.

Call it Deane or call it Nine Ounce. Either way it's great: photo by Jo Josephson

Some apples have two names that both stick. We feel compelled to say them both rather than settling on one or the other. The most famous of these is Fameuse. Sometimes we say Fameuse. Sometimes we say Snow. Usually we say some weird version of "Fameuse which is also called Snow" or "Snow-Fameuse," or "Fameuse-or-Snow." Might it be easier if the apple had ten names? Then we'd be forced to pick one.

Another one of these twofers is Deane, sometimes spelled Dean which was also called Nine Ounce. It's uncertain why Deane and/or Nine Ounce has two names. As far as I know, no one found any other Deans or Deanes out there in the mid-nineteenth century when the apple first appeared. I doubt there was another Nine Ounce.

Cyrus Deane farmed in Temple. The Deanes were Quakers, and the Quakers and the Deanes were abolitionists. Temple was on the underground railroad, seventy miles north of Brunswick and seventy miles south of the Canadian border. The Deanes' farm was a safe house. Richard Pierce writes in *A History of Temple*, "The Cyrus Dean family, in particular, were active in this work and a descendent (Flora Dean Weeks) recalls vague traditions of Negroes arriving under cover of darkness from the southern part of the state and the following night being on their way to the Canadian border."

Because of the extreme risk to those involved in assisting escaped slaves, little verifiable information about this episode in Maine's history exists. We do know that dozens or perhaps hundreds of escapees traveled through the state annually. Beyond that we try to piece together a narrative as best as we can.

Brunswick was an abolitionist stronghold and an important staging point for runaway slaves. Despite its proximity to the coast and its long history of ship building, by the mid-nineteenth century Brunswick had diversified into manufacturing. The town had the luxury of feeling un-beholden to the shipping and fishing industries. Shipping and fishing were dependent on interstate and international commerce, and that meant slavery. Most coastal towns of Maine couldn't help but support the slave trade. They built the ships that brought the slaves from Africa and they built the ships that delivered the New England salt cod and pickled mackerel that fed those same slaves down south. But not Brunswick. Brunswick was situated at the falls on the Androscoggin River. Falls meant manufacturing, and manufacturing seemed to be far enough removed from slavery to make Brunswick a haven for runaways. Brunswick also had Bowdoin College, which was well established and had become a focal point of abolitionist thinking. According to local lore, an elaborate web of secret tunnels was built connecting safe houses throughout Brunswick. Even the college was on the grid. Tunnels also connected much of Topsham on the other side of the river.

Heading for Coburn Gore

These were not run-of-the-mill, crawl-on-your-belly tunnels; they were tall enough to walk in, and some were even large enough to drive a wagon through.

Escaping slaves knew to get themselves to Brunswick. From there they could follow the river up through Lewiston and Auburn to Jay where they would veer inland to Temple and Phillips, and up what is now Route 142 to Kingfield. Their last push to Canada took them either northwest along the Carrabassett River to Eustis and Coburn Gore or east up the Kennebec to Jackman along Benedict Arnold's 1775 route. Making the last hundred-and-forty-mile trip from Brunswick to the border was tricky, especially in winter when snow was deep, travel difficult and tracks could be followed. Bounties were high. Slaves didn't have the luxury of determining when they'd arrive in Brunswick so they were often forced to winter-over until the weather broke. Then they ventured north, many of them sleeping over in the safety of the Deane farm at the halfway point.

In 1851 Harriet Beecher Stowe wrote *Uncle Tom's Cabin* in Brunswick. In 1858, at the height of the underground railroad and two years before Abraham Lincoln declared, "Government cannot endure permanently half slave, half free…" the Deane family's apple was first exhibited at the meeting of the Franklin County Board of Agriculture. That apple was labeled Nine Ounce. A year later it was exhibited again, this time as Deane. In the ensuing years Deane (or Nine Ounce) gained popularity. It became rather common in Franklin County and was trialed in various locations around the state, including Bath, Orono, and points in between.

One place the Deane apple had been growing for a hundred and fifty years or so was in the orchard behind Michael Rothschild's old, rambling farmhouse at the top of Tory Hill in Phillips, deep in the heart of Franklin County and fifteen miles north of Temple along the underground railroad.

Wendy and Michael with one of the four Deane trees, 2018

In the fall of 1999 Michael and his longtime partner, Wendy Fleming, contacted me about visiting them and identifying some apples. I don't recall how they'd heard about me. Most likely it was through Howard Wulf, a photographer, poet and orchardist who split his time between his parent's farm in New Portland and his own in Unity. Howard had done much for me over the years, and I knew that he visited the Rothschild's farm now and then.

I had a sense that Michael and Wendy would be interesting people to meet so I put together a plan to include Phillips on my next trip up that way. I like to combine my fruit exploring missions whenever I can. If your tree is only a few miles from here, or on my way to work, I'll probably drop by without much fanfare. If you're farther away, I'd rather visit three or four places in one blast. Phillips is a couple hours drive and not too far from Weld, so I combined my trip to Michael's with visiting Dave Hutchinson's Golden Ball trees. After saying goodbye to Dave I puttered my way down various dirt roads and emerged at the top of Tory Hill. There was the farm, with its unobstructed views to the north and south. You could just about see Brunswick one way and Coburn Gore the other.

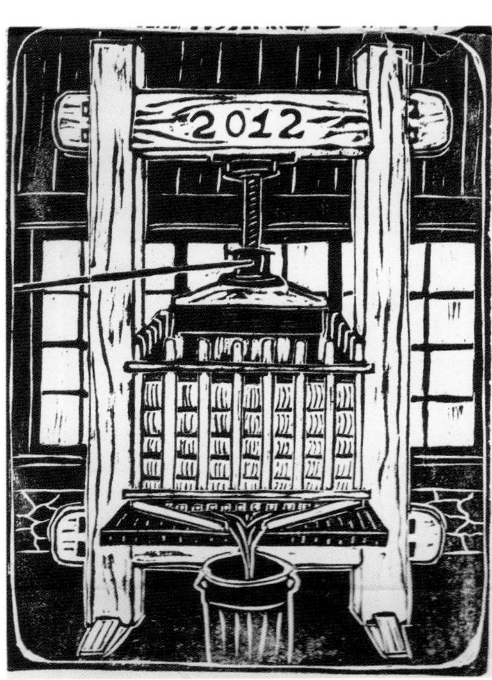

The Rothschild cider press: woodcut by Walker Fleming

Most people take some time to get to know. Others you never know, no matter how long you try. Sometimes you find out years later that the person you thought you knew was someone else altogether. Some people have a wall around them so impenetrable you don't even bother to try. Sometimes you meet someone, and you feel as though you've known them all your life despite the fact that you know you've never met, at least not this time around.

I pulled into the driveway, hopped out and found my way to the front porch. The door opened, and Wendy led me into a long, warm kitchen with a low plaster ceiling and a floor of wide pine boards with cracks in between. When Michael entered, we shook hands. He was one of those people I felt as though I'd known before. Maybe I had. Maybe it was long ago. Maybe he knew so too.

Michael is an artist, a writer and a farmer. He went to Colby, but graduated before my time. He rarely leaves Phillips although I learned that he likes Fenway Park. He says his friends don't understand why he lives on an old farm at the top of a hill in western Maine, two hundred miles north of Boston. Maybe he could tell that I did understand. There I was, and there we met, and off we went on a tour of the farm. He showed me his studio and office tucked into one corner of the old farm house. We visited his work shop in the barn where he was carving an enormous relief of a halibut on a piece of Monson slate. He was also in the process of building a huge oak cider press with a gigantic hand screw. We checked out his nursery where he lined out young grafted trees destined for their orchard. And then we visited the old orchard itself of about two hundred trees out back behind the barn.

About three quarters of the orchard was Wealthy, all set out in one large block. The other quarter was Baldwin. The Wealthy trees would have been planted towards the end of the nineteenth century, about the time Wealthy first came to Maine. The Baldwin block could easily date from fifty years earlier. They

Harry Rothschild, Wendy Fleming, Michael & Chee Rothschild all leaning on Cyrus Deane, 2018

were two distinct orchards side by side. Set into the Baldwin rows were four trees of another variety, three in one row and one in the next. All four appeared to be as old as the Baldwins, probably planted the same spring. That apple was why I was there. Could I identify the fruit?

As you know, I always ask myself first, "Is it a seedling or a grafted tree?" In this case there was no question. There were four trees of the same variety all within spitting distance of one another. These were grafted trees. They had to have a name. So then I ask myself, "Have I ever seen this apple before?" That's where visual memory kicks in. I see a painting in a book or on somebody's wall, and I know if I've seen it before. I might not know the name of the artist; I might not even like the painting; but I usually know if I've seen it before. It's the same with an apple. Have I seen it before? If I have seen it but I don't know its name, perhaps it's one of those perennial headaches. Some varieties have defining characteristics subtle enough to keep me confounded for years. I might have seen one of them a hundred times and know I'm seeing it again, yet still I don't know what it is. What's wrong with this picture?

Sometimes I look at the apples and something tells me that it's a variety I've never seen before. That was the case that October afternoon on the top of Tory Hill. If it's something new, it might be something local. I know I've never seen it before. I'm far from home. It's grafted. The trees are a hundred and fifty years old. Where are we? We're in Franklin County. What was growing here? What are the local apples in Franklin County? What are the varieties from up this way that I thought were lost? That I thought I'd never see again?

I collected a bag of fruit. By that time it was dark. I said my goodbyes and headed home. Did I let on what I was thinking? No, I just said goodbye. This would be great. Could this be the Nine Ounce apple of Franklin County, also known as Deane? Deane or Nine Ounce. Tonight I'll drive home; I'll weigh them tomorrow.

I arrived home and immediately got to work. Why go to bed? I pulled out my photocopy of W. M. Munson's 1907 report describing in detail a number of obscure Maine apple varieties, including Deane. Over the years I've relied heavily on the research and writing of Welton Marks Munson for an overview of the state's historic apples. Munson (1866-1910) of the University of Maine wrote and published extensively about plants, including vegetables, ornamentals, native perennials, orchard fruits, and, happily for me, apples. I had obtained a copy of his *Preliminary Notes on the Seedling Apples of Maine*. In thirteen pages of photos and text he describes thirty-eight apples originating in twelve counties. Not only does he describe each variety in great detail, he also pinpoints the location where most of them originated.

Typical Deane apple weighing in at about nine ounces: "A popular apple wherever known. Productive" Maine Agricultural Yearbook 1874

When I made up my first batch of WANTED posters, they all were based on Munson's descriptions. Many of the apples painted on the 2009 Common Ground Fair poster were also based on information from his report. I had created a Deane WANTED poster using Munson's description and even drew an "artist's composite" which turned out to look a lot like the apple that was sitting in front of me that October night. Deane is medium-large sized (and weighs about nine ounces). It's "oblate or roundish conical, sometimes a little angular, and flattened at the base; skin whitish, shaded and obscurely splashed and mottled with red, with numerous yellowish dots; stem short, small, inserted in a rather large, greenish cavity; calyx closed; basin medium, slightly corrugated; flesh white, fine grained, tender, juicy, with a sprightly vinous or subacid flavor. September and October." The description was a match. The apple originated on the Cyrus Deane farm in Temple sometime before 1874. Temple is two towns over from Phillips.

Deane is a beautiful apple. It looks like a cross between Northern Spy and Duchess. Maybe it is. The greenish-yellow ground color can be very light. The surface is partly to mostly shaded and rather obscurely splashed and mottled with light red and pink stripes and blush, then moderately sprinkled with those yellowish dots. The red stripes have a bit of that Spy-pinkishness. You can see large areas of the ground color here and there. The shape is slightly flattened with a deep basin. It is also slightly crowned, but only to the touch; it's not obvious like a Red Delicious. Like many of the old varieties, it gets no scab.

The white flesh of the Deane is fine textured and melting, with the consistency of a Mac. It's juicy and tart, and suitable for dessert and for cooking. Its season is said to be September and October, but we still cook with them into December. The sauce is tart and tangy with a fluffy texture and a pink-apricot color. In the 1889 Maine Pomological Society report, Deane was described as, "One of the most profitable, and one of the best autumn apples."

I contacted Michael and Wendy to let them know what they had. They were happy. I grafted it onto one of our trees here at home the following spring. That large branch has been bearing fruit for several years now. We grafted trees for the Fedco catalog. We put it in the orchard at MOFGA. I spoke at the grange in nearby Farmington one fall and gave a grafting class in Phillips. The trees are getting spread around Franklin County again. The Deane or Nine Ounce is back. This time it won't get lost.

There are other apples from up that way that remain on the loose. One is Franklin Sweet. I'm assuming that it's sweet and red. It might make good cider, and as far as I know it has only one name. I want to find it. And there are others in Franklin County, including Boardman, Sarah, Russell, Parker Sweet and Hoyt Sweet. We have the list. We have the Wanted Posters. We are on your trail. We will track you down.

## Twenty

# Getting Lucky

I call it luck, but it would not have come my way had I not been looking out for it.
Sherlock Holmes, *The Adventure of Wisteria Lodge*, p. 882

Cherryfield a.k.a. Collins

Unless you're on the way to eastern New Brunswick or Nova Scotia, you could live in Maine your whole life and never go to Washington County. It's a long way from everywhere else in Maine. It got stuck over in a corner off the beaten track. It's Downeast—way Downeast. When you go north up the Maine coast, you really don't go north at all. You go east. You could get in a boat in Portland and sail due east and hit Nova Scotia. From Palermo you drive half an hour east to get to Belfast on the coast. If you hang a left and continue along the water from there, you keep on going farther and farther east. And then you get to Washington County. Cherryfield is one of the first towns you come to. It's about a hundred miles up the coast from us. A hundred miles east but not more than ten miles north.

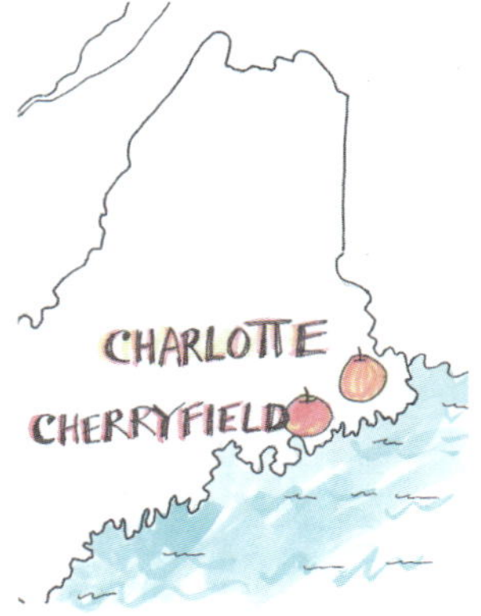


Washington County is not what you'd call the epicenter of apple growing in Maine. It is the epicenter of another important Maine crop: lowbush blueberries. It is, in fact, the world epicenter of lowbush blueberries. The low-growing ground cover, *Vaccinium angustifolium,* has covered thousands of acres of Washington County ever since the glacier receded ten thousand years ago. The blueberries grow wild almost everywhere across what looks like the heathered moors of Devon. Native Mainers have always eaten them. When the Europeans arrived, they immediately got into them too. In the fall the tiny leaves turn a dazzling combination of reds, oranges and purples. The visual effect is over-the-top amazing. Lowbush blueberries are delicious. Many Mainers don't consider the highbush blueberry (*Vaccinium corymbosum*) to be worth eating. You call those things blueberries?

But like everywhere else in Maine, Washington County also grows apples. Apples love all of Maine. Once the Europeans arrived, the apples were not far behind. Early on, the county had a vibrant apple economy growing an assortment of varieties that came from maritime Canada, southern New England and across the Atlantic. Apple seedlings soon popped up all over Downeast. Mostly the apple varieties that originated in "Sunrise County" were only locally known. Understandably so. The early state meetings of the pomological society and other agricultural organizations were held in the central and southern parts of the state, where most of the people lived. As a result, the majority of our documented Maine varieties come from areas where decent roads allowed growers to hop in the wagon or sleigh and get themselves to a meeting to show off what they'd discovered. It's not an accident that so many varieties originated along a thirty mile stretch on either side of the Kennebec River from Fairfield down past Gardiner. While it's true that some of the best soils in the state were along the river, perhaps more importantly, that was where the organized action was. It could be that nearly as many apple varieties originated Downeast. They just didn't make it out of Washington County.

Paul Molyneaux & Regina Grabrovac

My pomological connection with Washington County began in the early nineties when I met a fellow named Paul Molyneaux. Paul and I connected instantly. He is a stocky guy with fang-like teeth, an infectious smile and a permanent twinkle in his eyes. He's a fisherman, a storyteller and a writer. One of his books is about through-hiking the Appalachian Trail with his young son. He's married to a fruit exploring friend of mine, Regina Grabrovac. Regina is always on the lookout for old apples. She was the one who found the Bourassa apple not too far from their home in Washington County.

The first time we met, Paul told me a story about an apple from down his way. We met, we stood, he smiled and the story began. The apple was called Beauties of Wellington. He was emphatic about the plural. No singulars. Each apple was a Beauties of Wellington. I knew at some point I would have to make the trip to Washington County to see it myself.

The Wellington in Beauties of Wellington is not a place. It's a person named Wellington Johnson James, one of a select group who get to have a last name for a first name and a first name for a last name. I've known a few others. All males. One of them was a neighbor of ours, Kempton Tobey, much my senior who taught me a lot about Palermo history. Another married my mom's cousin. His name is Arsen Charles. How does that happen? It can wreak havoc on a young boy's life. "Is Charlie Arsen here?" "Tobey Kempton?" "Jimmy Wellington?"

Wellington Johnson James was born in December 1861, eight months after the attack on Fort Sumter and the first days of the War of the Rebellion. He died in 1945 bookending another war. He was spared fighting in either, being just a baby for the former and a man of ninety when Roosevelt declared war on Japan. Somewhere in between those two iconic dates he discovered a seedling apple that had sprung up in his pasture in Charlotte, Washington County, Maine, not far from the Atlantic Ocean, and thousands of miles from all that hell and conflict. He knew he'd found a good apple, and he kept track of it. Funny how life goes on. If he chose to name it, no one will ever know.

Sometime near the end of his life, Wellington clipped scionwood from his seedling apple tree, and gave it to Reuel Furlong, another Charlotte resident, with instructions to keep it going. He knew it was the now-or-never time to pass the baton. Otherwise his tree with the beautiful apples would someday die and be gone forever. Reuel gave the wood to his son, Damon, to graft. Damon was a local grafter. The month was June. The year was quite possibly 1945. Despite the fact that June is usually too late in the spring to do any topworking in Maine, according to the story, Reuel told Wellington James that "if anybody can graft these now, it's my boy." Although most of us finish our topworking by mid-May at the latest, Damon Furlong was just as capable as his father said he was. The grafts took, and the two trees are still standing on the Furlong family farm seventy years later.

Paul told me all this within minutes of our meeting. There he was with his gleaming eyes and his mesmerizing smile. We were standing together at a conference that had nothing to do with apples, surrounded by dozens of people milling around us, transported to our own beautiful world of the Beauties of Wellington.

It was Damon Furlong who named the fruit Beauties of Wellington, and he was the one who was emphatic about the plural of Beauties. When Paul once referred to the apple as a Beauty of Wellington, Damon was quick to correct him.

On October 4, 1996, I made the trip up to the Furlong farm in Charlotte to see the two trees. I met Paul at his place, and we drove the last few miles together. By that time a sprinkling of other Beauties of Wellingtons (oh those plurals) had been grafted in the area, but the apple was still exceedingly rare. The two Furlong trees were somewhat buried in a massive thicket of small hardwoods and American Cranberry, but both looked healthy and were producing. I finally got to see and eat the fruit. It was worth the four-hour drive Downeast. The apples lived up to the name. They were already looking magnificent despite the fact that they were not fully ripe. They had good color and good flavor.

Beauties of Wellington—always plural: photo by David Grima

Each Beauties of Wellington is perfectly symmetrical, round to conic in shape and medium in size. No ground color is visible as the apple is entirely red blushed and striped. The surface is covered with a mass of small white dots. Each apple is a beauties. The cavity is acute and deep, and the basin varies from being almost nonexistent to being medium in width and depth and abrupt on the sides. The stem is medium in thickness and length. The flesh is fine-grained and slightly off-white with a faint hint of a pink hue. The core is fairly large. Core lines are meeting or slightly clasping. I headed home with half a dozen fruit. The following winter, Paul sent me scionwood, and I grafted it onto one of the trees at home as well as on one over at MOFGA where it could begin to beautify central Maine.

Beauties of Wellington is one of those apples that applesauce makers love. I made a small pot of it this morning. I cut them up at 7:00 AM and by 7:15 they were ready to spin through the food mill. Sometimes I like to cut out the cores when I make sauce. That eliminates the need for the mill. More often I just cut up the apples and toss them into the pot with a half inch of water. That way I get all the apple, including the core, in the sauce. Running five or six apples through the mill takes only a minute or two anyway. I think it's worth it. Beauties of Wellington cooks down fast. The sauce is thick, creamy and apricot-orange in color. It's a little tart but not too. It's excellent. I'd call it a superior sauce apple.

Seedling apples even love to grow along the ocean's high tide line

Sometimes when I mention applesauce, you might chuckle. Can't you make sauce from any apple? Isn't that the way to use up the junk? Far from it. I make a small pan of sauce five or six days a week from September to May. I covet the good sauce apples. They are among the best of all varieties. It's barely eight o'clock in the morning, and I've already eaten four or five apples. Try to do that with raw fruit and see what happens. Good sauce apples cook quickly and thoroughly. Fifteen minutes in a pot brings out a rainbow of flavors. There may be no better meal in the world than apple sauce and oatmeal. Pretty cheap too, but fit for a king or queen. Occasionally I might throw in some dried plums, but please, never any sugar or spices in the sauce.

Fast forward to the third week of October 2006, nearly ten years from the day Washington County first entered my life, and I find myself headed all the way Downeast again. It had been too long a hiatus. The Sox had just completed a poor season two weeks earlier, the same day as a 4.2 earthquake hit Maine. Baseball season winds down, and apple season cranks up. The apples were flowing in from all over, and as fast as I could look, see and guess, the IDs were flowing back out. On the 14th a package arrived containing two green and red, long-stemmed, conic-shaped apples from Cherryfield, along with a note from someone I'd never met named Barbara Maurer: "These might be Collins/Cherryfield apples, but we're not sure. Please let us know. I got these from Margery Brown who also lives in Cherryfield."

A lot of packages arrive every fall. Each one holds within it a possibility. I never know what's inside. I try not to expect too much. I usually open the boxes when I'm alone so that if it's a McIntosh or, God forbid, another seedling, no one will see the disappointment descending over my face. Not another Black Oxford? But sometimes I get lucky. Just like on Christmas morning sixty years ago when I'd snap the ribbon with a hard tug, tear off the wrapping paper, open the box and discover it's not another pair of weird-looking slippers but that boxcar I've been dreaming about for months—the one with the blue plastic door that pops opens and the little man who swivels out and looks like he's tossing you a bundle of something important.

The package from Barbara Maurer was the apple equivalent of that Christmas gift with the boxcar inside. I'd been waiting for a lucky break in the Collins case. I had looked at a couple of other possible Collins apples in the past, but nothing matched. This was promising. As Thomas "Fats" Waller liked to say, "One never knows, do one?"

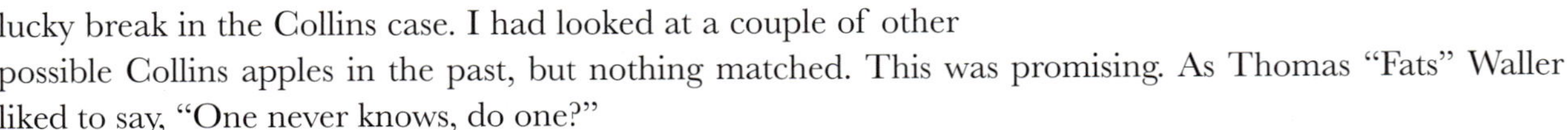

A couple of weeks later, I received a phone call from Barbara Maurer. She filled me in with more details. She believed that she had a historic apple from the Cherryfield area. It had been found by a handyman named Larry Brown and identified by Larry's mother, Margery, who was co-founder and president of the Cherryfield-Narraguagus Historical Society. Unfortunately, Mrs. Brown was too ill to meet or even to talk with me. She died the following month. Barbara went on to tell me that I would be welcome to talk to Margery's daughter, Kathy Upton, or her son, Larry Brown. Larry was the one who grafted Margery's tree. They would all be pleased to have me take scionwood.

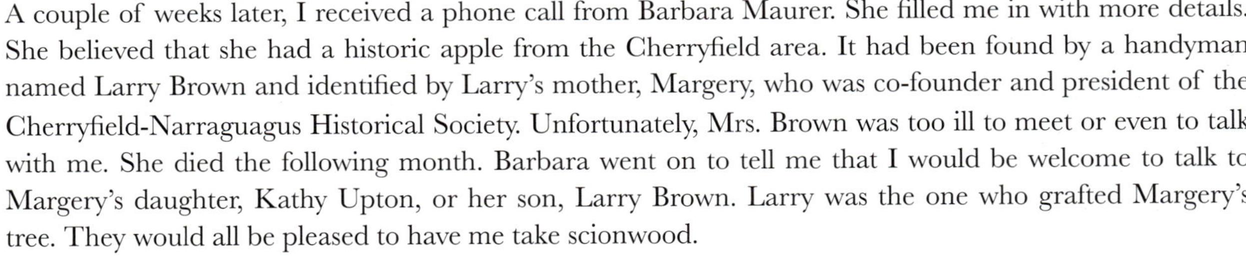

On March 17, 2007, I talked to Kathy Upton, who told me more. As far as anyone knew, the original tree was gone; but they suspected that other trees might still be around. The two apples I received were from

Larry Brown, Kathy's brother. They hoped it would bear fruit again that year though the tree was not in good shape. "We had no idea what it was," she continued.

Kathy Upton & Barbara Maurer, 2018

A couple days after I talked with Kathy, Larry called. It was he who discovered the apple. He had found the old tree with its delicious fruit while mowing the lawn on Persis Noddin's property. When the fruit was ripe and he was there, he would take a few apples home. "We got a rare antique here. They tasted good…Henry Nichols said we could save it." Mr. Nichols then taught Larry to cleft graft. "I watched him do it. I grafted a limb an inch big, put two in; one lived one didn't. It was right here in town, not quite a quarter mile from the old homestead. The tree, at the time…we took the cuttings, it got broken down by the ice storm [1998]. The original tree is gone. We saved it."

None of them knew the name of the old variety, so they nicknamed it The Smith Apple. Then, according to Kathy, "the limb grew and after a few years started producing apples. At the time, we had no idea what kind of apple it was only that it tasted good. When you inquired about the Cherryfield Apple, our mother Margery Brown remembered the grafted apples and sent one to you for verification… and the rest is history!"

Larry Brown with a young Cherryfield tree: photo by Kathy Upton

Larry sent me wood that spring, and I grafted trees. I also topworked a branch on one of our trees at home. It did well, and is now large and producing crops most years. It appears to be annually bearing.

The Collins/Cherryfield apple originated on the farm of Wyman B. Collins and Abigail Ray in Cherryfield in about 1850. According to Kathy Upton, Collins was born in 1807 and died in 1885. He worked as a "joiner and a laborer." A joiner was a skilled version of what we might call a cabinet maker today. The apple was more than likely a random seedling that he or Abigail or one of their eleven children discovered somewhere on the farm. Like the Brown family a hundred years later, they knew they'd found something good. Although Collins shared his first name with one of the area's most famous families, apparently he was not related. Wyman is an iconic name in Cherryfield, Milbridge, and the rest of Washington County. In 1874, Jasper and Edward Wyman started J. and E.A.Wyman Company, canning seafood and wild blueberries. What is now called "Wymans of Maine" is the largest producer of lowbush blueberries in the world, and it's based in the small town of Cherryfield. Sometime around 1907, David Wass Campbell, also of Cherryfield, sent apples to W. M. Munson. Campbell is another of the most prominent family names in Cherryfield, and David had to have been an important citizen. He was also an ancestor of Persis Noddin. The David W. Campbell house is on the National Registry of Historic Places. Campbell obviously thought the apple was good. Munson did too. He recognized the value of the apple and described it as follows in his Maine seedling apples notes:

The David Campbell house on the river in Cherryfield

> Mr. David W. Campbell of Cherryfield, Me., who sends this apple, writes that it has been extensively grafted into all kinds of apple trees in the vicinity of Cherryfield, and that it proves hardy, a good bearer, and of excellent quality. It is a favorite variety in that locality. Under ordinary conditions it keeps through January, and has been kept in good condition until April.

By the time Munson received fruit, a different seedling apple had been discovered near Fayetteville, Arkansas, and named Collins, so the Maine apple was renamed Cherryfield, at least in officialdom. In Cherryfield it probably remained Collins. In her 2006 letter to me Barbara Maurer called it Collins/Cherryfield. Another apple doomed to be saddled with two names.

Overall the Cherryfield apple resembles Golden Delicious in its shape and long stem. Technically it's roundish-conic but the impression it gives is conic. The basin is so shallow that it sometimes appears to have almost none at all. There are several distinct furrows radiating from the base of the apple, extending over the top. The coloring is a wash of light, reddish-pink blush and slightly darker red stripes. There are dots, but not many, and they are not prominent.

In April 2009 I sent three trees up to Cherryfield as a thank you. One of those trees is now growing and fruiting next to the barn of the Cherryfield-Narraguagus Historical Society just across the Narraguagus River from the David Campbell house. In 2010 we found another old Collins/Cherryfield tree in Ellsworth, not too far away in Hancock County. We've grafted and planted a tree in the Heritage Orchard at MOFGA. We've also got it well established on our farm. We've had enough fruit now over several years that we've gotten to know the apple pretty well. It begins to ripen about a week after Deane. It cooks up quickly into a tart, smooth-textured, golden apricot-colored sauce. It makes a good pie. It's a good dessert fruit. It keeps well into the new year. We love it.

When we don't understand why something happens, we often call it luck. Maybe I do understand why it happened. Maybe this thing with Collins wasn't about luck at all. Larry Brown had discovered an old tree with excellent fruit. Henry Nichols had taught him to graft it. Somehow Barbara Maurer had heard about what I do and sent me two apples. I had read Munson. Munson had corresponded with David Campbell. It all fit together rather logically. But what if I hadn't opened the box before the two apples rotted? It has happened. Or what if Barbara hadn't heard of me, or didn't think it was that important, or had gotten distracted by any number of other things life tossed her way? What if I had glanced at the apples, decided it was another Golden Delicious seedling and tossed them into the compost pail. What if I hadn't been looking out for it? Maybe I did get lucky.

## Twenty-One

# When the Wrong Clue Is the Right Clue

"I could not but admire the cunning with which my friend had inserted the wrong clue in the evening paper…" Dr.Watson
Sherlock Holmes, *The Adventure of the Six Napoleons*, p. 591

Fletcher Sweet

Sometimes I have unidentified apples before me like the ones that turned out to be Deane or Collins, and I have to put a name on them. Other times I have a name, and my task is to locate an apple to go with it. It's the mirror image of the apple ID. You have the name but you have no apple. This is frequently the case when tracking down lost varieties. It requires a different set of strategies. It begs for creativity and imagination. After all, you're looking for an apple that no one may be able to identify, including you. All you have is the name and whatever description or historical information you can muster. Sometimes, that's not much.

Many years ago when I had just begun to look for old apple varieties and Fedco Trees was still in its infancy, our customers picked up their fruit tree orders at the old Fedco Food Co-op Warehouse in Bob Plourde's chicken house in Winslow. The chicken house was one of the thousands of victims of the abrupt demise of the Maine poultry industry in the early eighties. For a number of years back when Fedco was mostly about food, not seeds and trees, Fedco rented the building.

Selling trees was more like a hobby in those days. I created a list on a manual typewriter, all in capital letters. Larry Dansinger printed the list for us in his basement, and we stapled the pages together ourselves. Two pages and an order sheet to be filled out in pencil or pen and submitted by mail. We included a second order sheet with each price list so customers could keep one for their records. We hand-wrote the vinyl tree tags with Sharpies. We had an endless supply of pallet wrap on hand at the warehouse to wrap up the boxes and bags of food destined for the various co-ops around the state, so it made sense to wrap up the tree roots with damp sawdust and pallet wrap that we hung in rolls from the ceiling. We pulled it out as we wrapped the trees like you would roll out paper towels by the sink.

When the customers came in the spring to pick up their orders, I sat at a small table and checked them off a list as they left. I met them all. A year or two later, CR Lawn took over that job. For several years, he met each customer as they came in, settled up with them and sent them off with their trees. It was a simple affair, and we got to meet and hang out with pretty much everyone. Some of them became good friends.

One in particular I'll never forget, though as far as I know I only met her once. I don't think I'd recognize her, and I don't remember her first name, but even after thirty years, I won't forget that her last name was Fletcher. As she picked up her order, she said to me, "Do you know where I can get a Fletcher Sweet apple tree? I'm a Fletcher. The apple originated in Lincolnville." Alas, though I wanted to impress young Ms. Fletcher, I had no Fletcher Sweet trees to sell to her, and, even worse, I hadn't even heard of the apple.

Topworking #14: Give the stub a good thorough coating

Over the years, I did find myself in Lincolnville from time to time. It's only forty minutes from home, over on the coast and still in Waldo County, more or less equidistant between Camden and Belfast. But I never had an opportunity to try to locate the Fletcher Sweet. The only description of the apple I ever read was a brief citation in Bradford's thesis. "This apple originated on the farm of Jonathan Fletcher, one of the first settlers of Lincolnville. As the name implies, it was a sweet apple and its season was late fall or early winter." Not much to go on.

Incorporated in 1802, Lincolnville was not named after our sixteenth President. It was named for a Revolutionary War general. The town was one of only three coastal Maine communities to vote in favor of separation from Massachusetts in 1807. Thirteen years later Maine became a state. Lincolnville has always been a little bit radical.

The name Fletcher Sweet has a bit of a ring to it, and I kept it tucked into the recesses of my brain. Lincolnville held a special place in my life since it was the summer residence of my first wife's family, and consequently, an important place in my daughter's life.

Also, for several years it was one of my apple collecting stomping grounds. I picked a lot of apples in Lincolnville back in my early cider pressing days when I pressed ten or twelve barrels every fall. What a way to make a living. I'd do two barrels at a shot and then sell the cider by the barrel to my friends. Sometimes it seemed as though they'd buy a barrel just to be nice to me. One year John Doan consented to buy a barrel. I brought it over, and we muscled it into his basement. I still can recall noticing that it felt rather warm down there. The barrel was all bunged up of course. We talked about air locks, but, alas, no airlock got installed. A month or so later he still hadn't put one in. One night the family heard a tremendous boom. A board on one end of the barrel had lifted up and separated itself from the rest. A large gush of cider had exploded out. An instant later, the piece of cracked oak was sucked back into the barrel, and no more cider escaped. A year or two later, John gave me back the empty barrel and I continued to use it for years. It was still tight but I always knew which barrel it was by that cracked board on one end, raised slightly higher than the rest.

Seth Yentes and JPB, 2017

Many years later, in the fall of 2002, I was invited to do an apple display in Lincolnville. The occasion was a daylong, fall festival at Kelmscott Farm. The farm was known for its rare breed conservancy. It seemed like a good idea so I said yes. Off I went to the Vancycle Road on October 19 with an apple display and some catalogs and handouts to give away. My young, fanatic, fruit exploring buddy, sixteen-year-old Seth Yentes, was with me. Seth was a cello-playing homeschooler who had fallen in love with apples. We had great fun together. Years later he became a grower for Fedco Trees and started his own farm in Monroe.

The day at Kelmscott was frigid. It was the first really cold day that fall, and the wind blew a gale. We stood and shivered in the large barn. The wind freight-trained though the narrow opening between the two huge sliding doors, and plumes of dust swirled around us. I was underdressed. Seth was a trooper. Amidst the dust and the brays of a donkey in the corral across from us, we chatted with folks and entertained them with samples of apples I thought would be good that day: Opalescent, Bailey Sweet, King David, Starkey and Gray Pearmain.

About mid-morning as the wind was picking up and it was beginning to rain, a tall slender fellow appeared at our table. He was wearing a baseball cap that was embroidered with an antique car and the words "Rosey's Restoration." His name was Rosey Gerry. He had a couple of others with him, including a woman named Diane O'Brien. They were from the Lincolnville Historical Society. For the next hour or so, they proceeded to educate Seth and me about the history of Lincolnville. At some point in the conversation it dawned on me that an opportunity was presenting itself, and that these people were going to be the reason why I would never forget this day.

As we chatted, Rosey told us he knew nothing about apples, but he did know every old farm site and every cellar hole in Lincolnville. He said he would be delighted to take us on a tour someday. Maybe there would be old apple trees. Meanwhile I'm thinking, I love historical societies. If anyone knows where the old places are, it's these people. These are the people I want as my allies in the search. Maine history is largely an agricultural history. Agricultural history is a big piece of fruit exploring. Everyone grew food. History people get that. They have a different view of the present. They see the present moment as a part of a continuum. All of a sudden for the first time in twenty years, I flash on the name Fletcher Sweet. All I can think of is that I have to find that apple and if I am ever going to track it down, these are the people who are going to help me.

King David originated in Arkansas but loves it in Maine

"I'm looking for an apple called Fletcher Sweet," I blurt out. "It originated in Lincolnville." They all stared at me. None of these Lincolnville historians knew the variety, but I could see the wheels turning somewhere under Rosey's Restoration Antique Car baseball hat. "I know a place called Fletcher Mountain. That must be where the apple still hangs, if it's anywhere in town."

And then they were gone. We had agreed that Rosey would find out where the old Fletcher farms were and, with any luck, I'd get a call sometime soon. I went home feeling pretty upbeat about the day despite the cold and wind and dust. That evening I wrote in my journal, "Went to Kelmscott Farm in Lincolnville for the day. Did a display of 106 Heirloom varieties—all Cortland or earlier. Cold and Windy but fun. Seth Yentes assisted me. Then picked a bucket of Briggs over in Waldo. Ate dinner and crashed early."

Since it was fall, I quickly got absorbed with other tasks: bring in firewood, harvest the cabbage, pick the Black Oxfords, plant the garlic. On October 22 Seth and I went out to Islesboro to look at old trees. That was an excellent day. On the 28th I did an apple tasting at the Maine State Prison. Who can forget the hollow clang of the door locking behind you followed by a second door and a second lock? The taste test winner that day? King David, a very flavorful apple.

Meanwhile, unbeknownst to me, Diane O'Brien had written a brief note in the Lincolnville section of the October 31 *Republican Journal* and *Camden Herald*. In the article she put out the word that Fedco was looking for the Fletcher apple. The hope was that someone would see the article and call Rosey, Diane or the paper. Diane inadvertently referred to the apple as "the Fletcher apple," not by its correct name, Fletcher Sweet. She was like Sherlock when he "inserted the wrong clue in the evening paper." A small, seemingly insignificant error turned out to have fortunate consequences.

Rosey and I did attempt to plan a trip together, but we were both too busy and nothing came of it until about two weeks later when I received a message on my machine. "This is Rosey Gerry. I found the Fletcher Sweet."

Diane O'Brien, Clarence Thurlow, & Rosey Gerry Nov. 5, 2002

I called him back. He told me about Diane's article. A day or two after the paper was published, a man named Jenness Eugley called Rosey and told him about a Fletcher Sweet tree at the farm of Clarence Thurlow in Lincolnville. Then Clarence called as well. While he had never heard of an apple called Fletcher, he assumed the article meant Fletcher Sweet. The tree at his home was dead, but he knew of another at the Four Corners. Clarence's father had worked for the Fletchers, and when Clarence was a boy they had gone every fall to collect Fletcher Sweet apples. He knew right where the tree should be.

Diane couldn't have put it better. Asking for a Fletcher apple was almost a perfect set up, even if she hadn't meant to do so. The fact that Jenness Eugley and Clarence Thurlow both knew the name Fletcher Sweet made me optimistic that this was the real deal.

A few days later, on November 5, I met up with Rosey, Diane and others at the Lincolnville Fire Station. We got in our vehicles and paraded off in a convoy, converging at the old picturesque farm of Clarence Thurlow, tucked away down a long dirt road through a field of goldenrod. Mr. Thurlow met us out front, and, after introductions, we all went to find the Fletcher Sweet.

We parked at the intersection of High Street and Moody Mountain Road. We were right in the center of what had been Fletcher Town, at the base of what had been Fletcher Mountain. The names had changed over the years. It was no longer Fletcher Town or Fletcher Mountain. Gone were the houses. Gone was the store. In fact, it was all gone. But it was still the same place. Maybe it's still the same apple too even if the name changes every few years or every few towns. A thick stand of sugar maples had overgrown the entire area. Other than the stone walls and the foundations, there was nothing left of Fletcher Town. Well, almost nothing.

We parked our various vehicles and walked a few feet up a chained driveway that led onto an open field to the northeast. This was once the driveway into one of the Fletcher farms. There, surrounded by sugar maples, was a very old, bark-less tree about sixteen inches in diameter. It was rather tall and had no branches. "This is Fletcher Sweet," Clarence pointed. "I used to eat off it when I was a boy." We stood and looked up. It was still alive. There was a narrow, four-inch strip of bark ascending from the ground up fifteen feet to one tiny branch with a few leaves. Sort of a tree version of Mr. Thurlow himself.

Yes, it was alive, and yes, it was an apple tree. About two feet off the ground was the graft itself, now almost entirely bark-less except for that one narrow strip. Without the bark the graft was much more visible. It was a large rippled bulge that wrapped its way around the tree.

Of course there was no fruit. But that was of no concern to me. We had found a Fletcher Sweet. I knew what to do next. We would save this thing. While I was elated with the discovery, I think my companions may have been a little perplexed by my delight with a dead tree.

Just one small branch was still alive, but that was all we needed to cut a stick of scionwood and save the Fletcher Sweet

I wrote that night, "Then to Lincolnville to meet up with Rosey, Karen, Diane and others and Clarence Thurlow and a trip to the Fletcher Sweet tree!!"

That winter I returned to Fletcher Town with my pole pruner and snipped off a three-inch piece of the small branch. I didn't want to kill the tree, so I removed just enough to do a few grafts. As I suspected, there was virtually no new growth so I would have to do the grafting with three or four-year-old wood. That would be tricky but doable. If I grafted a half dozen or so, even with this poor wood, I should be able to get at least one or two takes. A few months later I grafted trees, and they took. The Fletcher Sweet was saved. When I returned to Fletcher Town a year later, the tree was dead. Not long after that I received word that old Mr. Thurlow had died as well.

The beauty of grafting is that once you successfully graft a single tree, that tree will produce scionwood for thousands more. It's a bit like Mickey Mouse and the brooms in the Sorcerer's Apprentice. One begets another, which begets several, which begets many. Now that we'd grafted a couple of Fletcher Sweets, they would in turn grow and provide us with the wood to graft many more. Which they did, and which we did. On May 3, 2005 we planted a young Fletcher Sweet tree at Breezemere Park in Lincolnville in a small ceremony in memory of Clarence Thurlow. Since then, we've offered about a hundred more Fletcher Sweet trees through the Fedco catalog, many of them destined to be planted back in the Lincolnville area.

Rosey Gerry guarding the wrapped up Fletcher Sweet tree at the Lincolnville planting day in 2005: photo by Priscilla Pattison

A year or two later I tracked down a second Fletcher Sweet tree, just a couple miles away as the crow flies in the front yard of an old Cape on Church Street in Hope. The home and tree belong to Tom Hardy. Tom told me that his father Bill received a tree or scionwood from Clarence Thurlow in the early thirties. I've collected fruit and scionwood from that tree as well.

I also met and have gotten to know Lys Pike, whose family owns the land on which the old Fletcher Sweet tree stood. I've given her a young tree to replace the old one. The Pike family also has a rose they call the Fletcher rose. Lys wrote to me, "My mother said they came from the northeast corner at the Four Corners where the foundation of the Fletcher House was, or so I have always assumed."

As I poked around looking for more information on the Fletchers, I came across the *Timothy Fletcher House Expense Book, 1825-1833*, housed in the special collections at Virginia Polytechnic Institute and State University. Here's a bit of what I learned. The first Fletcher to come to Lincolnville was most likely Timothy Fletcher who was born in Acton Massachusetts on May 9, 1779:

> When he was just twenty-days old, his father died and he was bound out to a farmer for the rest of his minority. In 1792, Fletcher left Massachusetts for New York City to search for his long lost brother named Daniel. Fletcher made most of the trip alone and on foot, covering about fifty miles a day till he reached Bridgeport, Connecticut; from there, he sailed to New York.
>
> In 1803, Fletcher moved to Lincolnville, Maine, where he would spend the rest of his life. In that same year, he took up a teaching job, bought a farm, and married Mary L. Brown, a Maine native. The couple had several children together. Throughout his life, Fletcher was known as an upright man and stout Christian, faithful to the Fletcher family's puritan roots. When Fletcher was eighty-five, he wrote in his journal that he was, "[still] not too old to hate rum and slavery." He died on August 21, 1864, just nine months before the war's end.

It would be safe to bet that it was Tim Fletcher or one of his descendants who discovered the seedling that became known as Fletcher Sweet. That apple has now fruited for us for several years. It resembles Tolman Sweet in appearance and it would not be a stretch to guess that Tolman could have been one of its parents. After all, Tolman Sweet was grown throughout coastal Maine at that time. Some Fletcher

Fletcher Sweet is reborn in Lincolnville, 2005

specimens even have the suture line common to Tolman Sweet. The apples are often partly russeted and sometimes feature an orange blush.

Despite being called *Sweet*, Fletcher Sweet has enough acidity to make it a decent dessert apple. This totally makes sense since Clarence Thurlow told us he loved to eat the apples right off the tree as a boy. Fletcher Sweet also makes a nice, light yellow sauce, loose textured with a hint of blackberry. The flesh is white with a slightly green tinge. The skins do not breakdown in cooking. The fruit always seems to be scab-free. Like most yellow apples, though, it can get discolored by sooty blotch. It ripens on about the first of October and keeps for two months.

Can I ever know for certain that we found the Fletcher Sweet? No, of course I can't. Do I think we found it? Yes, I do. It's hard to discount two facts: that Janness Eugley and Clarence Thurlow knew of the apple, that Clarence knew the location of the tree and that he knew of no "Fletcher" apple but did know "Fletcher Sweet." It is unfortunate that the only written description of Fletcher Sweet really isn't much of a description. But that will have to be okay. Is the evidence good enough for now? I'd say it is. Did I ever meet Ms. Fletcher again? No, but if you see her, please tell her I found her apple.

Fletcher Sweet resembles Tolman Sweet in appearance only; not in taste!

# Twenty-Two

# Wanted Alive

…when the facts slowly evolve before your own eyes, and the mystery clears gradually away as each new discovery furnishes a step which leads on to the complete truth.
Sherlock Holmes, *The Adventure of the Engineer's Thumb*, p. 274

Marlboro

Fletcher Sweet is not the only name in search of an apple. There are hundreds or even thousands. In Bradford I found the names of dozens of apples that originated in Maine, most of which probably haven't been identified in many decades. Then I discovered the *Maine Agricultural Yearbooks* and that great article by W. M. Munson. Before I knew it, I was assembling lists of apples from every county in the state. I'd already found a few of them, including Black Oxford, Starkey, Somerset of Maine and Fletcher Sweet. Others, however, remained more elusive. Munson's list became my dream team of apples to locate. But how would I find them?

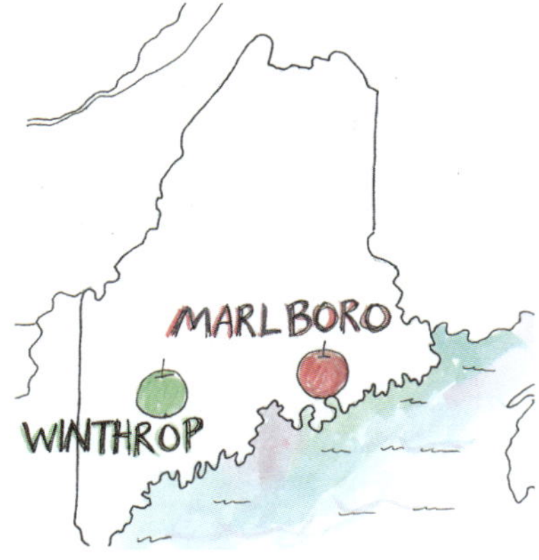


I don't think of myself as having original thoughts. I assemble my universe from the bits and pieces I see and hear around me, from what I observe. I like to pick and choose. Maybe all of us endlessly put together seemingly random things into infinitely novel combinations. Could invention be simply a rearrangement?

I don't remember where, when or how I came up with the idea of the Wanted Poster. I suppose it partly came from a love of Westerns. Or maybe it came from those summer mornings as a young boy waiting in line at the post office with my grandmother. I was fascinated with the old-fashioned, gold post office boxes with the square glass windows and the combination dials. I was a little taken aback when "Bombu" told me how she remembered the combination. "Jesus God Damn." J-G-D. It certainly worked. I still haven't forgotten the combination and that was probably sixty years ago. The post office isn't even there anymore. As she and I stood together waiting to buy stamps, I was forever scanning the old walls for interesting stuff. How could you not be curious about the ten most wanted list? Each of them got their own page. Where were they now? Should I be on the lookout?

Once I hit on it, the idea of the Wanted Poster was a no-brainer. In early September 2002 I made up my first "wanted alive" posters, each one featuring a Maine apple. Because these were all local varieties, it would be key to get the posters into the areas where the apples would most likely be found. I assumed that this would be where they originated. Thanks to Munson and others, I now had that information. Using bits of Munson's descriptions and photocopies of his photos, I put together the first set of a half dozen. It was essential to include information about the names of people associated with the apple, as well as any locations that might be relevant, such as where it originated. My hope was that someone in the area would recognize some name or farm or road and lead me to an old tree. Then I could worry about making the ID.

The posters were a less morbid variation of the old western version. It wouldn't do much good to locate a dead apple tree. They needed to be alive. Just barely alive was fine, as in Fletcher Sweet. But they had to be alive. So the word "dead" never made it into the headline.

Origin, farm of S.H. Remick, Marlboro [Hancock county] Me.
Fruit medium, oblate[slightly flattened] -spherical, yellowish green overlaid with rich crimson on the sunny side, with numerous small dark dots; cavity medium, flaring, regular, slightly russeted; stem slender, 1/2 to 3/4 inch; basin very wide, shallow, plaited; calyx partly closed; flesh white, crisp, fine grained, very firm, pleasant acid; core small. Good. January to May.
In March specimens of this variety were received from Mr. Remick and they were in prime condition with a rich aroma.

IF YOU KNOW THE WHEREABOUTS OF THIS APPLE
PLEASE CONTACT FEDCO BOX 520 WATERVILLE 04903 873-7333

I set out the first batch of posters on September 20 at the Common Ground Country Fair. I figured that the Fair would be a perfect place since people attend from all over the state. I printed twenty or thirty of each apple. These were my bait. I didn't necessarily think that they would work; but they were fun to make, and the people seemed to enjoy them. As it turned out, it didn't take long to get my first nibble.

Although I wasn't aware of it at the time, one of the posters was picked up by a fellow named Don Cooper. A few days later I got a phone call from Don's wife, Jo, the president of the local historical society. Her husband brought home the wanted poster for the Marlboro apple. She had read the information on the poster, and she knew the Seneca Remick farm in Marlboro where the apple originated. She wanted to help find the apple.

Marlboro, Maine, is one of several villages that make up the small coastal town of Lemoine, across the narrows from Mount Desert Island, Bar Harbor and Acadia National Park. Incorporated in 1870 but settled a hundred years earlier, Lemoine's early shipbuilding days are now long gone. These days you could call the town a bedroom community for Ellsworth and a beautiful spot for a summer home.

Always be on the lookout for two different apples on the same tree; but beware—one might be a grafted variety (L) and the other might be the seedling rootstock (R) sprouted below the old graft

During our phone conversation that afternoon, Jo offered to go to the farm, pick apples from all the old trees, map and number them and send me samples from each. How could I refuse? On October 8 she picked the apples, made a map, and sent it all to me.

On the 17th, a couple days before my trip to Kelmscott Farm in Lincolnville, I opened the box from Jo and pulled out seven plastic bags of samples from the trees in Seneca Remick's yard. I always try to discourage plastic bags when mailing apples. They get gross pretty quickly. But these hadn't been in there too long, and the specimens were in good shape.

One jumped out. It was clearly a Northern Spy. I got out my 1907 description of Marlboro and compared it to the others. There were two that vaguely resembled the description and one other that came close. Oddly enough, there were two very different apples in that bag. I called Jo Cooper back, thanked her for her amazing effort, and asked her about the two different apples in bag #1. She said she'd look into it. I made notes on the map she sent, but I had other projects to deal with, and I didn't think I'd be heading to Marlboro any time soon. Nothing happened again for three years.

The Marlboro tree in front of Seneca Remick's house

Seneca Higgins Remick—or Captain Remick as he was known—was born across the narrows from Lemoine in Eden near Bar Harbor on September 8,1834. He died on May 25, 1914. "Family stories have him going to sea at nine and being called Capt. Remick." He was married three times. Sedulia Hodgkins, Cornelia Haines and Annie Kingman were his three wives. He was a mover and shaker kind of a guy. The family farmhouse doubled as the post office. He built a church. He was an active grange member. He was probably a grafter, and he discovered and named an apple.

In October 2005, three years after that first phone call with Jo, I had a bit of time and decided to contact her again to see if there was any fruit on the Seneca Remick trees. She called work and left a message: "Jo Cooper's apple tree has apples on it now, but not much longer." When I got back in touch with her, I learned that the tree was still standing but that a subdivision was going in next door and Lemoine was getting rapidly developed. I decided I better go.

Off I went on October 10. We met just north of Ellsworth. She was driving a green minivan with the license plate, "Go Jo" and a bumper sticker: "Musical Kids are Noteworthy." My kind of bumper sticker. I followed her in my black Corolla for the last few miles into Marlboro. The Remicks had been gone for many years. The house was being rented. We knocked on the door. No one was home. I scurried around looking at the trees. Yes, that one was a Northern Spy. I was saving tree #1 for last, the tree directly in front of the house.

Marlboro looking very red with the small russet stem "splash"

Now came the time to examine tree #1. Jo looked at me and said, "This gives me the chills." Not only did the tree have two varieties on it, it had three. It became clear to me what Remick had done. If this was the Marlboro apple, he had topworked it onto the farm's most prominent tree, an old Tolman Sweet, just fifty feet from the front door. It made total sense that it was a Tolman Sweet. Tolman is one of the oldest and most common varieties planted around New England before 1900. No one knows of its origin. Some say Dorchester, Massachusetts, and others say Rhode Island. It's a medium-sized, round, yellow apple, occasionally with a very slight blush, often with a distinct thin green or brown "suture" line running from one end to the other, and an absolutely distinct flavor. Tolman was also used as a rootstock for generations, presumably because the grafts were compatible and it made a rugged, long-lived tree. I have run into Tolman trees all over the southern half of Maine, many of them exceedingly ancient. Often the trees are entirely hollow and still bearing regular crops.

Someone—almost certainly Seneca Remick himself—had topworked the Tolman Sweet tree in his front yard with two additional varieties. One was Alexander, that large, red, Russian apple brought over by the Massachusetts Horticultural Society in about 1835 and well-known throughout much of the state including, as you recall, on Len Alexander's farm in Chelsea. Two other branches of the tree were grafted with something else, a variety I'd never seen before. It had to be Marlboro!

I compared the fruit to Munson's description. It was a perfect match. I sliced one in half and laid it on top of the half-cut photo I had. Then I drew a pencil line around the fruit. When I lifted it off, the two profiles also matched perfectly. This was great.

We marked each of the two huge branches with a twist tie. I planned to come back to get scionwood when there would be no apples on the tree, and I didn't want someone to see flagging tape and remove it thinking it was trash. Even worse, I didn't want some zealous arborist crew to pull over and cut it down, thinking that the flagging was some sort of marker for destruction. So I used a discretely placed twist tie, one that I could find in a few months. I then made a drawing of the tree on a brown paper bag showing which branches were which. Insurance. I'd return in February or March to get the scionwood.

We collect piles and piles of scionwood every winter and spring, and 2006 was no exception. I kept delaying a trip to Marlboro. It was a long way to go. But on March 30, I hopped in the car and drove the two and a half hours. There was the tree. There were the twist ties. I cut the scionwood. That spring I grafted a branch onto a tree on our farm, and Delton Curtis, one of the Fedco grafters, grafted another twenty trees. Two were destined for the Marlboro Historical Society. The rest would find homes somewhere nearby, I was sure. A year and half later the trees would be ready to dig in the nursery. I returned to Marlboro on July 25, 2007, and gave a talk at the Lemoine Baptist Church not far from Seneca Remick's farm. Afterwards, on the long

Scionwood piles in the snow...is there a better way to spend the winter?
photo by Laura Sieger

drive home, I listened to the Sox lose a squeaker to the Indians—one- zip. The talk was fun and well attended. It was also the first time I met Todd Little-Siebold, the history professor from College of the Atlantic. He and I would meet again, and in more ways than one, he would change my life.

I've created about two dozen more Wanted Alive Posters over the years with occasional bites here and there. When Steve Miller, one of my fruit exploring buddies from northern Maine, contacted me one fall about an apple he found that might be called Hartford Sweet, I knew the name sounded familiar. I pulled out the posters. I had made one for Hayford Sweet, an apple originating in Maysville. Maysville is just a few miles north of Presque Isle, the home of a fellow named Norm Johnson who owned the tree he thought might be called Hartford Sweet. The names were too similar. The locations too close. Next fall when I went up to the County to visit Steve and his wife, Barb, we swung by to see the tree. The fruit matched Munson's description. It was a true old-fashioned, sweet variety: dry, low-acid and not an apple many Americans would go rushing out to buy at the supermarket. Hayford Sweet, however, keeps at least until January and not many winter storage apples are hardy enough to survive the winters up in the County.

I heard a few years later that Norm Johnson had cut down the tree. At the time Steve first discovered it, it may have been the last Hayford Sweet tree left on earth. Fortunately we got to it in time to take scionwood and keep it going. Now there are at least a couple dozen young Hayfords that we've grafted and disseminated. This past spring I received an email from someone up that way looking for scions. I cut some twigs from our tree and put them in the mail.

More recently the Wanted Posters worked their magic once again. In the fall of 2015, Maine had its best apple crop in many years, maybe even in decades. My intern, Laura Sieger, and I were visiting trees all over central and southern Maine every moment we could carve out the time. One of our jobs that fall was to identify a couple hundred trees in an old orchard in downtown Augusta. You'd never know it was there. Right in the middle of town, tucked in behind an old mansion, a Hannaford's and a shady residential street is a beautiful, small, commercial orchard, planted in two sections, the later dating from about 1925. Most of the trees are Spys and Cortlands, although we did find one Rhode Island Greening and a scattering of other odds and ends. We raced along row after row. I would snatch an apple out of the tree

with my pole picker and call to Laura, "Spy, Spy, Spy, Cortland, Cortland, Cortland, Spy…" I was channeling Francis. Laura was writing as fast as she could. The sun was getting low, and we wanted to finish. We had to get over to Winthrop before it got dark. We were hoping to find a Fairbanks tree, maybe the last Fairbanks tree alive.

A few days earlier I'd received an email from a Peter Nielsen of Winthrop: "Picked up a flyer at MOFGA apple event a week or so ago that said you are looking for a Fairbanks apple that originated in Winthrop. According to the town history book, we are living on the Elijah Fairbanks land. There's this old tree. I said, wouldn't it be something, so I got some apples this weekend. They match the description on your flyer pretty well, yellow fruit, some russetting, juicy. We made some excellent cider from some drops, and I ordered a fruit picker to get the ones still on the tree. Do you want to check this out? The town book mentions the Fairbanks apple, and the Moses Wood, whose land is also in the area. There may be another of these Fairbanks trees just up the road, on a neighbor's land. Not sure."

Peter Nielsen and the Fairbanks tree

Peter Nielsen's daughter had seen one of the Wanted Posters at the Great Maine Apple Day. She had noticed the word Fairbanks. She stopped and read the information below the drawing of the apple. She turned to her father and said, "Dad, we live on that farm!" I had been looking for Fairbanks off and on for years. This might be our best chance to find it. Augusta is about half way to Winthrop. We should go. We quit our romp through the Augusta orchard and headed for the truck. We pulled into the Nielsens' driveway as the sun was setting. It was October 30.

Peter met us at the door. "The tree's across the street behind the barn. We never paid much attention to it until my daughter saw your poster." We crossed the road and waded through the goldenrod over to the huge, ancient tree. We collected a few handfuls of fruit off the ground. They were not in great shape. Then we returned to the house where Peter had more apples in a bucket in the shed off the kitchen. These had been picked a few weeks before, and they looked pretty good. We got out the camera and took

Fairbanks

a lot of photos of the fruit, and of Peter and his wife, Mary Richards, who arrived a little while later. It was a celebration.

Elijah Fairbanks was born in 1756. He died at age ninety-seven in 1853. He was almost as old as Francis Fenton. He planted the apple that bears his name on June 9, 1779, the day that General Francis McLean and the British Royal Navy attacked and occupied Castine at the mouth of the Penobscot River. According to Elijah's grandson, Daniel Allen Fairbanks, "The native tree was killed by the hogs, which were kept in the orchard, stripping it of its bark. There are several trees on the Town Farm in this town, [Winthrop] engrafted with this variety by Dea. Williams, some years ago. Every person who knows these apples values them highly."

The description of the apple itself is not very detailed. Munson calls it "light yellow, obscurely striped, with patches of russet." There's a good chance that this could be it. This is the farm where Elijah Fairbanks lived. The tree is certainly old and could date from his time. Perhaps it's a replacement for the original tree that got girdled by the pigs. Was this the Fairbanks apple? Is there anyone who can say if this is the Fairbanks? We may never know if it's the true Fairbanks. It is a Fairbanks apple from the Fairbanks farm. Does that make it a Fairbanks? For now we will graft a few trees, and put it in the Maine Heritage Orchard.

By the time we finished celebrating, it was dark. Peter pulled out an old map of Winthrop. There was the Elijah Fairbanks farm, and there was the farm of Moses Wood just around the corner. These guys all knew each other. They were in this together. They were buddies who were fruit exploring and discovering seedlings and naming fruit over two hundred years ago. They were the grafters, the baton passers of the past.

We drove home giddy with discovery. The basement and both root cellars were already packed with dozens of bushels of apples to be pressed or sold, and hundreds of paper bags filled with mysteries to be identified. Laura carefully placed a bag labeled "Fairbanks" into the box marked "E-F."

# Twenty-Three

# Roy Slamm

Simply as a mental exercise, without any assertion that it is true, let me indicate a possible line of thought. It is, I admit, mere imagination; but how often is imagination the mother of truth?
Sherlock Holmes, *The Valley of Fear,* p. 802.

Canadian Strawberry
"Sweetest apple you'll
ever taste"
he says,
Roy Slamm
in a dream now
smiling-
sweet, sweet-

gary lawless

Canadian Strawberry

What do you do when you're presented with a named apple and you can't find a reference to the variety anywhere? Do you believe them when they tell you it's so?

At the 1995 Common Ground Fair, Roy Slamm was one of those who came to the booth. It was the first time I'd met him. He told me about an apple he called Canadian Strawberry and he told me that I should come see his old trees and taste the fruit. They were perfectly ripe right then, though he hadn't thought to bring one with him. He seemed nice. He was certainly enthusiastic. When I got back home that night, I checked through my old books for any citation of the apple. Finding none, I dismissed it as nothing too important and promptly forgot about it.

A year later Roy Slamm was back at the Fair, this time with his daughter, Amanda. Again, he told me about this Canadian Strawberry and insisted that I come to Solon to visit the trees. He was friendly, persistent and persuasive. Somewhat reluctantly, I said I would. The squeaky wheel and all that. A few days after the Fair, I drove a couple of hours north up to Solon. I knew the area, having spend a lot of time nearby in Athens years ago, although I'd never been down the South Solon Road where he and this supposed "Canadian Strawberry" both lived. I pulled into Roy's driveway, which was just a small parking area off the dirt road, a few feet from the front door of the old cedar-shingled Cape. The original barn was gone but Roy's two large rustic woodworking shops stood behind the house. Roy came out and we walked together a couple hundred feet into a small grassy meadow. There stood three ancient hollow apple trees. All three bore identical fruit so I knew right off that they were grafted. I also knew that I'd never seen this apple before. I tasted the fruit. It was perfectly ripe and perfectly delicious. That night I wrote in my journal, "Went to Roy Slamm's and got Canadian Strawberry apples. Yes!"

There's nothing like Canadian Strawberry for fresh eating in late September in Maine

That day marked the beginning of a fifteen-year friendship with Roy Slamm and a deep appreciation of the Canadian Strawberry apple. It's an apple I quickly learned to love. It's beautiful and it tastes great.

The fruit is roundish. It's ground color is a rich buttery yellow and it's about half to three quarters covered with "china" red stripes and blush. The blush is really just an optical illusion, as apple blushes often are. It appears like a wash of coloring over the surface of the fruit, but it is, in fact, made up of hundreds of small dots, often quite close to one another. Most Canadian Strawberrys also feature a splash of russet that fills the cavity and then spreads out over the side of the fruit.

The crew fast at work cutting up apples for the tasting, 2017

Because it is often at its peak on Common Ground Fair weekend, I picked apples at Roy's many times over subsequent years and brought them to display and taste at the Fair. On several occasions it's won our annual tasting contest, beating out much more famous varieties of the season.

The Fair apple tasting is a wild event. We do one on Friday and another on Saturday. We always have a good crowd. A hundred or even two hundred people gather. We cut up apples into very small bits and pass them around on paper plates. Everyone samples and yells out their opinions. Someone volunteers to be the scribe and writes down the comments for future reference. A team of three or four cutters slice up the next variety as fast as they can chop while I try to entertain the crowd with historical babble and other chitchat. We're restricted, of course, by what's ripe at the end of September but that's also part of the fun. Most years the selection includes Chestnut Crab, Egremont Russet, Cox's Orange Pippin and Honeycrisp as well as six or eight others that happen to be perfect just then. We bring whatever varieties we have from our place. For many years before he died, Don Johnson, an orchardist from Herman, Maine, with an excellent and interesting collection, always brought several. We sample ten or twelve varieties and then we vote. You get to vote twice, for your two top favorites. It's the apple version of ranked-choice voting. We count up the votes and declare a winner. It's always fun and always popular.

JPB & Jordan: always have a good scribe to take notes!

Invariably the winner is the apple that just happens to be the one at its best. You might think that a few days wouldn't matter with apples. Not so, especially with those early-fall dessert varieties. Although they're all usually quite good for at least a couple of weeks, in a winner-takes-all Common Ground Fair taste-off, you can be a day or two one way or the other and you'll lose. On several occasions,

Canadian Strawberry has hit that perfection meter on the nose, most recently in the fall of 2015 when they were exquisite. I was amazed and delighted the see them knock off both Chestnut and Cox's.

A few months after that first visit to Roy's farm, I returned to Solon to collect scionwood. I grafted a Canadian Strawberry tree in our orchard as well as trees for the nursery. We began to spread the apple around via the Fedco catalog. We planted a specimen at the Maine Heritage Orchard. Over the years I regularly visited Roy, often to collect apples in the fall or scionwood in the winter. Roy also came to visit us a number of times in Palermo.

Roy Slamm and Abby Shawn purchased the farm from two brothers, Elmer and Asher Davis in 1971. The farm was founded by the Davis family nearly two hundred years earlier. The two bachelor farmer brothers were the last in the family to inhabit the old cape. There were four Canadian Strawberry trees then, and it was Elmer and Asher who told Roy and Abby about the apple and its name. For years after the brothers were gone, elderly neighbors would stop by each year in late September to chat and to pick a few "Strawberry" apples. One of the four original trees died before I first visited in '96. Abby had moved away by then, though she still lived nearby. One more of the trees has since died but two ancient hollow Canadian Strawberrys still remain, as well as two young vibrant specimens, one grafted by local friends and the other, a contribution we made some years ago. There are other apples on the farm as well, most notably a large old Ben Davis.

How do you explain a variety, like Canadian Strawberry, that appears in no books but was known as such to generations of one family and all their neighbors? If this was a case of mistaken identity, it was a long-standing mistake.

Two of Roy's Canadian Strawberry trees in bloom: photo by Roy Slamm

No one may ever know the origin of Canadian Strawberry— or "Strawberries" as Roy's neighbors called them. It might be a synonym of some other apple. It's definitely not Chenango Strawberry, that pastel pink and soft yellow conic summer apple that bruises when you glance in its direction. And it's definitely not Early Strawberry, that exquisite late summer dessert apple that's not much larger than a Honeoye and tastes a whole lot better. Then there's Foundling, also known around central and Downeast Maine as Late Strawberry. Not that either. Too red.

One of Roy's neighbors who had been picking the apples for over fifty years suggested that the apple may have come from New York State where it was called New York Strawberry. There is also chance it may be another New York apple called Washington Strawberry. One of Washington Strawberry's synonyms is Washington of Maine. This is sounding close to home. I was able to compare Washington Strawberry and Canadian Strawberry some years ago and wasn't convinced.

Maybe I should try again. A few years ago I did locate one reference to a Canada Late Strawberry grown at the Michigan Agricultural College in 1887 but could locate no description of the fruit.

Roy wasn't an orchardist but he knew he had a great apple: photo by Jan Stringos

Is Canadian Strawberry a unique apple or is it a synonym for another? If I had to guess, I'd say that it probably is a synonym for something else. It may be another Briggs Auburn, finding itself a new persona as Naked Limbed Greening or another Williams Favorite finding itself re-dubbed as Thompson upon arrival in Maine. That being the case, of course, there would be no guarantee that the original name would even have "Strawberry" in it. It could just as easily be another name for Ho-dunk Nonsuch or Something-or-other Seek No Further.

The Canadian Strawberry could also be one of the multitude of apple varieties that originated on small diversified farms in small villages all over the eastern U.S. over the course of many generations of farmers, ever observing and selecting the best of what grew well on their own land, along their own stone walls, occasionally naming varieties and saving them locally for their friends and neighbors.

Perhaps the apple came up from Concord, Massachusetts, with the Davis family over two hundred years ago, either as a seed, a scion or a small tree. Isaac Davis came from what we now call Acton, just northwest of Concord. Back then it was still part of Concord and was long known as "the village." Isaac Davis was a gunsmith from the village. On April 19, 1775, he was killed at the Old North Bridge in Concord. He was the first American officer to die during the Revolution. He was thirty years old. Later he was memorialized for all time as the Minute Man in the Daniel Chester French statue at the rude bridge that arched the flood. He still stands there proudly today with his plow by his side, holding one of the muskets that he likely made himself.

Isaac Davis a.k.a. the Concord Minute Man: sculpture by Daniel Chester French

Around 1800, Isaac's son Ephraim moved up to Maine and started the farm in Solon. The Davis family continued to occupy the place for nearly a hundred and eighty years. It could be that Nathaniel and Dora Davis grafted the four Canadian Strawberry trees on the farm in about 1900. They would have been Elmer and Asher's parents. Or maybe it was Nathaniel's parents—Elmer and Asher's grandparents—who grafted the trees. It could be that the scionwood came from even older trees grafted by an earlier Davis generation. Or maybe they obtained the variety from one of the traveling grafters who were not uncommon long ago. We'll probably never know.

In this day and age of emails and text messages, Roy was an anomaly. He regularly wrote cards and notes the old-fashioned way. I don't believe I ever got an email from him. He always wrote in capital letters. In November 2007 he wrote to me, "It was a spectacular apple year. Our Canadian Strawberry apple trees outdid themselves in production. Most of the fruit was perfect and delicious. [Roy never sprayed a thing.] Lamentably we lost one of the triad of grand lady trees. The most venerable hollowed trunk. Fell in a strong windy day."

Roy was not a pomologist. He was not a fruit explorer. He was a fine woodworker, artist, good father, good neighbor and a homesteader. He was also a keen observer who took pleasure in life. And he was persistent. Were it not for Roy, the "Strawberry" apple might have been lost forever. He knew that he was the steward of something special, and he was determined to see that it was never lost. Roy died in February 2011.

One of the last cards I received from him read, "Hey John, A few frozen apples clinging to the gnarled branches with little snow hats above."

Fortunately, we've been able to spread the apple around New England and beyond via Fedco. It won't be lost even after Roy's last two trees finally pass on. New trees are now bearing in many locations. I love to read the occasional delighted email or note from a friend or a Fedco customer expounding on the merits of this wonderful apple. In the spring of 2015 we topworked a row of Canadian Strawberrys at the Apple Farm about twenty miles south of Solon. As far as I know, this event marked the Canadian Strawberry's first foray into the world of commercial orchards. I hope it doesn't go to its head. The orchard is owned by Steve and Marilyn Meyerhans. Roy was a longtime friend. Each spring a bunch of us go up and topwork for a day or two at the Apple Farm. As we were finishing up the row, I pulled out some homemade vinyl-siding tags and a #2 pencil and prepared to write. Steve was there with me, "Just write 'Roy Slamm' on the tags," he said.

Vinyl siding marked with pencil makes excellent long-lasting tags; don't use a Sharpie!

Some day, in a hundred years or so, a couple of friends will be eating the apples from these trees we grafted that spring. The trees will be a little broken down and a little bit hollow. "It's an old variety," one friend will say to the other. "It's called 'Roy Slamm.' Oh yes, we think it's a synonym for another apple, but we're not sure which one."

# Twenty-Four

# Blue and Gray Pearmain

Most people, if you describe a train of events to them, will tell you what the result would be. They can put those events together in their minds, and argue from them that something will come to pass. There are few people, however, who, if you told them a result, would be able to evolve from their own inner consciousness what the steps were which led up to that result. This power is what I mean when I talk of reasoning backward, or analytically.
Sherlock Holmes, *A Study in Scarlet*, p. 83

Gray Pearmain

Not only were Steve and Marilyn Meyerhans good friends with Roy Slamm, they have also become good friends of ours over many years. They purchased their orchard from Royal and Irene Wentworth about the time I moved to Palermo in the early seventies. By then, Steve had been working at the orchard for a couple of seasons. The old fellow wanted to retire. Marilyn had orchard blood in her family and it seemed like the perfect thing to do. So they bought the orchard. They changed the name from Wentworth's Orchard to the Apple Farm. That was forty-five years ago.

The Apple Farm has thrived and expanded under their care. Most of what they've grown have been the typical Macs, Cortlands and now Honeycrisp. But they also grow an impressive assortment of heirlooms, some of them planted years ago by Royal Wentworth, many others added over the decades by Steve and Marilyn. It's a good collection. We've used many of their apples in our displays. We've also collected a great deal of scionwood there. In recent years, they've let me have a row or two here and there to graft whatever I like. Small sections of the orchard have become mini-preservation blocks.

They had a lot to learn when they first took over the orchard. Royal was a good teacher, but some things got lost in the shuffle, such as the origin of two of his varieties, Gray Pearmain and Sweet Red. Unfortunately neither Steve nor Marilyn knew enough yet to ask the soft-spoken Wentworth about the apples. They were just two more of the many varieties growing on the magnificent hillsides looking east and west.

Purchasing Wentworth's Orchard on April 1, 1974: Steve and Marilyn Meyerhans, Royal and Irene Wentworth & Nick and Suzi Preston

I couldn't resist the challenge of taking on both apples. The first of the two would have to be the Gray Pearmain. With its beauty and its excellent flavor, it begged for a positive ID. It appeared that, like Canadian Strawberry, Gray Pearmain was another apple with no reference point. I looked and looked but couldn't find it anywhere. I assumed it must be a synonym. Every so often, I spent time scouring the old books for information about Gray Pearmain. But I could find nothing. There are a few similar names such as Great Pearmain, itself a synonym of Winter Pearmain. According to Beach, there are several Winter Pearmains that appear in the old literature. I've seen the one down at the Tower Hill Botanic Garden collection in Massachusetts, and it's definitely not Marilyn and Steve's. There's a Gray Apple, synonym of Pomme Grise, a small very tasty russet, delicious but nothing like Gray Pearmain. Another Gray Apple, synonym of McAfee, is a southern variety. No chance there. No Southern apples ever made it up to Maine, right? Autumn Pearmain, synonym of Long Island Pearmain, is a very large apple. I found myself traveling down a smattering of dead ends.

Like Canadian Strawberry, people try Gray Pearmain once or twice and they love it. The firm white flesh is pear-like, juicy and mildly tart. The apples are medium-sized, oblate, obscurely ribbed and muffin-shaped. They keep well into the winter. The skin is a soft opaque greenish-yellow with a rosy pink blush, a bit of a russet netting, and a grayish bloom. You can pick them late, even after the Spys. They store reasonably well in the root cellar although they may shrivel a bit like many of the Russets.

Jesse Stevens with a young Blue Pearmain tree planted in the middle of the remains of a huge ancient Blue Pearmain tree in Norway, ME, 2016

What is a Pearmain, anyway? Where does it come from? What does it mean? There are many Pearmains listed in the old books. Beach mentions twenty-four in *The Apples of New York*. Robert Hogg lists fifty-nine Pearmains in his 1851 *British Pomology*. According to him, the oldest known variety was a Winter Pearmain, dating back to the year 1200.

The one Pearmain typically grown in old New England orchards is Blue Pearmain, an old baking apple of uncertain origin, most likely from just west of Boston in Suffolk or Middlesex counties. One of Blue Pearmain's claims to fame has got to be its inclusion in Henry David Thoreau's *Wild Apples*. It's one of the few "modern" varieties that Thoreau speaks highly of. It's also one of our favorite apples. It's wonderful for baking.

Blue Pearmain with its signature dusty bloom was made to be stuffed and baked

The name Blue Pearmain seems to give everyone trouble up here in Maine, perhaps because no one knows what it means. I've heard quite the assortment of variations over the years. Blue Pear Maine is evidently a native son, perhaps exhibiting Bosc-like overtones in its flavor profile. Blue Pomade comes from Sedgwick, and probably makes a tasty summer cider. Blue Pearamell is from along the Kennebec River a few towns south of Len Alexander's farm. Maybe it refers to the hints of caramel when you bake it? Blue Pearamay from Readfield has a French flair to it. Maine Blue Pear was one that Russell Libby heard and passed along, another reference to Bartlett or Bosc, no doubt. My favorite though was one I heard some years ago: Painbear Bluemain.

The *Oxford English Dictionary* gives two definitions for Pearmain. Probably from old French, "An old variety of baking pear," or "Any of several varieties of apple with firm white flesh." The OED is usually spot-on with everything but when it comes to apples, I'm not impressed. None of the Pearmains are pears. They are all apples. Gray Pearmain does taste like a pear but the two other pearmains I know well, Adams Pearmain and Blue Pearmain, do not. And there are dozens of apple varieties with firm white flesh that are not called Pearmain.

Beach's explanation of the term is about as vague as you can get. He says that Pearmain "has been applied to very many different varieties of apples." To his credit, however, he then directs the reader to Robert Hogg. Hogg's explanation is worthy of Sherlock himself. Here it is in its entirety:

> Much doubt has existed as to the origin of this word...The earliest forms in which it was written, will be seen from the synonymes above, they were Pearemaine and Peare-maine. In some early historical works of the same period, I have seen Charlemagne written Charlemaine, the last portion of the word having the same termination as Pearemaine. Now, Charlemagne being derived from Carolus magnus there is every probability that Pearemaine is derived from Pyrus magnus. The signification therefore of Pearmain is the Great Pear Apple, in allusion, no doubt, to the varieties known by that name, bearing a resemblance to the form of a pear.

Let's pick up where Hogg left off. I think he was on the right track but he could have gone one step further. Until well into the twentieth century, apples and pears were lumped together in the same genus. The common apple was *Pyrus malus* and the common pear was *Pyrus pyrus*. A *Pyrus* could therefore be an apple as well as a pear. We should assume that the pear in Pearmain is about *Pyrus*, but not necessarily about pears. Pearmains don't need to resemble pears. A pearmain—*Pyrus magnus*—can simply be a great apple. Returning to the Oxford English Dictionary for a moment, they might not have been as wrong as they seem. If apples "with firm white flesh" were rare and considered superior, then maybe the OED is correct. Maybe firm, white fleshed varieties were wonderful, coveted *great* apples, worthy of the name "Great Apple," or Pearmain.

What about Gray Pearmain? While Gray Pearmain is not pear-shaped, it does taste remarkably like a pear. Given that it's safe to bet that Charlemagne or *Charlemaine* had been more or less forgotten in central Maine by the time this apple came along, it makes sense that whoever named it was free to use his or her own criteria. Let's say that the apple originated shortly after the Civil War. The trees were old in 1970 when Steve spent his first season at what was then Wentworth's Orchard. That would make the trees a hundred to a hundred and twenty years old in 1970. Split the difference. Call it a hundred and ten, which would date their planting to the spring of 1865, right at the end of the war. Blue Pearmain was a fixture in orchards throughout New England by then, including central Maine where it was commonly grown. A farmer in Fairfield discovers a new pear-tasting seedling apple. Everyone's got blues and grays on their minds. It's a great moment. It's a new day. The war is finally over. Everyone's got a Blue Pearmain. This new apple does have a sort of grayish tint to it. Let's call it Gray Pearmain!

Gray Pearmain is one Pearmain that really does taste like a pear; it's wonderful fresh eating

There were still five or six of the old Gray Pearmain trees when Steve and Marilyn took over the orchard. Although they didn't know anything about its origin, they liked the apple and grafted a couple dozen young trees before the old ones died. In more recent years, we've grafted several hundred for the Fedco catalog and it too—like Canadian Strawberry—is gaining a following. It is an excellent dessert fruit. Still, though, I could find no word on its origin. Until a couple of years ago.

The short-lived Maine Pomological and Horticultural Society was established in 1847. It was followed by the Maine State Pomological Society in 1873, which has been with us ever since. Maine began publishing its *Agriculture in Maine* annual reports in 1858. Beginning in 1873, these bound yearbooks included a report of the Pomological Society. The early years provide a treasure trove of information on fruit, vegetables, grains and livestock. No Maine fruit explorer should be without a complete set. You could spend years browsing the pages. No matter what your farming interest, they are fantastic.

A few years ago someone sent me a citation from the 1885 edition. A number of Maine growers were donating apples to be sent down to New Orleans to the World's Exposition, presumably as a promotion of Maine agriculture. Buried in the list of twenty-nine growers and dozens of apple varieties appears one line in reference to a C. A. Marston of Skowhegan:

> "C. A. Marston, Skowhegan, one barrel: Northern Spy, Tompkins King, Gray Pearmain, Black Gilliflower."

Just a trifle, but oh what a trifle! You can imagine my excitement when I read this line. It was, at last, confirmation that the Gray Pearmain is, more than likely, its very own variety. Not only that but it was grown—and most likely originated—within a few miles of the trees we'd been picking. Skowhegan is right next door to the Apple Farm. Recently Steve told me the Meyerhans' orchards actually were originally in Skowhegan until they moved the town line years ago.

Steve and Marilyn Meyerhans in 2006: photo by Martin Brown

Still we've found no other citations. But one is enough. The next thing to do was track down the Marstons and find out who they were. First off, we can presume that C. A. was an orchardist. His name was likely Charles. His family probably started, owned and operated The Marston Worsted Company of Skowhegan. He was an important figure in town, vice president of the Skowhegan and Norridgewock Railway and Power Co., member of the "committee of sewerage" as well as the school board. Most interesting, though, is that his home was at the intersection of Main Street and the Middle Road. This puts him on the south side of the Kennebec River within four miles of the Apple Farm. Maybe the apple originated in North Fairfield. Maybe it originated on Steve and Marilyn's farm. I suppose we'll never know.

Blue Pearmain

# Twenty-Five

# Eliminating the Impossible

"That process," said I, "starts upon the supposition that when you have eliminated all which is impossible, then whatever remains, however improbable, must be the truth. It may well be that several explanations remain, in which case one tries test after test until one or other of them has a convincing amount of support. We will now apply this principle to the case in point."

Sherlock Holmes, *The Adventure of the Blanched Soldier*, p. 1011

Sweet Red

With Gray Pearmain, it was pretty straightforward. It was just a waiting game. I waited around for twenty years, and finally someone found the citation and sent it to me. Coming to a resolution about Sweet Red, on the other hand, has required more effort. I began with the assumption that the name Sweet Red was a synonym for another variety. Sweet Red sounds too much like a made-up name for an unidentified apple. It's definitely low in acidity. It takes forever to cook into sauce. Low-acid apples are always slow to cook up on the stove top. That makes it a true, old-fashioned sweet variety in the tradition of Tolman Sweet, Pound Sweet and others. It's also solid red in coloring. It often has no stripes or ground color visible at all. Just red. So it's definitely sweet, and it's definitely red.

All classic apple books note whether a variety is "sweet" or "tart." These two terms describe the single most important taste distinction for the apple user and for the apple identifier. The terms refer to the relative levels of acidity and sugar. Neither describes an absolute. It's all about relativity. The more acid relative to the amount of sugar, the more tart the apple will taste. The more sugar relative to the amount of acidity, the more sweet. There are no entirely sweet apples, and there are no entirely tart apples. Every apple has its own unique combination of both. All apples lean one way or another. For the chef or the apple eater, the distinction is all about flavor and use. For the apple identifier, the distinction is all about reducing the pool of possibilities. Once you determine if it's sweet or tart, you eliminate a lot of the impossible.

The most desirable dessert apples, in other words, those for eating raw or fresh, lean slightly towards the tart side of the scale. They please our modern taste buds. These apples are often called "subacid." Honeycrisp, Gala, Fuji, Braeburn and all the rest of those apples in the grocery store these days have been selected for you in large part because of that balance. They're close to the middle, like politics used to be. You might think you like your apples sweet, but, more likely, you like them on the acid side of that balance point. Once they get a lot more to the left or right of center, apples can get pretty bizarre, also like politics. When it comes to fresh eating, it's all about that balance.

The apples themselves probably don't care if they're balanced or not. The vast majority of them are not. Most of these "unbalanced" varieties lean further into tart territory while the occasional renegade leans towards sweet. This may be the primary reason why so many of those wild apples out along the roadside are affectionately called "spitters." They can be hard to swallow. Hedrick calls those apples "austere." That's why lots of money is being spent trying to find the next Honeycrisp. Otherwise, you could simply grow your apples from seed and be delighted with whatever you got. But waste not, want not. Other than for fresh eating, those balanced Braeburns and Honeycrisps are mostly useless. The grocery store varieties are lousy in pies. They're lousy in sauce, and they're irrelevant in cider. Traditionally, the vast majority of apples were cooked or pressed, not eaten fresh. Those seemingly out-of-balance apples were searched for and coveted for generations. Our ancestors knew just what to do with them.

Topworking #15: Dab the scion tips with goo

Knowing if an apple is sweet or tart is essential to the pie-maker, the cider-maker, the sauce-maker, the molasses-maker, and anyone else who's using an apple for anything other than a snack. Tart apples cook more quickly than sweet apples. I've tried making sauce with many sweet apples. It takes forever. They don't cook down. The slices float around the sauce pan like pieces of limp leather. Cole's Quince, Kavanagh and Roxbury Russet are three of the best sauce apples. All three cook into sauce in a few minutes. It's worth growing the trees just for the sauce. Cole's ripens in late summer, Kavanagh in fall and Roxbury after New Year's. Their skins dissolve. There's no need to peel them. You wouldn't even know the skins were there.

John Paul Rietz stirring apple molasses at the farm, 2011

All the best pie apples are also tart, each to its own varying degree. Some, like Black Oxford and Spice Sweet, are only mildly tart, while others, like Moses Wood and Scott's Winter, are quite so. But they all live on the acidic side of center. They cook thoroughly in an hour. The slices stay firm, and the crust won't collapse. They're not crunchy. The acidity also brings out the flavors in those pies and crisps and sauces. One Thanksgiving I tried making a pie with a neighbor's Pound Sweets. Boy, was that a bummer. The texture was weird, and the pie had no flavor.

Sweet apples, on the other hand, have their own attraction. They make excellent baked apples. Set three or four sweet apples in a baking dish with a little water and put them in the oven whole. No need to add sugar, spices or anything else. Cook them for an hour or so. They will be delicious. Serve with hot milk or cold ice cream.

Emily Skrobis pouring off the sweet apple molasses, 2011

Use sweet apples to make molasses. Press them into cider, and then boil down the cider in a pan just like making maple syrup. You'll get a sweet molasses, better than any you'd ever find in the store. Particularly back when sugar was scarce, sweet-apple molasses was the best thing going. You can use any sweet varieties, although Tolman Sweet is one of the best. Apple molasses made with dessert varieties is okay but noticeably more acidic.

Sweet apples stewed in milk was considered a medicine for those with a multitude of illnesses. They are also ideal for those who want to minimize their acidic food intake in general. This might include you, and it might include your cow, sheep, horse or goat.

Cider makers usually want about half their apples to be sweets. As the cider ferments, the sugar turns into alcohol. If you ferment your cider to dryness, all the sugar is gone. If your apples are mostly acidic, you will have a really acidic cider, one that can be un-drinkable. By using some portion of low-acid varieties in your mix, your final fermented cider should develop into that delicious, well-balanced beverage that's every cider-maker's dream.

You can't train an apple to be sweet. You've got to go out and find them. Over the centuries, an assortment have been found and cultivated by cider-makers. That search continues today and is driving a Renaissance in orchard planting. While there are some famous acidic cider apples out there, many of the most coveted are sweet. Dabinett, Yarlington Mill, Harry Master's Jersey and Chisel Jersey are just a few of the increasingly well-known English bittersweet cider apples. Not only are they relatively low in acidity, they also have tannins, which impart bitterness and/or astringency to the cider. The bitterness in these apples plays a role roughly analogous to hops in beer.

After reading James Thatcher's 1825 *The American Orchardist*, you might never want to eat a well-balanced subacid apple again. Here's some of what "the honourable T. Pickering" says about sweet apples, as quoted by Thatcher on pp 13-14:

> After providing a due proportion of apples for the table and the ordinary purposes of cookery [both presumably sub-acid or tart varieties], I do not hesitate to express my opinion, that, for all other uses, sweet apples are entitled to the preference. The best cider I ever tasted, in this country, was made wholly of sweet apples. They afford also a nourishing food to man and all domestick animals. What furnishes a more delicious repast than a rich sweet apple baked and eaten in milk? I recollect the observation made to me by an observing farmer, before the American revolution, that nothing would fatten cattle faster than sweet apples. Mentioning this, a few years since, to a gentleman of my acquaintance in an adjoining state, he informed me, that he was once advised to give sweet apples to a sick horse. Happening then to have them in plenty, the horse was served with them, and he soon got well, and, continuing to be fed with them, he fattened faster than any other horse that he had ever owned that was fed with any other food.

I felt as though I owed it to Steve and Marilyn to identify their sweet apple. Not only have they become great friends of ours, some Octobers it feels as though I'm spending every other day in their orchard. After pondering Sweet Red for years, I finally put together this story. It might even be correct. It begins with Royal Wentworth or someone before him who owned the orchard. One of them was given scionwood for a sweet red apple, probably with a varietal name. Being an orchardist and not a preservationist, Wentworth topworked the scionwood onto a few of his trees, promptly forgot the name, and went about his orcharding life. By the time the trees fruited a few years later, memory had blurred and the name had evaporated away. "What was the name of that red apple?" But the fruit was good so he kept the grafts. It was sweet, and it was red so Sweet Red it became. And so it remained Sweet Red until today when all that will change right here before your very eyes when its past life will once and for all be revealed. I hope.

The envelope of a letter to my parents in 1977

So we begin with the premise that Sweet Red is a real variety but it's not a variety named Sweet Red. Now we just need to determine which one it really is. Of course, that also presumes that it is only one variety. With all the synonyms out there floating around, maybe it's got ten names. We'll cross that bridge later.

Sweet varieties should be easier to identify simply by virtue of the numbers. There are fewer of them. That means fewer to agonize over. There just aren't that many sweet heirloom varieties. There are even fewer red sweet heirloom varieties. It can't be that hard to ID this apple. God forbid if it's another of those rare locals, though something keeps telling me that this is a named variety.

On several occasions I spent the better part of a day scouring the sweet apples for a match. I was able to knock off a few impossibilities in the process, but ultimately I failed every time to settle on a name. I think I was making it too difficult for myself. I was doing too much speculation. I was considering obscure varieties with almost no descriptions. Too much theory and not enough attention to the details. Consider the possible ones until they become the impossible. Then dump the impossible and move on.

Eventually I did reduce the pool to a handful of suspects. I began to feel as though I was accomplishing something. I did this by looking at all the sweet varieties I could find that had any association at all with Maine. This was made possible in large part by historical apple name protocol. Most of the sweet apples include the word "sweet" in the name. What a gift. I suppose that's because the sweet apples were relatively rare and were coveted by those who appreciated their qualities. They wanted to know which ones were the sweets. Conversely, it made it far easier for those who did not want a sweet apple to avoid them.

Then there was the issue of the descriptions themselves. I assumed that I would need to have a little bit of blind faith in the accuracy of the descriptions. I suspected it might come down to one or two fine points. Once you arrive at that level of detail, one ill-chosen word can send you off on the wrong track.

I went to the basement and brought up a dozen Sweet Reds. We still had a bushel. It was December, and the fruit was in perfect shape. This is not an early apple. I spread them out on the dining room table and put on some music, Rahsaan Roland Kirk's Bright Moments. "Bright moments, right now!" I've got lots of Sweet Reds so I cut up several of them both ways. Now I can see them up close and personal. I'm going to nail this apple today. Right now in real time. In front of you. Right now!

A bushel of Sweet Red

I want it to be Franklin Sweet so I start there. That would be perfect. I've been looking for a Franklin Sweet for years. I could kill two apples with one stone. This would be great. Better think again. Or, maybe I should say, better think? All of a sudden I can hear Sherlock rolling around in his grave. I can hear him snarling about how that's not the way to approach any mystery. No theorizing before you have the facts. You have to let the facts dictate the way, not the whims and desires of the apple sleuth.

Franklin Sweet originated in Franklin County, the land of the Deane and only twenty-five miles from the Apple Farm. I also know that the apple was being grown in nearby Kennebec County from information Cammy dug out of the old *Maine Agriculture Yearbooks*. So what? Interesting, but let's look at the facts. Two details disqualify Franklin Sweet almost immediately. It's season is September, and the flesh is white. Sweet Red keeps all winter. The fruit is still in good shape in the spring. September apples don't keep that well. On top of that, the flesh is off-white or even yellowish. So reluctantly I say goodbye to Franklin Sweet. It was a nice thought. Or maybe it was a bad theory.

Next I go for even closer proximity. Parlin is another apple I've been looking for. It's from Norridgewock, even closer to the Apple Farm than Franklin County. Norridgewock is bisected by the Kennebec River, just west of Skowhegan. It shares a short stretch of its eastern border with Fairfield, a few miles from Steve and Marilyn's orchards.

I don't know if I've ever seen a Parlin. There's a detailed description in Beach. It almost exactly matches the Sweet Reds cut up on the dining room table, right down to the funnel-shaped calyx tube. Remember the calyx tube? Every so often it can be funnel-shaped. The seasons are the same. The color is described as "nearly as dark red as Jonathan or Gano." The flesh is described as "whitish with a yellow tinge, moderately firm, tender, moderately fine-grained, not crisp, moderately juicy, sweet or very mildly subacid, slightly aromatic, good or sometimes very good." That's nearly Sweet Red all the way. Alas, there are three minor inconvenient snafus. Beach says Parlin can be sub-acid. That's a stretch. Sweet Red is really a sweet apple. Parlin's seed cells are supposed to be closed, or nearly so. Sweet Red's are what I would call open. Beach says that the skin is "often irregularly veined with russet." No russet on Sweet Red except around the stem. Are any of these deal-breakers? I'm not sure. It's so close! I like Parlin. I want Parlin. I wish it was Parlin. But of course, Sherlock doesn't care about my wish list.

Pumpkin Sweet of Orland

Next I look at Parks Sweet from Skowhegan, home of the Gray Pearmain. That would be as close geographically as Parlin. Parks Sweet might also be the Parker Sweet of the USDA watercolors. Parker Sweet was grown in Franklin County. Sadly, there's almost no description to speak of. The fruit in the watercolor looks too large and too elongated-ovate. Something doesn't ring true. Bradford doesn't have much to say about the apple although he does mention that its season is "early autumn." Cross that one off too. Sweet Red keeps until spring. So no go for Parks Sweet even though it originated right around the corner from the Apple Farm.

So I abandon Parks Sweet and move onto the most likely suspect. This is the one I've been coming back to over and over again. Ramsdell Sweet is one of the most famous sweet apples. It was grown in Maine, though it's unclear exactly where. "Popular in some sections," reads the 1875 *Catalog of Fruit for the State of Maine*. It originated in West Thompson, Connecticut, on the farm of the Rev. Hezekiah Stocking Ramsdell and dates back to the early decades of the nineteenth century. Steve Meyerhans told me long ago that when he ate Sweet Red for the first time, he knew that he was eating something he remembered from his childhood in Woodbury, Connecticut. Ramsdell? How far is it from Thompson to Woodbury? Ramsdell was certainly famous enough to travel.

Pumpkin Sweet of Orland (a.k.a. Wardwell Sweet?)

I pull out Beach and read. Ramsdell was also known by many other names, including English Sweet, Ramsdell's Pumpkin Sweet, Ramsdell's Red Sweet, Red Pumpkin Sweet, and a few other variations. Why didn't I ever notice that before? Red Pumpkin Sweet and Ramsdell's Pumpkin Sweet are both synonyms for Ramsdell? How had I missed that? I've had another mysterious sweet red apple from Orland for twenty years that was introduced to me by Bob Dow as "Pumpkin Sweet." I've struggled with that one ever since. It's not the large yellow Pumpkin Sweet of Connecticut, nor is it the orange russety Pumpkin Sweet of southwestern Maine, which may also be known as Pumpkin Russet. Not knowing what else to call it, I've been referring to the apple as Pumpkin Sweet of Orland Maine or sometimes Wardwell Sweet after the 1912 USDA watercolor lookalike, painted from an apple sent to the USDA by Charles Atkins of East Orland. Bob Dow's Pumpkin Sweet/Wardwell Sweet is a red Pumpkin Sweet if there ever was one. If it's *the* Red Pumpkin Sweet, then that would also make it Ramsdell Sweet.

I met Bob Dow at the Common Ground Fair in the mid-nineties. Bob was an "old Mainer." He loved gardening and trees and visited me at the Fair for several years. On more than one occasion I went to visit him at his place in Orland. One year I gave a talk at one of his grange meetings. He introduced me to the apple he called Pumpkin Sweet from an old tree at his next door neighbor's. He knew it was rare. Bob's Pumpkin Sweet is a true, low acid, sweet apple. It's large, roundish-conic-ovate, entirely red, and, as Bob put it, "even ribbed like a pumpkin." I topworked scionwood onto a tree at home. When Bob's wife died, he moved, and he and I lost contact.

Lyman's Pumpkin Sweet may be what Mainer's call Pound Sweet

I liked Bob's Pumpkin Sweet apple, and wanted to save it. I grafted it for Francis. When his tree fruited a few years later, he was incensed. "That's not Pumpkin Sweet!" he needled me. In one sense, of course, he was correct. The most widely accepted Pumpkin Sweet is the Pumpkin Sweet of Beach (a.k.a. Lyman's Pumpkin Sweet) which is a large, yellow-green apple that originated in Connecticut. But Pumpkin Sweet is a great name, and other Pumpkin Sweets emerged here and there in the nineteenth century. I featured Bob's Orland apple in my Pumpkin Sweet Fair display. Pumpkin Sweet of Orland Maine is little long, but the name allowed me to keep track all my various Pumpkins. I don't want to mix them up.

Then along came Wardwell Sweet. I was looking for apples on the USDA website from the Orland area primarily because of the work Pete Jenkins and Todd Little-Siebold had been doing. I also had it in the back of my mind that maybe Bob Dow's Pumpkin Sweet was a synonym for another variety. Since there is no written description of Wardwell Sweet, I could only go by what I could see in the painting. The apple and the painting were remarkably similar. You could say identical. Not only that but both came from East Orland. Maybe the apple in the painting came from the tree in Bob Dow's neighbor's yard. The tree was certainly old enough.

Then one day Todd sent me photocopies of correspondence from late 1911 and early 1912 between Charles Atkins and G. B. Brackett, a USDA Washington DC pomologist. Among the letters is one in which Atkins lists the varieties he has sent to the USDA to be described and painted. Included in the list is Wardwell Sweet: "I know of only one tree of this sort. It is in the orchard of a Mrs. Wardwell of East Orland, who says her father planted the tree and as he sometimes bought trees from N.Y. Nurseries, this may be a known sort." New York nurseries? Could they have sent him a Ramsdell Sweet? In a later trip to Orland I found that the Wardwell Road is less than a mile from Bob Dow's tree. Surprised?

Ramsdell Sweet (a.k.a. Sweet Red?)

Back down into the root cellar I go to find a Wardwell Sweet, a.k.a. Pumpkin Sweet of Orland Maine. There's only one left, so I only have one chance to cut it up. I have to treat my Wardwell well. I sit, and I stare. Just as I fear, the Wardwell red Pumpkin Sweet and the Sweet Reds look to be identical. Why didn't I think of that ten years ago? I see these apples every fall. You'd think I might have noticed. I set the Wardwell off to the side so I didn't mix it up with my ten Sweet Reds. That would be easy to do.

I now know Wardwell Sweet was another "provisional" name. That explains the lack of any printed information on the apple. There is none because there was none. In fact, there is no variety Wardwell Sweet. It too is a synonym for something else. I'm putting some things together now. Wardwell is probably the same as Bob Dowe's Pumpkin Sweet of Orland Maine. Pumpkin Sweet is an old synonym for Ramsdell Sweet. Wardwell resembles Sweet Red. What does that say about the relationship between Sweet Red and Ramsdell Sweet? Now we're cooking.

Back to Ramsdell. I read on through Beach's description. Ramsdell and Sweet Red are nearly a perfect match. There are slight discrepancies in the flesh color, the stem color and the dots. Ramsdell's flesh is yellow. Sweet Red is really light yellow. When does light yellow become yellow? Were these guys pomologists or artists? Ramsdell's stems are "often red." Sweet Red's only sometimes. When does sometimes become often? Sometimes I stop by the bar for a drink on the way home from work. How many days a week would that be? Oh, sometimes, but not often. Ramsdell's dots can be "rather large" and "often submerged." Sweet Red certainly has the dots. They are distinct and prominent. But could I call some of them large? I'm not sure. I pulled up twenty or thirty photos I've taken of Sweet Reds over the years, and it's clear that sometimes the dots are larger on some of the fruit. I can live with that. The one difficulty is the basin, which Beach calls "rather small" and "faintly furrowed." I'd call Sweet Red more furrowed than faintly and more medium than small. In his cosmo equations, Sun Ra often flipped two-word or two-number combinations. After all, three twos are always equal to two threes. Red Sweet should therefore equal Sweet Red.

Lastly I consider a relatively obscure apple called Sweet Pearmain. I decided to consider it because of its inclusion in Bradford. Apparently Sweet Pearmain was shown by the Bangor Horticultural Society in 1852. That would be about the right time frame and about fifty miles from Fairfield. And who's to say who brought it to the Bangor show? Could it have been an orchardist from Fairfield? Sweet Pearmain is said to have originated in England before the Revolution. It was brought over and grown here and there in southern New England and also evidently somewhere near Bangor here in Maine. Its description sounds identical to Ramsdell Sweet. I can't find a single difference between the two. Can anyone tell these apples apart? One of Ramsdell's synonyms was English Sweet, and Sweet Pearmain is said to have originated in England.

Sweet Red channeling its inner Ramsdell Sweet

This requires another detour. I get out *Elliotts Fruit Book: or, the American Fruit-Growers Guide in Orchard and Garden*, written by F. R. Elliott and published in 1854. Elliott quotes his various contemporaries including Downing, Thomas and Robert Manning. Manning was one of the founders and an early officer of the Massachusetts Horticultural Society. Downing and Thomas call Sweet Pearmain synonymous with English Sweet. Manning calls English Sweet and Ramsdell Sweeting synonymous.

Next I go back to Bradford. What does he have to say about Ramsdell? "Ramsdell has been known mostly in Maine as Ramsdell's Red Sweet, although it is catalogued by the Maine Pomological Society as English Sweet." Bradford also lists another synonym as Ramsdell's Pumpkin Sweet. I think we found our apple. Drop the Ramsdell and you have Red Sweet. Flip the two words around and you have Sweet Red.

Back to Royal Wentworth. He grafts a few trees and they fruit a few years later. What was the name of the apple? Something or other red and sweet. Was it Red Sweet? Sweet Red? We'll call it Sweet Red. Back to Bob Dow's apple. It's an easy slide from Ramsdell's Pumpkin Sweet to Pumpkin Sweet. It's like someone who says, Oh that's Frank's Northern Spy. What's the name of the apple? not "Frank's Northern Spy," just Northern Spy. So Ramsdell's Pumpkin Sweet is really just Pumpkin Sweet. (Who is Ramsdell, anyway?) and Ramsdell's Red Sweet is really just Red Sweet. Red Sweet is a bit odd sounding so we'll call it Sweet Red. That sounds better.

Lastly I get out my copy of *The Fruit Manual: A Guide to The Fruits and Fruit Trees of Great Britain* by Robert Hogg. I flip through it. No Sweet Pearmain. No Pumpkin Sweet. No Red Pumpkin Sweet. Then I turned to Pearmain on page 169, and I read that Pearmain was also called Old Pearmain and English Pearmain. That last one sounds like a scary combination of English Sweet and Sweet Pearmain. Remove the sweet from the end of one and the beginning of the other and what have I got—English Pearmain. This is like grade school algebra. According to Hogg, not only is this the oldest Pearmain on record, it's the oldest English apple on record. The description sounded just like Ramsdell. Help! Hogg even calls the flesh "yellowish, tender, juicy, sweet, and highly flavored." The basin sounded more like Sweet Red than Ramsdell: "plaited." Alas, I was saved by the stem which sounded too short, the cavity which was too shallow and the dots which were too large. I wasn't quite ready to say the Meyerhans have a row of the oldest trees recorded in English pomological history. Not yet at least.

I made an executive decision. I decided to say that Sweet Pearmain and Ramsdell Sweet are the same apple, whether Sherlock likes it or not. Because the description of Sweet Pearmain includes no conflicting details with Ramsdell, and both were grown in central Maine, there is probably no way to consider them independently anyway. So I'll call them the same for now.

Now it's time to cut my only Wardwell Sweet in half and compare it to Sweet Red. The one Wardwell Sweet I have is axile and open-cored. The core-lines are clasping, and the stamens are placed just as they are on Sweet Red. It appears to be a perfect match, although I know I should cut up a few more Wardwell Sweets next fall to be sure. One apple is not a good data set. But until that time I decide to assume that they are the same apple.

Parlin and Ramsdell Sweet, however similar, are almost certainly not the same apple. Wardwell and Sweet Red probably are the same apple. I was first introduced to Wardwell as Pumpkin Sweet. Not only that, it is undeniably a red Pumpkin Sweet. Since Ramsdell is also called Red Pumpkin Sweet, I suppose that makes Wardwell equal to Ramsdell, at least by name. And if Sweet Red and Wardwell are synonymous, then Sweet Red becomes Ramsdell.

One more time through Ramsdell and Parlin. I tried to focus on every detail. Sherlock would have been proud of me. "Data! data! data!" I liked Parlin because of the proximity. But there were still two things that worked against Parlin. One is the veins of russet that sometimes cover the dark red skin. I don't see them on any of the Sweet Reds. The second is the closed seed cells. Sweet Red's cells are not closed. I read through Ramsdell Sweet one last time. Where's that fine-toothed comb? Everything seems to match up well, if not perfectly. The only troubling trait is the basin features. Beach calls it faintly furrowed and wrinkled. While Sweet Red and Wardwell are not deeply furrowed, they seem to be more than faintly.

I'm exhausted. I need dinner. It's dark. I'm alone. It's snowing. Fortunately there's a pot of stew on the cookstove. Am I satisfied? I don't think I found Parlin even though it would be fun if I had. I do think there's an excellent chance that I have Ramsdell.

Identifying apples is not easy, even the sweet ones. Apple aliases are worse than those on the ten most wanted cheat sheets on my grandmother's post office wall. Too many. The Sweet Pearmain and the Ramsdell Sweet are probably the same apple. The red Pumpkin Sweet (Wardwell Sweet) of Orland and the Sweet Red of the Apple Farm are also probably the same apple. All four are probably the same variety as well. What we should call that variety is a toss up. Ramsdell Sweet probably? There's a vague chance that all four are also synonymous with the Old Pearmain of England, making them one of the oldest varieties on record. That's a bit more of a stretch.

When I gave a talk in Blue Hill a couple of years ago, not far from Orland, to my great surprise and total delight, Bob Dow showed up for the event. He had reappeared and his new wife was with him. She was his high school sweetheart. For now I'll continue to call Bob's Orland apple Pumpkin Sweet like he did or maybe Wardwell Sweet if I feel like it. And I'll call the apple from the Apple Farm Sweet Red. But until new information appears and send us off in another direction I'm willing to argue that both of them are Ramsdell Sweet a.k.a. Red Pumpkin Sweet. Now, please, get me out of here!

Sweet Red
an equation in names

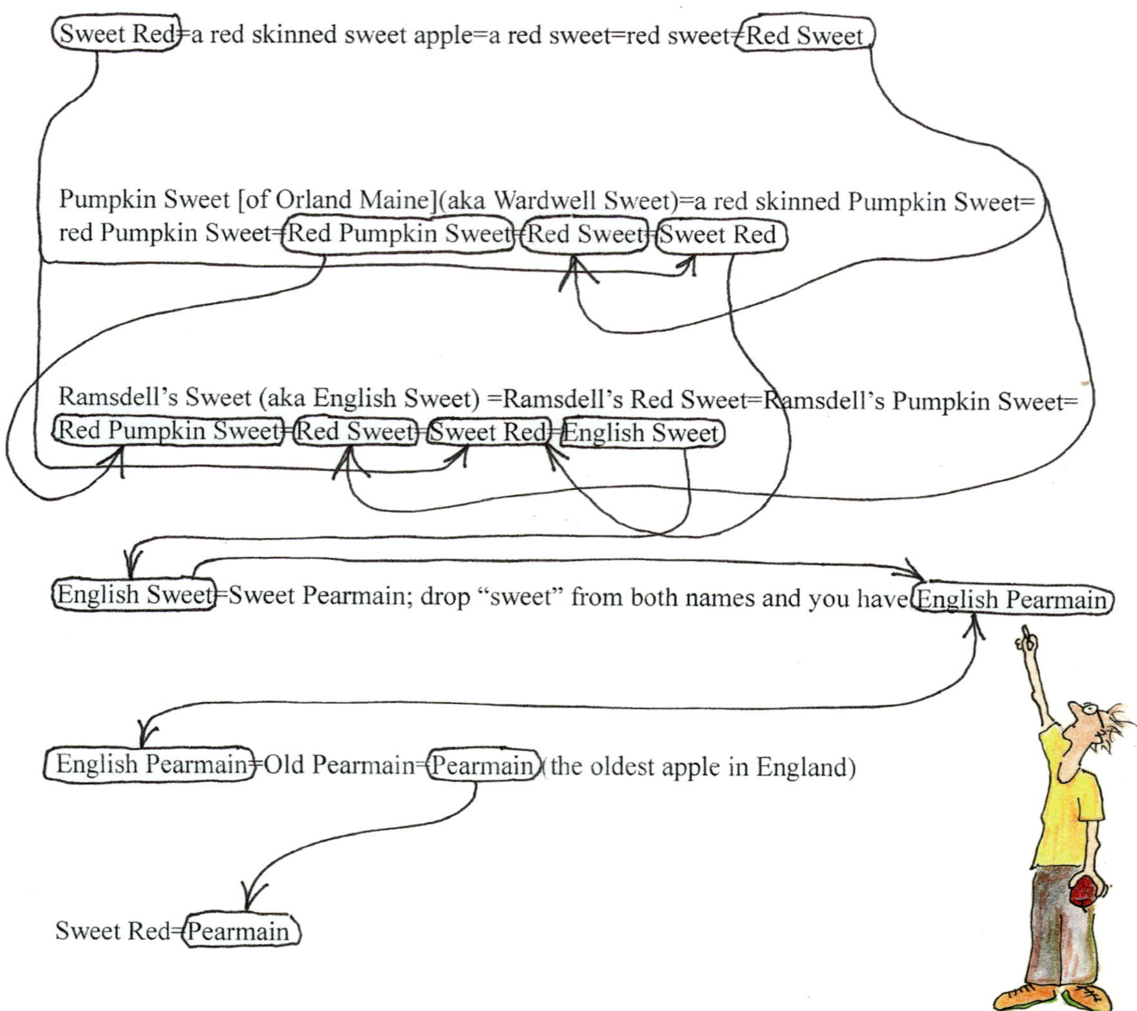

# Twenty-Six

# Greenings and Pippins

Ah! my dear Watson, there we come into those realms of conjecture, where the most logical mind may be at fault.

Sherlock Holmes, *The Adventure of the Empty House*, p. 495

Rhode Island Greening

Mr. A. ordered a Clingstone peach because he thought it would be more durable in his family than a Freestone, and when it came to bear it bore Rhode Island Greenings.

*Agriculture of Maine*, 1892 p. 52

Gros Frequin

For the apple identifier it's a blessing that apple namers of the past nearly always used the word Sweet in the names of the sweet varieties. It is equally convenient that many of the green-colored apples include the word Greening in the name. During the later half of the nineteenth century there were close to twenty different Greenings being grown in Maine alone. How did that happen?

For centuries there was rarely anyone dictating anything about how apples were named or what words could or could not be used. And yet customs, conventions and protocols did develop over time and over a wide geographical area, despite what we might think of as the relative isolation in which people lived. How and why did so many English and American apples latch onto the word Pearmain, for example? Where did the word Frequin originate, and why are there so many different Frequin apple varieties in the orchards of northern France? According to French etymologists, a frequin is a quarter cask. I suppose that would be about twelve gallons. But why would multiple apples be named for a small barrel? Could it be that all the various Frequin apples make such excellent cider that you only need a little bit? Who needs a barrelful of cider when a quarter barrel of Frequin Audievre or Frequin Blanc or Gros Frequin will do the trick?

Pierre Bourdon identifying one of his cider apples for us in Normandy; Sharpie is perfect for writing directly on the skin; Could it be a Frequin?

In English we have the word frequent, lifted from the French, meaning numerous or repeated as an adjective, and meaning to visit as a verb. We also have frequently, frequency, frequenter, frequentation and even frequentative. Undoubtedly all these variations refer to the Frequin apple tree that we can assume bears annually and productively and makes a delicious cider that's so good it warrants repeated trips to the quarter cask. In other words, the frequentation of the frequentative fermentation of Frequins inspires the farmer to become a frequenter who frequently frequents the frequin with great frequency.

At least we *think* we know what sweet means. Don't we? We can imagine that the convention of using the word "sweet" with the sweet apples began in England or France, and is also directly connected to cider. Cider makers want to know the relative acidity of the apples they press. So the convention of noting sweet apples in the name probably began long ago, and was simply continued in North America by the growers who were all cider makers early on.

René Magritte's dilemma

But what about the word "Greening," which keeps showing up practically any time we have a green-skinned apple? While we do have a generous number of varietal names that include the word "Red," the names of most red apples give you no clues about their skin color. Baldwin, Stark, Esopus Spitzenburg, and Mother are just a few of the well known historic red apples that reveal no color in their name. Same is true with yellow. We do have Yellow Transparent and Yellow Bellflower but what about Porter, Primate, Peck Pleasant and Ortley? They're all yellow but you'd never know it from the name. So few Yellows; so few Reds; so many Greenings.

A reasonable guess would be that the convention of using the word Greening began over here in the colonies. Greening is a word rarely, if ever, found in English and French apple names. It probably makes sense, therefore, to turn our attention to the earliest American varieties and see what we find.

Rhode Island Greening is America's best known Greening. It's also one of our oldest and most important varieties. There are several published histories of the apple. One claims that it originated in Foster, Rhode Island, west of Providence on the Connecticut border. Another claims it was introduced by a Richard Smith in Smithfield, just north of Providence. A third claims that the original apple tree was a gift from the King of Persia to a Captain Green Chanson who saved the king's son in a shipwreck. In that version, the apple is a direct descendent of the Tree of Knowledge from that famous garden in Eden. The first two origin-stories have been disproven while the third remains inadequately explored. The most accepted story mentions an inn just north of Newport and a fellow named Mr. Green. I decided I would dig a little deeper and see if I could fabricate a plausible account of the history of the apple and how the word greening entered into the apple lexicon.

Rhode Island Greening is green even when ripe; it often has a reddish-brown blush

It appears that in about 1635 a Puritan named John Green (or Greene) left England with two brothers and headed across the Atlantic, possibly first to the

These Maiden Blush are green enough to be Greenings

British West Indies and then north to Aquidneck Island, a.k.a. Rhode Island, down near the mouth of Narraganset Bay. Green arrived on Aquidneck in 1637, a year after Roger Williams had been kicked out of Massachusetts and came south. This was during that era of draconian religious intolerance and misogyny; Anne Hutchinson and a bunch of others were banished from the Commonwealth and headed off to what became the state of Rhode Island. So much for religious freedom!

John Green—or possibly his brother William—settled in what became Portsmouth and eventually Middletown. It was all wilderness back then. There he established a tavern and began to farm on the shore of what is still called Green End Pond, a beautiful body of fresh water that serves as the modern-day reservoir for the city of Newport. Not long ago, I drove the length of Green End Avenue across the southern edge of Middletown, stopping to take in the view of the pond. It's not hard to imagine the site as the perfect location for a farm and a trading post. There was ample fresh water and easy access to the bay, not much more than an apple-throw away. Trading posts were multi-use back then, doubling as forts, taverns and inns. Green's farm became known as Green's End and the tavern became Green's Inn. Like many other colonists back then, John Green grew apples from seed. Sometime around 1650, one of his seedings grew up to produce green-skinned apples. It was a green apple from Green's Inn at Green End.

Green End Avenue and Green End Pond, 2018, near the site of Green's Inn and the first Greening in America

All apples are green until they ripen. Most of them eventually acquire blushes and stripes of red as they mature, others turn entirely yellow or red or a combination of the two, but a select group of apples never lose that green color. They remain green right into maturity. They might have a red or pink blush and they might have a russet splash, but predominately they remain green. Not only was Green's apple green, it was also an excellent variety. It still is, three and a half centuries later. It ripens late, keeps in common storage until spring and is an exceptional cooking apple. It was one of the most important of all American varieties for nearly three hundred years.

Good apples were already in demand in 1650, especially top quality storage and cooking varieties, so the Green End apple began to spread. Before long it became known as the apple at Green End, the apple at Green Inn or the apple at Green's Inn. Beach says, "The scions of this tree were in such demand by the people who stopped there as guests, that the tree died from excessive cutting and exhaustion." Don't cut too much wood off your trees.

You can only say the "Green-End apple" or the "Green-Inn apple" so many times before words begin to blend together, especially during a time when not much was being written down. Who cared what the words actually were? It was just a name that blurred together. Before long it morphed into the "Green-en apple," or the "Green-in apple." As the variety's reputation spread beyond the state, people wanted to know where it came from. This led to the addition of Rhode Island to the name. Out of state it became known as the "Green's Inn apple from Rhode Island." Kind of a mouthful. Nevertheless, swap a few words around, and it's only a short stretch to Rhode Island Green's Inn or Green-en or Green-in—and, voilà! You have Rhode Island Greening.

Northwestern Greening is roundish, light green in skin color and the blush is usually a soft orange-pink—never brown like Rhode Island Greening

Then something unforeseen happened. Before too long, it seems as though every other green apple in America became a Greening. What began as a contraction of the location of the origin of one apple turned into a nomenclatural convention for dozens of varieties. This wasn't a regional fad either. One of the most famous Greenings is Northwestern Greening, which originated in Wisconsin, twelve hundred miles west of Green's Inn. Northwestern Greening did not sprout up next to a tavern or motel, as far as I know. Another fairly well known Greening, Patten Greening, originated in Iowa, even farther than Wisconsin from Aquidneck. Up here in Maine, not only did we have the rather famous Winthrop Greening, we also had Naked Limbed Greening of course, as well as Northern Greening, Palmer Greening, Prospect Greening, Mammoth Greening and even an Extra Large Greening. Bottle Greening originated in Vermont. Oconee Greening was discovered on the banks of the Oconee River not far from Athens, Georgia. There's even a Bossier Greening in Louisiana. Maybe the orchardists of the past weren't so isolated after all.

Barry Rodrigue with the Sasanoa tree in Phippsburg

What about Granny Smith? Granny Smith is not an American apple, but rather an Australian immigrant from New South Wales. The name refers to a Maria Ann Smith, who planted the seed about two hundred years ago that became the tree that produced the world's most widely grown green apple. I think we can call it an unofficial greening.

One of my favorite green apples is one that so far remains unidentified. I've found it in two coastal Maine locations over the past ten years, one in Boothbay and another in Phippsburg. Boothbay is at the mouth of the Sheepscot River, and Phippsburg is at the mouth of the Kennebec. I received the emerald green apples of what I refer to as Boothbay Greening from Peter Larson in 2007. Peter is a senior research scientist at Bigelow Laboratory for Ocean Sciences. Eight years later I opened a box in the mail with what had to be the same apple, this time sent from a professor named Barry Rodrigue from the University of Southern Maine. Barry came across the ancient tree while researching old abandoned farm sites on the Basin in Phippsburg. He calls the apple Sasanoa after the short river that connects the Kennebec with the Sheepscot, a fitting name for an apple that we've found only in those two locations. Maybe the grafter of both trees paddled the fourteen miles between the two locations on the Sasanoa two hundred years ago. I haven't visited Peter Larson's tree, but in early April 2017 I drove down to the Basin with Barry and collected scionwood from his discovery. If ever there was an apple that deserved to be called Greening, this is it.

A note of caution: in many locations, any number of very different green-skinned varieties were all simply called Greenings much in the same way that many different russets have been simply called Russets. Convenient and simple, but not easy for the identifier. Beware when someone tells you, "It's a Greening!"

What about pippin? What is a pippin? What makes an apple a pippin? We have a number of apple varieties that include that word in their name. The most well known of these is Cox's Orange Pippin, the famous and beloved apple of England and a parent of over one hundred of the world's most popular apples among connoisseurs. Another of England's most famous apples is Ribston Pippin, itself one of the parents of Cox. Of North American varieties, the most well-known Pippin is Newtown Pippin, or, as it's commonly called down South, Albemarle Pippin. But there are many others, hundreds in fact. Some Pippins are the accepted names, such as Fall Pippin, Golden Pippin and American Pippin. Many others are obsolete synonyms, once upon a time attached to some of our most iconic varieties. Ben Davis was sometimes called Baltimore Pippin and New York Pippin. The Vermont apple, Bethel, was also called

Shaker Pippin. The New Jersey apple, Yellow Bellflower, was also known as Bishop's Pippin of Nova Scotia. Why?

Cox Orange Pippin, photographed during our trip to England: one of the most famous apples in the world

We reluctantly turn to the Oxford England Dictionary. I say reluctantly because the OED sometimes strikes out when it comes to fruit. Still, it's worth checking. It defines pippin as "The seed of a fruit," or "Any of numerous varieties of fine-flavoured eating apple." That's not a lot of help. I also would call it a bit misleading. Not that I want to quibble with the most revered dictionary in the world, but there are dozens of other delicious apples for every pippin out there. And, if a pippin is the seed of a fruit, what does that make a pip? Back to the OED. There are multiple definitions of the word pip, most of which don't relate to fruit at all. The one of interest would be "Any of the seeds in certain edible fruits (as an apple, citrus, grape, etc.)." That sounds like a synonym of pippin.

Let's go elsewhere and see what we can find. Can we get any help from Dickens? Philip Pirrip, known as Pip, is the protagonist of Charles Dickens' *Great Expectations.* We can assume that all English teachers would agree that Dickens chose the name Pip because it means seed. It's a story about growing up, like trees do. It's a good choice for a name, maybe even better than Dickens knew. Like the countless, anonymous apple seedlings that have popped up over much of the temperate world's landscapes, Pip was also an orphan. Like all those apple seedlings, he too never knew his parents. And like that one-in-a-thousand exceptional seedling apple tree, Pip was able to defy adversity and reach maturity after many twist and turns. That's the seedling apple, all the way. That's a pippin from a pip.

After coming over from England, Ribston Pippin became famous in Maine up and down the Kennebec River and along the coast

I've also learned to consult the English pomologist Robert Hogg on all things apple. In his 1884 *The Fruit Manual*, he tells us that the "word Pippin is derived from the French Pepin, the seed of an apple, and in its earliest signification meant an apple tree raised from seed in contradistinction to one raised by grafting or from cuttings."

So, we'll be safe in saying that the use of Pippin in many apple names is all about the seed. Pippins are all apple trees from seed. When we look through the old books, and other publications, we can find ample evidence to back that up. Many of these Pippins were clearly ungrafted seedlings. From Bradford, for example, we find Britton's Pippin. You can bet that Britton found a seedling apple tree somewhere down on the farm, liked the fruit for one reason or another, and then brought it to the 1851 Kennebec Agricultural Society exhibition. Not wanting to spend time inventing a catchy name, they just called it Britton's Pippin. And it literally was. No one was trying to fool anyone. According to the November 11, 1847, *Maine Farmer*, Barnum Hodges of Winslow in Kennebec County sent in to the Maine Pomological Society an apple he called Hodges Pippin. "From seed sown sixteen years ago, he has received bountiful crops for several years." Hodges Pippin was definitely a Pippin. But why would Ben Davis have been called Baltimore Pippin and New York Pippin, and why did Bethel get called Shaker Pippin?

Ribston Pippin in Rockland; the original tree laid down decades ago, then two branches rose and became two new trunks: photo courtesy of Ilmi Carter

Unlike the word Greening, the use of the word Pippin was well established by the time the colonists planted their first pips over here in New England. So, rather than dwell on American apples, let's take a trip back to England. We'll look at one English Pippin and see what it tells us.

Blenheim Palace is located in Woodstock, Oxfordshire, England. The palace was built following the War of the Spanish Succession, a huge thirteen-year European blitz for the control of much of the world, instigated by a power vacuum caused by the death of the Spanish King, Charles II. This was back when most of the European leaders were related by blood or marriage or both. Most of them hated each other. Though the war dragged on for an additional eleven years, one of the turning points in the slaughter was a battle fought in 1804 near a small town in southern Germany called Blindheim. The hero of the battle was John Churchill, the first Duke of Marlborough. As a thank you to the Duke, the palace was constructed.

The War of the Spanish Succession is not generally on the radar screen for many of us over here in the New World, and who knew that the palace would be the birthplace and ancestral home of another English war hero with the name of Churchill two hundred and forty years later? No one seems to know how the name got changed from Blindheim to Blenheim. As they do with apples, names change. And wars will come and wars will go, and no one much cares about the King of Spain or the grand old Duke of Marlborough. However, for an apple explorer on both sides of the Atlantic, the name Blenheim is iconic.

Blenheim Pippin was one of the first English varieties to be grown in northern New England and the Maritimes. It's one of the few English apples that grew well north of Boston. Most of the English apples failed over here. Old Blenheim trees can still be found now and then in small Maine farm orchards. Hogg writes extensively about Blenheim Pippin in 1884. Blenheim (as Beach calls it) is a large, beautiful, oblate, red, yellow, orange and russetted dessert apple of excellent quality. The tree sprouted from a seed near the wall of Blenheim Palace. Three early Blenheim synonyms were Woodstock Pippin, Northwick Pippin and Kempster's Pippin. One of its other synonyms is Blenheim Orange. That's the name we know it by today.

Blenheim Orange a.k.a. Blenheim Pippin photographed by JPB, 2011 not far from the original Blenheim Palace

The name Blenheim Pippin suggests that the apple originated from a seed near the palace. It did. The palace is located in Woodstock. That covers Woodstock Pippin. The fellow who planted the pip that grew into the tree that bore the fruit was named Kempster. That covers Kempster's Pippin. What about Northwick Pippin? Northwick is fifty miles northwest of Blenheim Palace. We'll have to think about that one. Even after it was established in America, the apple was still primarily referred to as Blenheim Pippin, as A. J. Downing described it in 1848 in his *Fruits and Fruit Trees of America*, although Blenheim Orange was listed as a synonym. By 1875, John Thomas lists Blenheim Orange first and Blenheim Pippin as a synonym. Thirty years after Thomas, S. A. Beach adhered to the one-word protocol of his contemporary, W.H. Ragan, and got rid of both Pippin and Orange, stripping the name down to simply Blenheim. These days it's generally Blenheim Orange here and in the U.K.

So where's the monkey wrench in all this? It seems so straightforward. The apple tree comes from a seed. It's called a Pippin. Two monkey wrenches come to mind. First, all apple varieties originated from seeds. That would make them all pippins. If they're all Pippins, why call any of them Pippins? Secondly, all replicated named apples are from grafts. That would mean that no varieties are pippins. Now what?

For centuries apples were passed around from farm to farm and town to town and beyond. We know that there were thousands of varieties. The habit of planting apples from seed was common. New seedlings with desirable qualities were being discovered with great regularity. In the days of a largely preliterate civilization, most communication was by word of mouth. Catalogs and books were scarce. Consistency in names was unimaginable and unnecessary. But, keeping track of what

You too can start your own pippins

you had and where it came from would still have been important. Where did that apple come from?

When you hand me scionwood from your pippin, or Bob's Pippin or Sally's Pippin or Blenheim Pippin, the name tells a certain kind of story. In the case of your pippin, you're not giving me something you grafted from wood you received elsewhere. You're giving me scionwood from something you started from seed. It's from your farm. This could be important, especially if I want to grow apples from a certain terroir. I know where this apple originated. It's in the name. In the case of Bob's Pippin or Sally's Pippin, there is implicit information not found in the name Ben Davis. This is the apple Bob started from seed. We know where Bob lives. That's the apple Sally started herself, not something she grafted from who knows where.

The fact that so many of the synonyms include the word Pippin, further suggests that these various pippins were early "provisional" names, used when passing around newly discovered seedlings as scionwood to friends, neighbors and beyond. Our beloved Starkey was once known as Starkie's Pippin. Undoubtedly, that would have been before the apple's eventual reputation was established. It was the seedling apple Starkey discovered. It was Starkey's Pippin. Only years later did it become Starkey.

I have an orchard and you have an orchard. I receive scionwood from you to topwork a few trees. What is it you're giving me? If you're in the neighborhood, you probably know Kempster. After all, old man Kempster planted the seed. So, it that case, it's Kempster's Pippin. If you don't know Kempster, but you know the Palace, then, it's the "apple from Blenheim." "But there's many apples at Blenheim." "Well, this is the Blenheim Pippin." And if you live a bit farther down the road, the apple becomes Woodstock Pippin. Don't care much for the duke, but I do know the town. And if you live fifty miles away and never even heard of the duke and his palace or even the town, then it becomes Northwick Pippin, probably erroneously. "I got it from someone in Northwick. I think it's from seed." And so it becomes Northwick Pippin. Downing does not include Kempster Pippin or Northwick Pippin in his description of Blenheim Orange, which also makes sense. People in the states might have heard of Woodstock. Kempster would been long gone and Northwick might just be confusing. Where did you say it was from? It popped up from a seed. It's a pippin.

The recently revitalized American cider industry loves to use the word pippin. It has a nice ring to it. There's Pippin cider, and there's Big Pippin cider and Lost Pippin cider and Pippin Black, and Henry's Pippin Hard Cider, and undoubtedly many others. What do the cider makers mean when they use the word Pippin in their label? Do they even know? It probably should mean that the cider was made exclusively with apples from ungrafted seedling trees. Wouldn't you agree? This would have been the norm in America until about 1830. Up till then, seedling cider orchards covered many tens of thousands of rocky hillsides. But as modern commercial orchards came in, the seedling orchards went out. The only seedling apples in most of those pippin ciders are a few tokens here and there. But those times, too, are changing. Eric Shatt of Redbyrd Cider in New York State makes a Wild Pippin, entirely of locally collected seedling fruit. Andy Brennan of Aaron Burr Cider, also in New York, and Steve Gougeon of Bear Swamp Cider in Massachusetts have been scouring the countryside for new seedlings with cider potential. Others are following suit. The pippins are coming back!

Steve Gougeon and Andy Brennan with drinkable pippins in a bottle, 2018

Although most modern cider-makers covet the bittersweet and bittersharp cider varieties, true cider apples are hard to find. In many cases they're not available at all. Until the supply meets the demand, most modern cider makers are forced to use dessert fruit in their press. While many would be reluctant to admit that their cider is made from Honeycrisp or Macs or Red Delicious, they may be revealing that very fact by their choice of names. Back to Hogg who quotes Sir Paul Neile from his 1662 *Discourse on Cider*. How Neile came to the following conclusion we'll never be certain. According to Hogg, it had something to do with Shakespeare's *Henry IV*, but even Hogg seems a bit perplexed. In any event, concerning "pippin cider" Neile writes "For by that name I shall generally call all sorts of cider that is made from apples good to be eaten raw." In other words, Neile's Pippin Cider is cider made from dessert fruit. A travesty!

Back to Ben Davis, Bethel and Yellow Bellflower. Here's where we plunge into pure conjecture. We'll declare that Bethel was called Shaker Pippin because of the Shaker community in Enfield, New Hampshire, only forty miles from Bethel, Vermont. The Shakers were famous for orchards and innovation. It would have been a new variety and the Shakers would have been growing and disseminating it. Perhaps it was originally theirs. It was the Shaker Pippin. As for Ben Davis, we know that everybody and his daddy claimed to have originated the first Ben Davis, so why not some guy from Baltimore or New York who headed west with a variety from Maine, or was it Tennessee? And what about Bishop's Pippin of Nova Scotia? Yellow Bellflower is one of the first of the American varieties. Historians seem to agree that it originated in New Jersey, not Canada. Maybe the good Bishop of Nova Scotia was pulling the wool over the eyes of his parishioners. Probably not the first time, or the last.

A few years ago, my intern Laura and I were out for a marathon scion collecting trip. It was March. Prime time for the scionwood collector. Other people hate March. It lasts forever. But I love it. It's not nearly long enough. I'd be happy with at least two or three more weeks of March this year, last year and every year. We visited seventeen ancient trees that day and still got home before it was totally dark. Our first stop was the home of Tim Nason, long time MOFGA publications guy, and coincidently a resident of the Alexander Road in Dresden, where Len Alexander grew up. We had intended to stop by just long enough

Cammy, Emily & Kelsey: enjoying Reinettes!

to pick up a painting I'd done for a different MOFGA poster several years ago. Tim had put the poster together and still had the painting.

After a cup of tea and a bit of leisurely conversation, I knew we ought to get rolling. As we got up to go, Tim reminded me of an odd apple he'd sent to me some years ago from an old dying tree down in the woods. "It tastes terrible but it makes a great pie." Amazingly, I actually remembered the fruit. It was pink and yellow, elongated and deeply ribbed. At the time I guessed it was a seedling. Tim agreed. "We should graft it." One thing led to another, and ten minutes later the three of us were down the road standing in a snowy gully looking at a row of old seedlings. The one we wanted was broken nearly in half and mostly dead, but it still had several nice sticks of scionwood. We took two or three and headed for the truck. "We'll graft you a tree." Tim smiled. We said our goodbyes and Tim turned to walk back home. Laura took out the duct tape and taped up the bundle of sticks. "What shall we call it?" "How about, Tim's Pippin?"

While we're defining Greening and Pippin, we should also knock off "Reinette." As far as I know, there are no apples originating in the U.S. called Reinettes, but we have many of them growing here. One of the favorite varieties among dessert apple connoisseurs these days is called Reine des Reinettes. Another, a green winter variety gaining popularity, is called Reinette Simirenko. There many others here and in Europe. In December 2017 I was wandering through the narrow stone streets of Villaviciosa in northern Spain when I came across a true corner market with baskets of fruits and nuts set outside on the street. On one side of the corner, next to the door, was a basket of medium-large oblate red and yellow apples labeled "Reineta Canada." Around the corner, not more than ten feet away, was a basket of slightly smaller (medium-sized) apples, some oblate and some round-conic, labeled "Reineta Pinta." They looked

A few of the Reinetas from Villaviciosa

pretty similar at first glance. I decided to purchase a few of each, take them back to my hotel room, and see if I could find any difference. Cammy had gone back to the states and this exercise would provide a bit of entertainment before dinner.

My Spanish being limited, I picked five of the Canadas and handed them to the two grocers, who appeared to be husband and wife, and who understood my intentions and held out a small bag. When I took three or four steps around the corner and picked out five of the Pintas, the couple clearly wanted me to put them all in the same bag. With hand gestures, I tried to explain that I wanted two bags. *Igual* they said. I wasn't convinced. Finally they gave in and allowed me to put the Pintas in a separate bag.

Topworking #16: A vinyl tag and a #2 pencil and you're all set.

As I paid my two euros for the ten apples, I noticed a third basket of apples, this one inside on the counter and also labeled "Reineta Pinta." These apples were large, blocky and green with a few faint stripes, bearing no resemblance to the Reinetas outside. I bought one of them too. What was I seeing? What did I have? What is a Reineta?

Having none of my apple books with me I resorted to my Spanish dictionary. No Reineta there. Pinto means spotted or speckled, as in the pony. Pinta is the feminine form. Then to the iPhone. Reineta is grown throughout Leon, south of Villaviciosa. Since 2002 it has been a protected Denomination of Origin product. It's described as "squatty [oblate] wider than it is tall, with a short petiole [stem]...pale green, not shiny, with a characteristic rust [russet] over much of the skin surface...may be consumed fresh and baked." Which Reineta might that be?

With the exception of Reineta Pinta #2, these were not green apples. And Reineta Pinta #2 was neither russeted nor oblate. Reineta #1 was not green but was russeted and oblate. Then I found a website that called the Reineta an old French variety, known as Reinette or Canada Gris. Perhaps that would be my Reineta Canada. The web photo, however, showed an entirely russetted fruit unlike any of those in my corner market.

I searched for Reineta Pinta and Pinto but to no avail. I did get directed over to a site that brought up another Reineta, Reinette de Caux which is somehow related to the famous Dutch apple, Belle de Boskoop. That took me to another French website which featured four different Reine des Reinettes, two Reinets, one Reinetille and a hundred and eleven Reinette this and Reinette that. Alas, no Reineta (or Reinette) Canada or Pinta or Pinto, although I did find a Reinette Gris du Canada. About that time I got sick of staring at my iPhone and was about to head out to one of Villaviciosa's twenty sidrerias to test out some Reinettes in liquid form, when Todd showed up with a bag of apples an old fellow had given him from his tree along the street just down from the hotel. The man called them Reineta Pintadas. Unlike the Pintas scattered across my bed, at least these were spotted. Next night we were in another small grocery store where they sold Manzana [apple] Reineta. These were large and russeted. We later found them labeled as Reineta Canada. Maybe these were the protected ones.

Towards the end of the trip we visited with Francisco Figueroa, a fruit explorer and historian in Madrid who has devoted much time to restoring historic Spanish orchards. We spent two delightful days with Francisco mostly visiting his favorite trees in Madrid, out near El Escorial. One of the varieties he's looking for might be one we have in New England, or even in Maine. Over there it's called Camuesa, Camuesar, Camoise, Camoise Blanche, Reinette Camoesa or Reinette Blanche d'Espagne (White Spanish Reinette). In the U.S. it was probably known as American Fall Pippin which means it was also probably called Fall Pippin, along with a few dozen other yellow fall apples. My basic Spanish book translates camuesa as "pippin." Does this mean that a reinette and a camuesa are both pippins—in other words, seedlings?

When I got home a few weeks later, I opened up Robert Hogg. From what he says, the pippin and the reinette are not one and the same. You might say, first comes the pippin, then comes the reinette. Quoting Hogg:

> There are various opinions respecting the derivation of this word [reinette]. At first sight it appears to have a French origin, and supposing it to be so, some have translated it Little Queen, though there is no such definition in any French dictionary I have consulted. Others say it is derived from Rainette, a kind of frog, because Reinettes are always, or ought to be, spotted with freckles, like the belly of the frog.
>
> Thomas Fuller, the eminent historian and divine, says, "When a pepin is planted (*i.e.* grafted) on a pepin stock, the fruit growing thence is called Renate." This, I think, is the origin of the word, Reinette being derived from Renatus—renewed or reproduced. A Reinette is therefore a grafted apple, and a Pippin is a seedling.

One of the synonyms of Reine des Reinettes is Queen Pippin; which comes first, the pippin or the reinette?

The names Renato, Renata, Renate and Rene are also derived from the latin Renatus. We don't have many Renatos in the U.S. but we do have quite a few Renés, both men and women. If a Reinette is a grafted tree, I suppose that means René should be a grafter. The University of Maine's associate professor of pomology is Renae Moran. Renae and I have become good friends. We occasionally collaborate on some project or another. And, of course, she's an excellent grafter. Lastly, I suppose that Hogg would declare Reinette Camoesa a "grafted seedling"!

## Twenty-Seven

# May-Less or Mal-Us

"One's ideas must be as broad as Nature if they are to interpret Nature," he answered.
Sherlock Holmes, *Our Advertisement Brings a Visitor*, p 37.

Cole's Quince

The orchard apple, as we know it, has its origins in the mountains of Kazakhstan, west of China and north of Afghanistan. It traveled along the Silk Road with camel caravans of traders, eventually arriving in Europe and then jumping across to America more than four hundred years ago. I've said a lot about varieties and seedling apples. But what about species? Sometimes I'm asked about species of apples as though McIntosh and Black Oxford are different species. McIntosh, Black Oxford, Honeycrisp and Cole's Quince are varieties, not species. There are many thousands of varieties of apples. On the other hand, there are roughly twenty-five to forty species, depending on who you ask.

For many centuries the apple was called both *Malus* and *Pyrus*. Then, in 1753, Carl Linnaeus, in his famous *Species Plantarum*, took a risk, dumped the word *Malus*, and combined apples and pears into the same genus, *Pyrus*. A year later, the British botanist Philip Miller, rejecting Linnaeus, placed the apple in its own genus, *Malus*. Miller wrote persuasively in *The Gardeners Dictionary* (1754) that "the Apple should be separated from the Pear: and this Distinction is founded in Nature; for these Fruits will not take by budding or grafting upon each other, tho' it be performed with the utmost Care. Indeed I have sometimes succeeded so far as to have the bud or graft shoot [i.e. take]; but they soon decayed, notwithstanding all possible Care taken of them; therefore I shall beg leave to continue the Separation of the Apple from the Pear, as hath been always practised by the Botanists before this time."

JPB with Thomas Chao, Geneva NY, 2016

The contest between *Malus* and *Pyrus* was not resolved for nearly three hundred years. It was a long battle, but eventually Miller won out. The Harvard taxonomist, Kanchi Gandhi, who I visited in the fall of 2018, recalled having professors, not so long ago, who still referred to the apple as *Pyrus malus*. These days, however, the genus of apples is *Malus*, a simple two-syllable word that has at least two pronunciations. Most people pronounce it with a short "a," like "malice." Maybe we like that pronunciation because it reminds us of the trouble caused by the first fruit in the Garden of Eden. Others pronounce *Malus* with a long "a," to rhyme with "pay-less." Kanchi told me that either is fine. Within the genus there is a subset of species and within each of the species there is a subset of varieties. All apple species are native to the Northern Hemisphere. In North America there are four to nine of them, depending on who you ask. In Europe, the number is said to be five. The other various species are native to an assortment of locations in Asia.

What is an apple species? A workable, though not very scientific, definition of an apple species is an apple that will come true to type when grown from seed. While this does not mean that all the trees and fruit will be identical, it does mean that they will all exhibit certain similar botanically identifiable characteristics. Each of the different species grew up in relative isolation over many thousands of years, untampered with by humans, propagating themselves all the while by seed. This allowed them to develop their own unique characteristics over huge amounts of time.

Why the discrepancy over the agreed upon number of species? Given the opportunity, apple species will cross with other apple species and create new hybrids. Some of these natural hybrids, or interspecific apple blends, have been around for long periods of time. Generally, taxonomists call these "hybrid species." While some botanists would maintain that these are not true species, others have said they are. Another reason for the confusion is that some Asian species have only been found growing in cultivation and have never been found in the wild. Because these apples have been grown for so long in isolation where they have exhibited unique traits, they have been awarded species status, despite the fact that it may have been human intervention that was responsible for their existence.

The disagreement in the number of North American apple species may also be due to the ease with with some of the American apples cross with one another. It's the apple version of the American melting pot. According to Thomas Chao, curator of the USDA ARS apple collection in Geneva New York, three of those species, *Malus angustifolia*, *M. coronaria* and *M. ioensis*, may be the oldest apples in the world, dating back to when the continents collided a hundred million years ago. A fourth species, *M. fusca*, is thought to be a more recent import, migrating from eastern Asia during the last ice age over the Alaskan land bridge in the belly of an adventurous wooly mammoth or reindeer.

So how many species are there? At the low end, John Fiala proposes twenty-five in his 1994 *Flowering Crabapples, the Genus Malus*. Barrie Juniper and David Mabberley list forty-one in their 2006 *The Story of the Apple*. Fiala's book, while not focused on the orchard apple, is a comprehensive resource, recommended for anyone interested in apples and apple species. Juniper and Mabberley is an engaging and scholarly examination of the origin of the apple. For those who want to dig deep, these are two books you'll want to read. Currently, the source accepted by most taxonomists is the USDA Germplasm Resources Information Network (GRIN). Their list can be found on the GRIN website. It consists of one hundred and forty-four entries including species, hybrid species and synonyms. Of those, forty are now considered to be unique species.

What is a variety? The well-respected plant writer, Michael Dirr, calls a variety a "subdivision of a species having a distinct though often inconspicuous difference..." In the world of ornamental plants, a variety comes mostly true-to-type from seed. Dirr defines a cultivar as a "cultivated variety," differing from the former in that only a "few woody plant cultivars are seed-produced, but these are the exception rather than the rule." A cultivar, in other words, is a plant selection that usually must be propagated asexually, from cuttings or grafting. Although some apples come relatively true to type from seeds, it's safe to say, in most cases, the only way to replicate a variety is by grafting. That makes Black Oxford and Honeycrisp cultivars by Dirr's definition. On the other hand, practically all apple texts use the word variety, not cultivar, so I guess it's okay to call Black Oxford and Honeycrisp varieties.

What about crabapples? Most apple species are what we generally call crabapples. The fruit is small and not something you would choose to eat. Fiala's definition is the best one I've found. He calls a crabapple any apple under two inches in diameter at maturity. By that definition, nearly all apple species around the world would be crabs; the wild *Malus sieversii* of Kazakhstan and most of our domesticated apples being the exceptions.

*Malus orientalis* (PI 633822.61) Geneva NY

What about the apple in the orchard? The apple we eat, cook and drink has been known by a few different species names over time. These days, *domestica* is the preferred epithet, although some call it *Malus pumila* or *M. sieversii*. Because our beloved apple is domesticated to the nth degree, you'd think that *domestica* would be a perfect name. But to some, the term "domestica" suggests hybridization and there's where the disagreement lies. For generations it was assumed that the orchard apple is a blend of species. That was challenged when Juniper and Mabberley and others, came to believe that the orchard apple was entirely or almost entirely one species, the wild apple from Kazakhstan, *M. sieversii*.

Despite Juniper and Mabberley's assertions, intuitively I preferred the mish-mash-blend theory. It made sense that as the apple traveled west along the Silk Road over the centuries, it mated with the locals and made hybrid seedling *Malus* children along the way. That would make the orchard apple a mix of apple species.

In 2012, "using rapidly evolving genetic markers to make inferences about the recent evolutionary history of the domesticated apple," a group of research scientists from six countries, led by Amandine Cornille, agreed with my hunch. They found that the domesticated apple is in fact a blend of species. First off, they verified that the Kazakh *Malus sieversii*—sometimes referred to as *Malus pumila*—is the original "Central Asian progenitor." In other words, *M. sieversii* is where it all began. But they also determined that the European crab, *Malus sylvestris*, played a large and defining role in the apple's evolution. "*Malus sylvestris* thus appears to have made a significant contribution to the *M. domestica* gene pool through recent introgression, building on the more ancient contribution of the Asian wild species *M. sieversii*." Not only that, but it also appears that at least two other species, *M. baccata* and *M. orientalis*, may be a small but notable part of the blend: "Some authors have also

*Malus baccata* (PI 589838.09) Geneva NY

suggested possible contributions of additional wild species present along the Silk Route: *M. baccata*...which is native to Siberia, *M.orientalis*...a Caucasian species present along western sections of the ancient trade routes..."

Even more recently, in an on-line *Nature Communications* August 2017 article entitled "Genome, re-sequencing reveals the history of the apple and supports a two-stage model for fruit enlargement," a group of forty-three collaborators—a bit more than one per *Malus* species—using the most advanced genetic testing, came to the conclusion that "Cultivated apples likely originate from *Malus sieversii* in Kazakhstan, followed by intensive introgressions from *M. sylvestris*...Notably, the recent introgression from *M. sylvestris* into *M. domestica* has been so intensive that the cultivated apples now appear to be closer to European crabapple *M. sylvestris* than to their progenitor *M. sieversii*..."

*Malus sylvestris* (PI 589382.13) Geneva NY

In other words, the apple we eat, cook and drink actually is a domesticated blend of multiple species, in particular the wild apple from Kazakhstan and the wild European crab. *Malus domestica* is the perfect name. When you have to write it down, *Malus x domestica* Borkh is technically correct. The "x" in *Malus x domestica* indicates that it's a hybrid species, in other words, a blend of species. The "Borkh" refers to the taxonomist Moritz Borkhausen who, in 1803, coined the name, *Malus domestica*.

Cider makers, in particular, are familiar with *Malus sylvestris*. The bitter, astringent wild European crab has played a key role in the cider-making traditions of England, northern France and other areas of the continent. According to Cornille, et al., cider apples are, like the dessert and culinary apple, primarily a blend of *M. slyvestris* and *M. sieversii*. Surprisingly, cider apples do not appear to have more of the bitter European crab in their family tree than do the dessert and culinary varieties. They just picked up the bitterness by the luck of the draw.

What are the other thirty-eight apple species good for? Many of the small-fruited species are highly ornamental. A few of these species have been providing the genetics for the thousands of flowering crabapples planted along the highways, in parks and in your backyard. Many are covered with persistent fruit for the chickadees, jays, juncos and waxwings during the white of winter and the brown of early spring. Many of these crabapples tend to be extremely rich in phytonutrients and antioxidants. Cider makers are seeking them out for their astringency, their high sugar content and their nutrient value. As I've become more familiar with the European cider varieties, it's interesting to see that many of the most famous are small enough to be called crabs. Some cider makers are now fermenting juice from apples the size of grapes or blueberries. Some crabapples have broken loose from the chains of our lawns and have seeded themselves out in the wild. Sometimes they cross with the orchard apples to create new sorts of hybrids. Remember that American melting pot. Or as they shout out in *Hamilton*, "Immigrants, we get the job done!" You can find those immigrant escapees along nearly every road, stone wall and field edge in rural New England.

What's the deal with Antonovka, Duchess of Oldenburg, and Fameuse? A few apple varieties, including these three, are famous for coming relatively true-to-type from seed. Does this mean that they should be considered their own species? Seedlings of Antonovka and Duchess have been used for generations as rootstock, in large part because of the predictably uniform trees they make from seed. When you graft onto them, you know what you're getting. Still, none of these are considered to be their own species. They are selections of *Malus x domestica*. Antonovka and Duchess are ancient varieties from Russia that may have been grown from seed in relative isolation for hundreds of years. As I've mentioned, we have a Fameuse tree down Finley Lane. Despite the tendency for the seedlings of these three to resemble their parents, as far as I know, botanists aren't ready to declare *Malus duchessii* a new species anytime soon.

Where can you see species apples? If you live near Boston, you can go to the Arnold Arboretum. The USDA Agricultural Research Service (ARS) orchards in Geneva, New York, are also worth visiting. Both have extensive collections of species apples from around the world. Go in early May for the flowers or in early fall for the fruit. You won't be disappointed.

*Malus sieversii* in all its diversity at a market in Kazakhstan: photo by Richard Ossolinski

# Twenty-Eight

# In Love With a Duchess

We approached the case, you remember, with an absolutely blank mind, which is always an advantage. We had formed no theories. We were simply there to observe and to draw inferences from our observations.

Sherlock Holmes, *The Cardboard Box*, p. 895

Duchess of Oldenburg

Wherever apples hitched a ride with travelers or traders, be it across Asia or across America, their discarded seeds sprouted here and there along the way, vying to become the next generation of trees. The hardy ones lived, the others died, and after long periods of self selection, as well as human selection, locally adapted populations emerged. Across central and western Russia, for example, hardy apples established themselves hundreds of years ago in the frigid north. Four of these varieties would one day come to America where they would alter fruit growing forever. All northern fruit identifiers need to be familiar with Alexander, Duchess of Oldenburg, Red Astrachan, and Tetofsky. In the early nineteenth century these four made their way first to England and then over to Massachusetts. All four names are

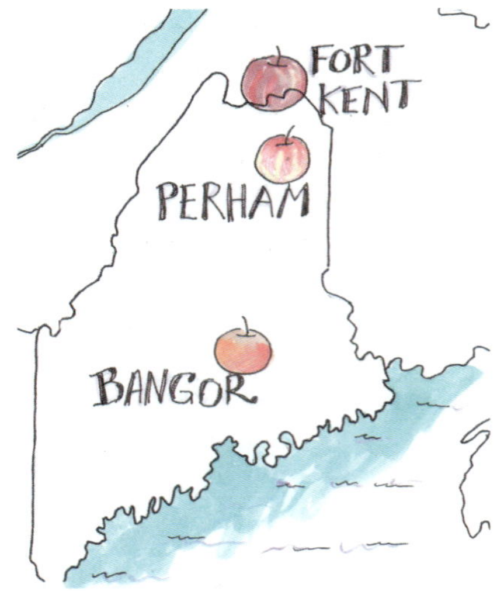


almost certainly not their original ones. Rather, they were fabricated somewhere along the way between the Volga and the young country of the United States, although no one knows by whom.

By the 1830's, many Americans were leaving southern New England and heading to Ohio, Michigan, Wisconsin, Minnesota and other points west. Others in Maine were looking for farmland due north in what is now Aroostook County. "The County" as it's called by almost everyone in Maine, was assembled from pieces of Penobscot and Washington Counties in 1839. It's also known as the "Crown of Maine," being the northernmost part of the state, the part that is almost entirely surrounded by Canada. It even looks like a crown on the map. It's a huge area, about the size of Connecticut and Rhode Island combined. Farming in the cold climes of the County posed many challenges to the early settlers, one of which was the lack of hardy apples. Farmers moving into northern New Hampshire, Vermont, New York and the upper midwest also needed fruit that could survive the long, tough winters. The varieties that had originated from the seed of European apples were not hardy enough.

There was no governmental support for farmers in those days. The USDA wouldn't get started until 1862. There would be no Maine Pomological Society until 1873. The slack was largely taken up by the Massachusetts Horticultural Society. Founded in 1829, the Society played an essential role in establishing agriculture in America. Members of "Mass Hort" collected, tested, described and disseminated new varieties of ornamental plants, vegetables and fruits. While most of their attention went to domestic introductions, in about 1835 the Society engaged in a major international project. They imported four apple varieties from England in collaboration with the London Horticultural Society. The varieties were Alexander, Duchess of Oldenburg, Red Astrachan, and Tetofsky. With this one action, apple growing became possible throughout the coldest regions of the U.S. It was a key moment in American agricultural history.

It is unclear how the four varieties were obtained by the London Horticultural Society or where they came from. The exact year they arrived in America is also uncertain, although we know it was about 1835. One way or another, each of the four got themselves to England and then over to the States.

Red Astrachan looks like a mini-Wolf River with its russet cavity

Red Astrachan was likely the first to arrive in England, supposedly in 1816 from Sweden. Based on its name, it is thought to have come from somewhere near the trading city of that name at the mouth of the Volga River on the

Alexander and its child, Wolf River, with the basin ends up; Alexander is always more conic and red; Wolf River is always more oblate (flattened) and pink

north shore of the Caspian Sea. There is also a White Astrachan, presumably from that same area. The Vikings were great traders and were instrumental in developing the Volga River route. Perhaps they brought the apple back home with them to Scandinavia. Red Astrachan has many synonyms, nearly all of which include the name Astrachan or Astrakan. None suggest a different point of origin. After arriving in Maine it became the go-to summer pie apple for several generations. Visually, it looks like a medium-sized version of a Wolf River. It's oblate, red and has a big splash of russet around the stem. It ripens in mid-August in Maine, and old trees are still sprinkled around the state, especially along the coast.

A year later, in 1817, Alexander became the second cold hardy introduction to arrive in England, and it may have been the first to travel on to Massachusetts. It is assumed that Alexander originated in Russia where it was sometimes known as Aport, Aporta or Oporto. These may or may not all be the same apple. The Arnold Arboretum has an accession in their collection they call Oporto that is definitely not the same apple as Alexander. It is small, roundish and red-fleshed. Among Alexander's other various synonyms are Alexander the First, Emperor Alexander, and Russian Emperor. Alexander I, the person, was born in 1777 and ruled as Czar from 1801 until his death in December 1825. It's a fitting name for a huge red apple. Alexander is a general cooking apple, suitable for pies, baking and sauce. Old Alexander trees can still be found throughout Aroostook County.

Tetofsky is now extremely rare but not entirely lost

No one knows where Tetofsky came from, nor is it known when it arrived in England. Perhaps it originated in the orchards outside the ancient city of Tetovo along the River Pena, in what is now the Republic of Macedonia, north of Greece and along the old trading route between Constantinople (Istanbul) and Ragusa (Dubrovnik). Maybe its parent was an apple brought across Asia from the Tien Shan mountains whose core was tossed by the roadside outside town. There is a Tetovo apple that looks suspiciously like the Tetofsky that eventually made its way to America. Macedonian tourist literature declares the Tetovo apple to be "very possibly—the best apple in the world."

*The* original Duchess of Oldenburg, Catherine Pavlovna

Despite the fact that the fruit isn't small and may not have originated in Russia, one of Tetofsky's English language synonyms is "Russian Crab." Technically it's a medium-small, roundish, red-striped, mid-summer cooking apple. Although it never gained the recognition of Red Astrachan or two other Russian summer cooking apples that were imported a few decades later: Liveland Raspberry (a.k.a. Lowland Raspberry) and Yellow Transparent, Tetofsky does live on as the parent of one of our favorite modern summer apples, Mantet, and as an ancestor of three highly touted University of Minnesota introductions, State Fair, Zestar! and SweeTango. (Who comes up with those modern apple names, anyway?)

The last and most important of the four imports is Duchess of Oldenburg. The iconic apple's name almost certainly has nothing to do with where it came from. More likely the name was chosen to make the funders happy. Oldenburg was the capital city of the Duchy of Oldenburg in what is now northern Germany. Although occasionally controlled by Russians, Oldenburg is not all that close to Russia. The city was settled along a trading route on the banks of the Hunte River. The Duchy bordered the mouth of the Weser River and the North Sea not far from Bremen. It played a key role in the history of that area for several hundred years. Duke George of Oldenburg ruled the Duchy from 1784 to 1812. His wife was the Grand Duchess Catherine Pavlovna of Russia, sister of Emperor Alexander. How nice to have two of the four imported apples named for a very rich and powerful brother and sister.

Like the czar's sister, the Duchess of Oldenburg apple came from Russia. Either the variety, or a similar variety, was commonly grown in the area surrounding the city of Tula, a hundred miles south of Moscow. This apple has multiple synonyms and multiple variations, complicating the search for the original. It doesn't help that it was never called Duchess of Oldenburg in Russia. There is some chance that it may be synonymous with the Russian apples, Borovinka/Borowitsky and/or Charlamoff/Charlamofsky. The name Borovitsky possibly refers to the Borovinyh family of Tula. Perhaps they played a role in the apple's discovery or promotion or both. There appears to be no record of who the Charlamoffs were. Duchess may also be the same as a slew of other apples. The *National Apple Register of the United Kingdom* (M. W. G. Smith, 1971) lists seventy-eight synonyms for Duchess. Sorting them out has been a pomological nightmare for nearly two hundred years, made worse by the fact that the Russian alphabet does not translate well into English.

Two of those who made Duchess a lifetime priority were born about the time it arrived in the Bay State. J. L. Budd was born in New York in 1835, and nine years later Charles Gibb was born in Montreal. Budd and Gibb grew up to be great fruit explorers and champions of Russian apples. Both devoted much of their lives to tracking down hardy fruits for the cold north. In 1882 the two men went together on an epic trip to Russia, in large part hoping to clear up confusion around Russian apple nomenclature. In 1870, eight years after its founding, the United States Department of Agriculture imported scionwood from two hundred and fifty-two additional Russian varieties. The importance of Russian apples in cold districts was, by then, well established. Orchardists wanted more. The wood was grafted, and the trees were tested in Washington, D.C., and then disseminated throughout northern apple growing regions. Unfortunately the translation of varietal names was poorly done. Scions were mixed up, and as one Agriculture Report put it, the nomenclature was "found to be in a hopeless muddle."

Budd and Gibb embarked on their trip, beginning in Poland and continuing on through Lithuania, Latvia, Russia, Ukraine and back to Poland. In 1884 Gibb published *On the Russian Apples Imported by the US Department of Agriculture in 1870,* in which he systematically described all two hundred and fifty-two apples. Budd eventually became Professor of Horticulture at the State Agricultural College in Iowa and went on to import more Russian apples; he spent much of his career testing and promoting them. Gibb returned alone to Russia in 1887 in a further attempt to sort out one variety from another.

Duchess of Oldenburg from Aroostook County; striped but not as pronounced as St. Lawrence

"In my report on Russian fruits," Gibb wrote in 1884, "I had spoken of Borovinka as the [apple] family of which the Duchess of Oldenburg is a member. In this catalogue the Duchess appears under all sorts of names, yet we did not see the Duchess in Russia, neither have I been able to find out the Russian name for it." Borovinka "is named after the family of Borovinyh, in the Province of Tula. In *Dutche Pomologie*, by W. Lauche, of the Pomological Gardens at Berlin, there is a beautiful colored print of Charlamovskoe [Charlamoff], which one can hardly believe to be other than Duchess. The description too is Duchess, and among the synonyms are Borovitsky and Duchess of Oldenburg. I think I have thrown enough light on this subject to make the darkness visible."

Controversy about the Russian varieties continued on, however, and on August 30 and 31, 1898, a committee of ten of the leading pomologists of the day gathered in La Crosse, Wisconsin, to make yet another attempt to sort out the names of the Russian apples. They brought their apples with them, spread them out on tables and began to compare. I wish I had been there! The report they issued states that "among the great number of Russian apples are found well-defined groups or families. By this is meant that some varieties so closely resemble each other as to be nearly quite identical." They went on to describe the various groups, including a Duchess group and a Charlamoff group. The great Niels E. Hansen, student of J. L. Budd, professor of Horticulture in South Dakota, and member of the committee, may have had the last word. He stated, "In European nurseries, Charlamofsky is a synonym for Duchess."

A couple of years ago I did a search of the apple collection at Geneva, New York. In the list of many hundreds of apples I found four different varieties listed: Borovinka, Charlamoff, Duchess, and Oldenburg. I sent for and received scionwood from all four and then topworked them all onto one tree: a Duchess seedling. As of this writing, we're still waiting for the fruit. In 2015 Cammy and I visited Pauline and Steve Hayer in southern Aroostook County to see their old trees. Two bore identical apples, similar to Duchess but different. The overall color was a purply red—a spitting image of the Charlamoff in the USDA watercolor collection. This would suggest that Charlamoff and Duchess of Oldenburg are distinct varieties, at least in the Unites States.

Charlamoff from Aroostook County with its faint lavender hue

Compounding identification challenges is the unfortunate fact that some old books call the apple Duchess, others call it Duchess of Oldenburg, and still others refer to it as Oldenburg. Some list Charlamoff and Borovitsky as distinct varieties while others list them as synonyms. Some of this confusion in names can be attributed to the work of T.T. Lyon and W.H. Regan in the late nineteenth century. As a result of their efforts to simplify nomenclature, many multi-word names became one-word names. Beach dropped the "Duchess" and called the apple Oldenburg. A nice thought, but he may have made things worse.

In 1833, even before the tree had arrived in America, William Kenrick, a founder and member of the Committee on Fruits at Massachusetts Horticulture, listed Borovitsky and Duchess of Oldenburg as two separate varieties in *The New American Orchardist*. In the 1870 USDA report *Russian Apples,* Charles Gibb lists Borovitsky, Charlamowsky, and Duchess of Oldenburg as three different varieties. Downing in *Fruits and Fruit Trees of America* (1886) includes only Borovitsky and Duchess of Oldenburgh, although he suspects they are one and the same. John Thomas (*The American Fruit Culturist,* 1908) lists Borovinka, Borovitsky, and Oldenburgh as three different varieties. He doesn't use the name Duchess at all. Beach considers them all to be one variety, but like Thomas, he calls it Oldenburg (though without the "h" at the end). A sample of the other Duchess synonyms includes Augustapfel, Early Joe, Queen Mary, and Smith's Beauty of Newark.

Why all the fuss? Duchess is a remarkable apple. Not only did it turn out to be the most important of the four varieties imported by Massachusetts Horticulture in 1835, one might argue that Duchess has been one of the most important of all apples. The Duchess tree is extremely hardy. In northern Maine winter temperatures routinely dip to -30° F. Many spots drop to -40° or -50°. The growing season is short, and winters are extremely long. Duchess thrives in these conditions. It ripens over a period of weeks, so it's perfect for those who don't want to harvest and process their fruit all at once. It drops its foliage and shuts down early before the cold weather sets in. It's widely adapted to differing growing conditions, including soils of poor fertility. It bears young, regularly and annually, with at least a small crop on the off year. It's a medium-sized tree of medium vigor, which means that it can be grown as a standard, full-sized tree and still be manageable in stature. It is scab resistant, if not immune. I rarely, if ever, see any scab at all on Duchess fruit, even on unsprayed trees.

The fruit is medium-sized, round, red-striped, irregularly splashed and mottled with crimson. Its yellowish, firm, fine-grained flesh is crisp, tender, juicy, subacid and aromatic. While it's too tart for some lunch boxes, it is a great culinary apple for sauce, drying and pies. At our annual September Pie Taste-Offs, the Duchess pies always have the most zip. You know you're eating an apple pie when it's made with Duchess.

In 1922 U. P. Hedrick, in his *Cyclopedia of Hardy Fruits,* wrote that Duchess, "still one of the best general purpose apples of its season, was the first of the Russian apples, and inspired interest in a group of varieties which has made fruit-growing possible in the colder parts of America." According to Bradford, "It was probably the hardiest apple grown in Maine...the farther north it was grown, the better the quality, the longer the season and the more it was liked…It's good points were its well-known hardiness, its early and prolific bearing, comparative immunity from many apple insect pests and the showiness of the fruit." New Brunswicker Eddy Dugas, in his 1992 *La Culture de la Pomme dans le Nord,* wrote that Duchess is "considered the benchmark used to evaluate new varieties" (Translation JPB). Lewis Hill wrote in a 1985 *Yankee Magazine* article that Duchess "was probably the all-time favorite with early growers. It's still one of my favorites today, unsurpassed for pies, sauces, pickles and cider. When I was a boy, if you knew of an old orchard with a Duchess growing in it, you didn't tell your best friend." Beach says of Duchess, "it was the extreme hardiness of this variety in the early test winters that kept up the hopes of prairie orchardists in time of great discouragement and led to the importations of more varieties from Russia."

Duchess seedlings are used as a uniform, hardy, standard (full-sized) rootstock called Borowinka. Presumably this makes Borowinka another Duchess synonym. I learned from a phone call to Michael Johnson at Lawyer's Nursery in Montana that the seed for the Duchess rootstock they raise and sell comes from Borowinka orchards in Russia. "The fruit is used for processing and the seed is sent to us," Michael told me. It was starting to make sense. A tree that produces uniform seedlings for rootstock in Russia might also produce uniform seedlings in Aroostook County, Maine. Maybe many of those seedling apples all over the County as well as the trees in peoples' orchards and backyards have this apple somewhere in their heritage. Maybe all the Boros and Charlas and Duchesses and Oldens are closely related seedlings, the result of hundreds of years of growing them from seed in western Russia. Could they be sort of synonyms, but not quite?

On September 30, 1998, I took my first fruit exploration trip to Aroostook County. My goal was to locate apple varieties that would be hardy additions to the Fedco catalog. I planned to visit the James Nutting farm in Perham where Nutting Bumpus originated. I also hoped to track down a Dudley Winter tree in its birthplace, Castle Hill. Both apples are thought to be Duchess seedlings. Other fruit explorers were flying to Kazakhstan while I was heading north up the highway in my beat-up old Corolla.

Steve & Barb Miller, and John Meader in Westmandland, September 2005

I had read about Nutting Bumpus a few years earlier in the Bradford thesis, and my Aroostook friend, Steve Miller, had brought me fruit in 1996. Every fall our own home in Palermo becomes one seemingly continuous apple tasting party. The Nutting apples that Steve brought down one year were a hit with the dozen in attendance one cold, rainy September night.

My first experience with the Dudley Winter apple had been through a Vermonter. Kenneth Parr of East Burke had been testing apples for many years and had a collection of over two hundred varieties. I read about Ken, who was about forty years my senior, in *Pomona*, the quarterly magazine of the North American Fruit Explorers. We began a correspondence. At that time I was looking for Dudley Winter. I had heard that the apple was extremely hardy, and I hoped that offering it in the Fedco Trees catalog would mark the beginning of a collection of apples for growers up north.

In a 1995 letter Ken recommended that I also propagate Duchess of Oldenburg, long assumed to be the parent of Dudley and also very hardy. "As for Duchess," he wrote, "I prefer the original but I can not give any reason. There are Duchess strains and mutations out there somewhere in other orchards and to my knowledge no one has ever put the whole story together. The national germplasm list I have does not have much." That spring Ken drove over to Palermo from Vermont for the day and brought me a fistful of scions of both varieties. I grafted them onto existing trees in my orchard and also grafted enough Dudley nursery stock to begin to sell them in the catalog. I kept his Duchess growing in my orchard but decided not to offer it for sale. In the back of my mind I was wondering about the apparent confusion

The hollow shell of a dead Duchess tree in the center surrounded by mature fruiting rootstock sprouts (not Duchess!) in Palermo: photo by Laura Sieger

around the different strains of this apple. I also wondered about "his" Dudley that was growing far from its place of origin in northern Maine. Maybe I should look for a Dudley up in Aroostook County.

After introducing me to Nutting Bumpus, Steve Miller encouraged me to come up to The County. Our friendship grew rapidly in large part through a mutual love of plants. He's a foot taller than I am, wears an assortment of rather floppy hats that vary according to the season, and always has astute observations to make about plants and life. Although he has a Ph.D. in Meteorology, he's spent his adulthood living on a small farm in Westmanland in the heart of Aroostook County where he and his wife, Barb, raise bees and made their living for many years selling honey and honey butters from their blueberries, raspberries, strawberries and currants.

JPB with Gloria Seigars, 2018

I arrived at Steve and Barb's, got a tour of their farm, met their goats and settled in. The next day we headed off to Fort Kent, at the junction of the St. John and Fish River, as far north as you can go in Maine. We drove up with another fruit explorer, Gloria Seigars of nearby New Sweden. Of the dozens of fruit explorers I have met over the decades, none has been more enthusiastic than Gloria. She and I hit it off instantly. You might accuse her of being a little nutty; she probably wouldn't mind. She loves to tell stories, and always has one ready to tell. She also does her homework. Over the years and my many trips to The County, Gloria always had suggestions for new places to go and new trees to check out. She would often join me on my apple adventures. She and her husband, Dick, cooked me many dinners, and she and I have become close friends.

One of Gloria's greatest discoveries was Garfield King. Garfield was not an apple. He was a retired teacher and superintendent of schools and a bit of a nut himself. One of the first things he said to me the afternoon we met was "When you start shaking the family tree, you never know what kind of nuts will fall out!" Garfield was born in Caribou and grew up in that area of northern Maine. He was twenty-five years older than I. He went to the University of Maine in Presque Isle and Orono, served in the Marines, and had a long and distinguished career as an educator in Aroostook County. With his crewcut, he looked like a Marine sergeant, even at age seventy. He had a lifelong passion for plants. That became our connection. That and a love of life. Introductions were brief, and within a few minutes we were off on a plant exploration trip for the rest of the day. For the next few years, every September I would take a three or four day excursion north to explore the gardens and orchards of Aroostook County. I stayed with Steve and Barb. Gloria usually played a big role in the trip, which always included a visit to Garfield and his wife, Venette, at their Page Street home in downtown Fort Kent. Most years Garfield would take us all on an adventure into the surrounding countryside or over the border into Québec. He was the consummate plant explorer, always on the lookout for something to bring back to his garden and share with his friends.

Garfield King and the Charette tree, 1998

It was during a superintendents conference in Bangor many years before I knew him that Garfield first learned about a hardy blackberry. As he told it to me, he was in a motel hallway when he heard several forest rangers chatting. When one of them said the words "hardy blackberry," he told me, "I pretended I was examining the wallpaper." The next spring when the blackberries were in bloom, he drove to the area they had been talking about as best as he could remember. He hoped he could find the plants by looking for the flowers. In an old abandoned settlement in Knowles Corner, he scored. Of course he always had a shovel in the car. Within a few years the blackberry became well known all around Fort Kent. Later that fall after he and I met for the first time, a box of roots arrived for me in the mail. That was the beginning of our own hardy blackberry patch. After several years of trials, we offered it in the Fedco Trees catalog. We call it Fort Kent King.

Garfield also introduced me to several extremely hardy and unusual apples he had discovered over the years. Two have become favorites. That first visit, he took us to an enormous four-trunked tree on Charette Hill just east of Fort Kent. It's thought to be over two hundred years old, dating back to just after the Revolution when both sides of the St. John River were settled by French speaking farmers and trappers, many of them Acadians who had fled the wrath of the British a couple of decades earlier. Garfield had two names for the fruit. The first was "the Donut Apple" because of the deeply sunken calyx. When you slice the apples, they sometimes look like donuts just off the cutter. The second name was simply Charette.

Garfield then took us to Sly Brook where we met with Elwidge Michaud, her cousin Phil Roy and the Pomme d'Or, the apple also known in the area as the Golden Pear. It has a large, often rectangular fruit, translucent yellow skin, and large hollow core that is sometimes filled with a golden nectar you can actually drink. Elwidge's grandfather had brought the apple down from Canada in 1870. She thought that his trees may have been started from seed but that the apple was originally from France. Of course I grafted both Charette and Pomme d'Or at our place. They've been growing here now for many years. I've also grafted two Charette trees over at Francis Fenton's and at the Maine Heritage Orchard.

Not only is Charette huge, it is also way more ribbed and lumpy than Alexander and Wolf River

Elwidge Michaud & Phil Roy holding a Pomme d'Or, 1998

Sometimes when I'd visit Garfield, we would go no farther than his yard. You could easily spend hours right there on Page Street despite the fact that his lot was barely half an acre. Subtract the house and the driveway and the lawn and that leaves about a third of an acre or less for planting. A year before he died, I took a partial inventory of the plants he'd been growing. Beside the driveway was the vegetable garden, about 30' x 125'. Interplanted with the vegetables were rhubarb, asparagus, blackberries, raspberries, highbush blueberries, grapes, various ornamentals and a dozen fruit trees. More blackberries (Fort Kent King!) and fruit trees peppered the backyard. Sweet potatoes grew beside the garage door. There was a small nut collection featuring two large butternuts, a large black walnut, some hazelnuts —squirrel food—and numerous seedlings of various sizes in pots. There were mountain ash and hawthorn and lilacs and crabapples and more. Practically every plant came with a story, including the fifty-year-old red oak on the other side of the house that Garfield planted from an acorn he collected off the sidewalk the day he graduated from the University of Maine. He and Venette lived together in a mini-paradise in the middle of town. Who would have thought you could do so much in such a tiny space?

On Friday, October 2, the day after meeting Garfield and Gloria, Steve and I headed off together for Perham. Garfield went about his life in Fort Kent. Gloria and Barb had other things to do. We drove up the Nutting Road and parked in the driveway of the immaculate Nutting farm with its magnificent easterly view of New Brunswick. The farm was by then occupied by Henry Nutting, grandson of the late James Nutting, the originator of the curiously named apple.

Pomme d'Or is later-ripening, larger and more barrel-shaped than Yellow Transparent

James Nutting settled in Perham in 1861 where he and his wife, Emma Huston, raised their four children. After fighting in the Civil War, James returned home where he ran a printing business, worked in public service, including terms in both the Maine House of Representatives and the Maine State Senate, and conducted years of horticultural experiments. At one time his orchard included over six hundred bearing trees of thirty-five apple varieties and ten different plums. In his nursery, he had between 3,000 and 4,000 young trees. In the 1880's he selected, named and introduced a large, red-striped apple. He named it Nutting Bumpus, and it was described by Munson in the 1907 *Maine Agricultural Experiment Report* as being "Hardy, vigorous, very productive. Highly prized by the originator." The description also noted that the apple was a "seedling of Oldenburg." Would that be Duchess, Borovitsky, Charlamoff?

Henry Nutting and his Nutting Bumpus tree,1998

Henry Nutting met us at the door. He had a welcoming and pensive smile. Although easily old enough to be Steve's dad or mine, Henry was all decked out in blue work pants, a heavy brown plaid shirt and a gray baseball hat, ready for adventure. He was as nimble on his feet as anyone our age. He took us to see his lush vegetable gardens and his trees. The original nursery and the large orchard were long gone. Only a few old trees remained. Two Nutting Bumpus trees stood next to a Duchess of Oldenburg just to the east of a plantation of pine where the orchard had been. I asked him about the origin of the apple's odd name, but he professed ignorance. The three of us then spent most of that day exploring the fields and abandoned orchards along a two-mile stretch of the Nutting Road. If you pick the right spot up there, you don't need to go far. We must have eaten fruit from three hundred different apple trees. Some were ancient hulks of the past, but most were young seedlings that had turned the old fields into forests of apples.

One of the oldest trees we ate from that day was an Alexander from the old "Paul" farm just down the road from Henry's. Not realizing its true identity, I called it Paul's Red for a number of years. Oh those synonyms! As we parted company late that afternoon, I had the feeling that Henry was still not ready to quit. Steve and I were bushed and, not only that, full. I had brought along a box of metal tags to mark any interesting trees I might want to return to later for scionwood. Surprisingly enough, I found that I had tagged nearly a dozen trees that were obviously seedlings. The chances of finding delicious fruit among ungrafted trees is said to be slim, but many of these seedlings up there in Aroostook County were great right off the tree. It was also interesting that the seedling fruit in Perham all looked strangely similar to the Nutting Bumpus in Henry Nutting's yard. There was something going on there, but I wasn't sure yet what it was.

Pomme d'Or

The next day I began to explore the backroads of the area looking for a Dudley Winter tree. Again I found many decent seedlings, all with a similar look. Wherever I went, I heard about one variety so often that it appeared to be a household word in The County. Every orchard supposedly had this variety, although I was never sure if I was seeing the same apple at each farm; it was more like a theme with variations. The variety was Duchess of Oldenburg, or as it's known locally, Duchess.

Steve and I talked quite a bit on the phone that winter, as we like to do, and we decided that not only would we offer Dudley Winter and Nutting Bumpus in the Fedco Trees catalog, we would also offer Duchess. I remembered what Ken Parr had written to me years earlier about its strains. It was beginning to look as though there were many Duchesses out there, maybe even a different one in every orchard. But how could that be? Varieties are grafted and therefore identical, aren't they? Could some or all of these be seedlings? I still didn't have enough scionwood downstate for us to propagate Duchess, so I was resigned to the fact that I would have to buy trees from a wholesaler. That would be okay for now, but I decided that the following year I'd have to track down the true Duchess in Aroostook County and use that tree as our scionwood source. If there was such a thing as the true Duchess.

Charette

I also continued to ponder the odd name, Nutting Bumpus. Where did it come from? My first hypothesis was that it was a play on the name, Natty Bumppo, the fictional hero of James Fenimore Cooper's five *Leatherstocking Tales* written between 1823 and 1841. Plausible, but not all that likely. Then I started to poke around looking for someone named Bumpus. That led me to Harry E. Bumpus, founder of the feldspar quarry in Albany Township in Oxford County in 1927, famous for its rose quartz and giant beryl crystals, some of them over thirty feet long. Maybe Bumpus and Nutting were buddies. Maybe a few bushels would be worth a thirty-foot beryl crystal. I'll keep you posted.

The old Everett Cunningham seedling tree in Hibbert's Gore: its rangy unpruned shape, its location growing out of the foundation and the lack of a visible graft line are all indicative of a seedling tree

And, by the way, who said seedling apples are ugly, anyway? Not only is the Everett Cunningham beautiful, it also makes great cider!

# Twenty-Nine

# Dudley Winter

The world is full of obvious things which nobody by any chance ever observes.
Sherlock Holmes, *The Hound of the Baskervilles*, p. 683

Dudley Winter

In 1858 we came to this county from the town of China, Kennebec County, there we had plenty of fruit. We boys missed that more than anything else. They told us then that we never could raise apples here. At that time there were a very few native crabs-trees and once in a while you would find a hardy seedling that would make a pig squeal if he ate one. But in a few years there was a change. They told us we could raise Duchess of Oldenburg. We tried a few and they stood our winters first rate. Some other kinds did quite well. The Fameuse and Alexander and Tetofsky, and they were shortly followed by the ironclad Wealthy from Minnesota, also the Yellow Transparent and Montreal Peach; then came the Dudley apple, which keeps nice until April. The tree is an ironclad and a very prolific bearer. I have several other varieties that I am testing, among them is a sweet russet, an apple of very fine quality and a good keeper. There are seedlings in Northern Maine and New Brunswick that are worthy of propagation and I believe if thoroughly tested would give us fruit the year round and of fine quality, good enough for a king.

J. W. Dudley, *Castle Hill, Aroostook County*, 1889

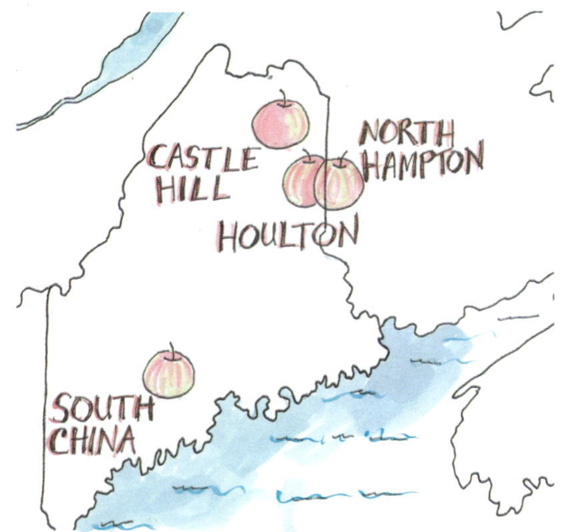


On September 8, 1999, I returned to The County with the intention of finding Duchess and taking another shot at locating Dudley Winter. I made the trip three weeks earlier in hopes of catching more apples at their peak of ripeness. Northern apples ripen earlier as a rule. They also drop their leaves earlier than the apples farther south and go dormant well before the cold weather sets in. Some of the worst cold damage happens in November when it comes off frigid after a warm fall. Most varieties get caught with their pants down. It can kill a tree. Not so Duchess and other hardy Russian types. They've been aware of the danger of late fall cold snaps for a thousand years. Not only that, but 1999 was a droughty summer in Aroostook, and everything was fruiting two or three weeks early.

Although I was pleased with the Dudley scionwood I had received from Vermont, I knew we should be offering true Dudley from its birthplace. Gloria Seigars, who was hot on the trail of the apple, had heard about a possible Dudley just across the town line from Castle Hill, in Ashland. Off I went.

In the West Midlands, eight miles from Birmingham, in the center of England is the small city of Dudley. The town sits on top of valuable deposits of limestone, coal and iron ore. In the center of Dudley sits the Dudley Castle, atop Castle Hill. No one knows when the Dudley Castle was first built, perhaps as early as the eighth century by Lord Dud, but more likely it was built several decades after 1066 and the Norman invasion. The Dudley castle was destroyed and rebuilt off and on over the centuries, as castles were. Today beautiful partially restored ruins survive.

The Dudleys were an illustrious bunch. Going way back, their ancestors include not only William the Conquerer but also Alfred the Great, the English king who ruled Wessex from 871 to 899. Other Dudley forebears include several kings of France, as well as associates of English royalty at least one of whom lost his head on the royal chopping block.

Dudley Castle 1993: photo courtesy of the Lena Dudley Kenney family

In 1576 Thomas Dudley was born in Yarling,

The Eliot burying ground Roxbury, MA

Hastings, England. A direct descendent of the Castle Hill Dudleys, Thomas grew up to become a staunch Puritan and eventually one of the founders of the Massachusetts Bay Company. The Puritans were unhappy with the brand of religion sanctioned by Charles I, which was far too Catholic for them. While most of the Puritans chose to stay in England and fight to "purify" the church, others left for America with the expectation of religious freedom. In 1630, at age fifty-four, Dudley and his wife and six children sailed to Salem Massachusetts on the Arabella, as part of a fleet of ships carrying seven hundred Puritans to what became Boston, Dorchester and Roxbury. Dudley was probably one of the founders of Cambridge, later moving to Ipswich and finally Roxbury, where he died in 1653.

The Puritans in New England wanted religious freedom for themselves but never pretended to believe in any form of religious tolerance when it came to anyone else. They knew what was right and wrong. They could be brutal and often were. Thomas Dudley himself was not a very liberal guy. The history books have not been kind to him. In T*he Life and Work of Thomas Dudley*, Augustine Jones refers to him at the outset as a bigot. But this was four hundred years ago. The world was a different place. We can wish that Dudley and John Winthrop and their Puritan friends had never graced the shores of Boston Harbor. We can wish a lot of things. We can also thank Thomas if we like. He almost certainly was one of the founders of Cambridge. He served as the second governor of the Massachusetts Bay Colony. His son Joseph signed the charter for Harvard College. Without Thomas Dudley we might not even have the diverse and vibrant community of twenty-first century Roxbury. And we definitely wouldn't have a very good apple.

Thomas, his son Joseph and several other Dudleys are buried in a tomb at the Eliot Burying Ground at the corner of Washington and Eustis streets in downtown Roxbury, not far from the original Roxbury Russet tree and not far from Dudley Street, Dudley Square, Dudley Station and dozens of stores, shops, and a wide assortment of other establishments all called Dudley this and Dudley that. In 1995 Michael Womble, Jenny Chen, Juan Moscat, Tiffany Clayton, Eliud Ferrer and Christina Knudson created a mural entitled "Faces of Dudley" that stretches across a long brick and cement wall in Dudley Square at the intersections of Washington Street, Dudley Street and Malcolm X Boulevard. Maybe Thomas would be rolling over in his grave if he could see that the mural that bears his name does not include a single white Puritan face. Or maybe he'd be proud to see his name associated with others who came to Roxbury seeking their own refuge and freedom.

Two hundred years after Thomas Dudley's passing, John Wesley Dudley was born near China Lake in China, Maine, in Kennebec County on July 30, 1850, the youngest of Micajah [my-CAGE-ah] and Olive's seven children. China is only a few miles from our farm. By mid-century the old seedling-orchard culture of central Maine was being transformed into the grafted orchards of the future. Good farm land had become difficult to find in that part of the state and the Dudleys had seven kids. Generous land grants were being offered to families who were willing to move to Aroostook County.

Faces of Dudley: Dudley Square, Roxbury, Massachusetts, 2018

In 1858 the Dudleys took the bait, left China Lake and headed north. They moved to Ashland and soon thereafter to Township 12, 4th Range, along the stage line about halfway between Ashland and Presque Isle. A few years later the plantation of Castle Hill and then the town of the same name were established. Although it could be that the Castle Hill name stems from a large log structure, resembling a castle, long ago constructed on one of the town hills, I suspect that the Dudley family also knew about an ancient English castle on Castle Hill in Dudley, England.

Hislop [sic] "Subacid - Good for Cider" fruit plate courtesy of Jim Arsenault

There in the new Castle Hill, the Micajah Dudley family founded their farm on a square mile block. They built the Dudley Road through the wilderness. Three of Macajah's four sons located their farms on that same block, including John Wesley. Micajah Dudley's 1849 house is still standing. It wouldn't be until almost fifty years later, in 1896, that the Bangor and Aroostook Railroad arrived ten miles away in Ashland, ushering in the enormous commodity potato industry still with us today. So, back then in the 1850s and 60s, the Dudleys planted apple trees.

Micajah Dudley had a farm on one side of the Dudley Road. John Wesley and his wife, Mary, began to farm across the road. They farmed ninety-five acres of open land, as well as three hundred and twenty-five additional acres of woodland. According to Gloria, "His house was very close to his father's. We found the cellar hole just in the clump of trees to the west of the homestead. Lena Dudley Kenney, his great grandniece, told me that the big orchard was behind the houses and she thinks the original Dudley apple was there. The family cemetery was within the orchard."

The story goes that in about 1875, John and Mary's daughter, Grace, planted seeds from a Duchess apple, supposedly pollinated by a Hyslop crab. Hyslop was one of the most widely known and cultivated crabs of the day. The Pomological Society report from the 1892 *Agriculture of Maine Yearbook* stated, "From the orchardists that are attracting special attention is Dudley's Winter, an apple originating from a Duchess seed fertilized by a Hyslop Crab. Mr. J. W. Dudley, Castle Hill, is the exhibitor." In a letter from January 25, 1907, John Wesley wrote to W. M. Munson, "The original tree bore its first apples in 1880, and has borne a full crop every year since."

Dudley Winter usually features more red patches and blush than Duchess

John Wesley Dudley went on to be a Maine State Senator. He was "deeply interested in agricultural pursuits and the development of 'the garden of Maine.' " He was master of the Grange and President of the State Board of Agriculture. "Always a Republican."

By 1895 the Dudleys had two hundred and seventy-five full grown trees in their orchard, including two hundred Dudley Winter apples. By the turn of the century, however, his brothers and everyone else in The County were turning their attention to potatoes. John and his family wanted to continue to grow apples, so they left. They headed west to where a new orchard industry was just getting off the ground. They settled in Ephrata, a hundred miles northwest of Yakima, Washington. The Castle Hill orchard was cut down, and the site has been in potatoes ever since.

But the Dudley Winter apple didn't die. It caught on and became popular in northern districts. Under the catchier name North Star, Dudley Winter apple trees were grafted and sold by one of the many huge Rochester, New York, nurseries of the day until it was discovered that the name North Star had previously been taken by an apple in Iowa. A century earlier two apples of the same name would not have been a problem; no one would have even known. But by the late 1800's commerce was driving agricultural practices. The railroad had connected the east and west coasts in 1880. Although you can occasionally still find it listed as North Star today, generally the apple has reverted to its original name, Dudley Winter.

In a May 2002 letter, Dudley's granddaughter, Jean Humpfrey, wrote to me, "I never met my grandfather as he passed away before I was born in 1924. My grandmother was a very aristocratic lady and we didn't see much of her. I think that the brick house is still there but the orchard is probably gone [in Washington State]."

Apples that keep well into the winter were at a premium throughout New England in the nineteenth century. There were many good storage apples, but few that were hardy. Baldwin and most of the others could not survive an Aroostook County winter. Dudley Winter was hardy. Farther south in central Maine, it is a fall apple, but up north it kept better than most.

Dudley Winter remained a staple of many Aroostook farm orchards and went on to gain popularity in Canada and in the upper midwest. It may be the most widely planted apple variety ever to originate in Maine. In 2012 I received an email from Alan Teach, a commercial orchardist from Sunrise Orchards in Gays Mills, Wisconsin. It read in part: "My family grows 250 acres of apples in southwestern Wisconsin, mostly McIntosh, Honeycrisp, and Cortland, but we still produce about 200 bushels of Dudley per year. Our orchard was planted in about 1915 and I believe Dudley was part of the original acreage. As late as the sixties our local growing area probably produced over 20,000 bushels of them mainly sold as an early pie apple in the Milwaukee and Chicago markets." In 2005 and 2006 I received three emails from Joe Dudley, great grandson of John Wesley. He wrote:

> John W.'s brothers went big into potatoes during the early decades of the 1900s, ultimately acquired a farm in Florida where they grew their own seed stock in the wintertime but ultimately went bust sometime in the 50s (because of a blight outbreak in FL and increased competition from cheaper and bigger Idaho potatoes flooding the New York/Boston markets ??)...and possibly partly because the surviving son and heir to the Castle Hill homestead farming operation was a pilot in WWII who came back to help with the family potato farming business but whose heart was never in the farming business and it ultimately broke him down (so the story goes)...

Steve Miller and I drove to Ashland on September 10, 1999, looking for old Dudley Winter trees. There we found two candidates behind a trailer. We asked if we could have fruit and ran out in the pouring rain to collect a few specimens. Although practically everything was ripening early that year, the fruit on those two trees seemed to be on the ground way too early to be called Dudley Winter. Still I collected some fruit and hoped for the best.

Earlier that day before the rains came, we had visited Warren Moody who also lives in Castle Hill. Warren's farm was a sort of unofficial fruit experiment station. He had dozens of trees planted across what was once a large expanse of lawn, many of them thriving considerably north of their natural range: oaks, chestnuts, nut pines, lindens and much more. Surrounding his vegetable garden was a small, grafted orchard of unusual, mostly unnamed apples he'd collected in his own explorations over the years. One even blooms in the fall. A forty foot hop vine decorated the front of his barn, and an equally large kiwi covered the entire back end of the farm house.

In his mid-eighties at that time, and dressed in cowboy boots, a turquoise T-shirt and a faded blue hat with the inscription "Naturally Potatoes," Warren gave us the full tour. His soft white hair and wispy beard perfectly matched the long white hair of his old dog, Jo Jo. When he talked about his years sampling apples, he said he'd always bring a stick along with him to "pry out the puckery ones." Finally he took us out back to his three small, quite old Duchess trees. They were set about fifteen feet from one another.

There had been four, but one year the state came by to exterminate his wild plums that were thought to be harboring Green Peach Aphids, a vector for Potato Leaf Roller Virus. In doing so, they poisoned the roots of his Yellow Transparent and the fourth Duchess. "I told them, 'Next time you come back, you better bring the sheriff.' They never returned."

The Duchess herself: fruit plate courtesy of Jim Arsenault

From the way that Warren Moody talked about his trees, it sounded as though each of the three Duchess was unique. We sampled the fruit. They were remarkably similar, although one did seem to be the best of the lot. I found myself wondering if maybe all the Duchess trees in Maine are seedlings. Maybe this is the exception to the rule of apple genetics. This is a tree that produces incredibly hardy, locally adapted, generally true-to-type seedlings—the perfect tree for the coldest climates. I left Warren's with an increased appreciation for this apple, but I was still a little confused. If we graft nursery stock from one of his Duchess trees, are we actually replicating the variety or just one of its progeny. I also found myself wondering how the apple ever found its way to Aroostook in the first place.

It was almost exactly a year later when I returned again to The County. I still hoped to find a Dudley Winter. 1999 had been a bust in that regard. I also hoped to learn more about Duchess. What was this remarkable fruit that seemed to have created an apple culture in northern Maine all on its own? Where did it really come from, and how did it get to Aroostook County? It occurred to me that all those seedlings in Perham along the Nutting Road not only looked like Nutting Bumpus, they looked like Duchess too. It was beginning to appear as though Duchess was practically synonymous with the word apple in The County.

On September 18, 2000, my apprentice David Herter and I made the long drive north. After a quick stop at Steve and Barb's, we picked up Gloria and Garfield, Garfield's wife, Venette, and Venette's sister, Burnette Bowker, and headed to the plum orchards of Maison de la prune over the border in Saint-André de Kamouraska, Québec. It was a great plum year, and all along the way we stopped and ate from Garfield's favorite trees. He seemed to know them all. One was up a tiny dirt road at the top of a hill with no farms in sight. The plums were the size of ping pong balls and delicious. At Maison de la prune the six of us received a wonderful tour of the orchards and the kitchens where a wide array of jellies and preserves are made by the Martin family.

Paul-Louis Martin was a college professor and agricultural history buff who had revitalized the heritage farm. On the tour of the place, he showed us his small and ancient apple orchard up at the top of a steep hill overlooking the farm and a thousand plum trees below. Of course the orchard included Duchess. I told him of my interest in the apple, and that I wondered how it could have gotten from Russia to northern Maine, New Brunswick and Québec.

The old Massachusetts Horticultural Society Building on Massachusetts Avenue in Boston; what a beautiful building!

The story of Duchess coming over from England to "Mass Hort" in about 1835 is pretty well known. But Massachusetts is a long way from northern Maine, especially in the mid-nineteeth century. Even though back then the Maine coast was a "Maine Turnpike" of schooners, it would not have been easy to get the apples from the coast up to Aroostook. According to Paul-Louis, two members of the Québec Parliament were also members of the Massachusetts Horticultural Society at the time. One of them was even Vice President of the Society. Could they have played a role?

Next day we found ourselves back at Warren Moody's in Castle Hill. While Jo Jo looked on, we ate Duchess from each of Warren's trees. Again we heard the story of the demise of the fourth tree, and as I listened closely, it seemed clear to me that at least two and perhaps all three of his remaining Duchess were in fact seedlings. But definitely not "natural fruit." Natural fruit is the Maine expression for a chance apple seedling. It is slightly, though not entirely, pejorative. Duchess seedlings, I was now convinced, are in a class entirely of their own in The County—they are not quite identical, but hardly worthless seedlings. They are simply Duchess in all their subtle and magnificent variations.

Crimson Beauty ripens on the last day of July in central Maine and makes beautiful red sauce

We then visited Warren's son, Frank, in Perham, who took us for a ride in his truck through fields of goldenrod and multiflora rose to sample his favorite seedlings—all of which, once again, had a suspicious similarity to Duchess. As we left, he asked us to try to find him an apple called Crimson Beauty. His father had it years ago, but lost it. I told him I would look for it, though I was not familiar with the variety. We also returned to the Nutting Road to further explore the burgeoning forests of apple seedlings. Beautiful, nearly flawless, tasty, tart, light yellow, red-striped Duchess lookalikes by the hundreds. Later we returned to the supposed Dudley Winter tree in Ashland, but we were disappointed. The fruit was of poor quality, and most of it was on the ground despite the fact that this year fruit was ripening late.

That night at the Millers, I received a return call from Daryl Hunter of New Brunswick. I had been trying to reach Daryl for several days in hopes of visiting the orchards of Francis Peabody Sharp on our way back south. I had read an article a year earlier about Sharp in the Canadian magazine *Rural Delivery*. The article had reported that Dudley Winter was a Sharp introduction. Since I assumed that it was John Dudley's apple, I fired off a correction to the editor, which was printed a few months later with little fanfare. Then I put the article away and forgot about it. But since I was going to Aroostook County anyway that fall, I thought I'd see if there was more to this fellow Sharp who was said to be New Brunswick's "father of fruit culture."

It's impossible to spend much time in Aroostook County without becoming aware of the intimate connection between The County and New Brunswick. They share a very long border. The people at King's Landing Historical Settlement in Prince William, New Brunswick, had directed me to Daryl, a consultant in a project to create an orchard of Sharp's apple varieties at King's Landing. He and I talked on the phone for over an hour that evening at Steve and Barb's; and though I never made it to New Brunswick that trip, I learned several more pieces of the Duchess story.

Francis Peabody Sharp (1823-1903) played one of the key roles—perhaps the key role—in bringing Duchess and other apples to northern Maine. After purchasing his father's farm in Northampton, New Brunswick, in 1844 at the age of twenty-one, Sharp established over a hundred acres of nurseries and orchards there and in nearby Upper Woodstock. He may have had the largest commercial orchard in North America at the time. He trialed over three hundred varieties of apples, including many imported from Russia. He developed a huge nursery business, lining out as many as 180,000 nursery trees in one season. He made over two thousand crosses in hopes of developing new varieties for the north. One of his introductions was a seedling of unknown parentage that he named New Brunswick a.k.a. New Brunswicker.

By 1849 Duchess was recommended for growing in Maine. As early as 1850 it was being exhibited at the Bangor Horticultural Society exhibitions. In 1856 Duchess was recommended by the Maine Pomological Society as being "of high promise and worthy of extensive trial." By 1863 it was well known throughout Aroostook County.

In 1870 fifty-six Swedes founded New Sweden, Maine, thirty miles south of Fort Kent. They planted orchards that all included Duchess. By then Sharp had also established nurseries on the U.S. side of the border in Houlton. It was probably his trees that were planted in those New Sweden orchards. In about 1877 the Dudley Winter apple was discovered, and in about 1885 Nutting Bumpus came into being. By the turn of the century, Duchess was probably the most common apple in Aroostook County. A lot was happening in a short period of time.

Francis Peabody Sharp introduced thirty-one apple varieties. While most have since disappeared, New Brunswicker remains a favorite in both New Brunswick and northern Maine. Evidently, New Brunswicker came from seed that Sharp received from Albert Emerson in Bangor. Emerson also provided the seed that Peter Gideon planted in Minnesota. (One of Gideon's seedlings became another much more famous variety, Wealthy.) Many pomologists think that New Brunswicker is a Duchess seedling, but no one knows for sure. Unfortunately for Sharp, the apple was often called New Brunswick Duchess, almost certainly because the two varieties are difficult to tell apart. Some literature assumes that the two are synonymous. When I received the Estonian book on apples, *Eesti pomoloogia*, from my friend Raivo Vihman, it was interesting to see that even the Estonians consider New Brunswick and Duchess to be the same apple.

New Brunswicker looks a lot like Duchess

Is it possible to tell a Duchess from a New Brunswicker? Since Duchess comes relatively true to type from seed and since Duchess seedlings were planted all over northern Maine, I suspected that New Brunswicker was a Duchess seedling or maybe even a synonym for Duchess. I was eventually able to obtain New Brunswicker scionwood and graft it in our orchard. That tree has been fruiting for a number of years now. If our Duchess really is Duchess, and if our New Brunswicker really is New Brunswicker, two big "ifs," then it's true the two are nearly impossible to tell apart. We have Nutting Bumpus down here as well, and Nutting is also nearly impossible to tell apart from the other two.

All three apples ripen for us from late August to mid-September. All three are roundish-oblate to sometimes a little oblong and blocky. There appear to be no consistent differences in stem, cavity, basin or internal structure. All three are colored with delicate, short to long, light-red stripes over a light-yellow ground color. The fruit qualities are also similar. They are excellent in sauce, good fresh eating and nearly unequaled in a pie. That being said, our New Brunswicker does appear to peak a week later than our Duchess. This could be a true difference. It could also be the result of the placement of our two trees. The New Brunswicker gets less sun. Maybe that delays ripening for a few days.

When I hear from old-timers that they have a New Brunswicker, I assume that they are probably correct as long as the apples have that Duchess-New Brunswicker-Nutting look. Where are the DNA testers when we need them? Wouldn't this be a great use of your hard-earned tax dollars? We could test all three of ours plus several dozen other candidates from northern New England. I'm guessing we'd be in for a surprise or two.

Sophie Olson and Walfred Jacobson: courtesy of the family

That night I also learned from Daryl Hunter that another of Sharp's introductions was called Crimson Beauty. I heard the name Crimson Beauty for the first time that afternoon at Frank Moody's. Now I was hearing about it again three hours later. Here was another connection between Sharp and The County. It turns out that Crimson Beauty was grown downstate as Early Redbird. How can I keep all this straight? It's mostly red, and doesn't resemble Duchess at all. Later I obtained scionwood, and now Crimson Beauty is the earliest apple we grow. It ripens for us at the end of July, makes excellent sauce, and keeps for about a week.

My big question for Daryl was, how could Sharp be credited with the introduction of Dudley Winter? Well, according to Daryl, John Dudley and Francis Sharp were friends. Dudley's daughter planted seeds from New Brunswicker apples that the Dudleys had received from Sharp. One of the seeds produced the seedling that became the Dudley Winter. So does that make Dudley Winter a Duchess offspring?

Later that evening, just for curiosity's sake, Steve pulled out a small book entitled *Looking Back on Westmanland (settled 1879)* by The Gustaf Johnsons. Steve thought he remembered some references to local orchards. He began to read aloud. "Sophie Olson came from Westmanland…She married Gustaf Johnson, and moved to lot 26 in Westmanland, where they started to clear land and built a home…" Steve continued. Sophie's husband died about 1899 and she "later married Walfred Jacobson." She and her new husband "moved the house down nearer to the spring…adding a wing to the original structures at this time. More land was cleared and an orchard of Dutchess [sic], North Star and Yellow Transparent was planted." We looked at each other around the kitchen table. Steve had seen the orchard some years ago but hadn't remembered anything special. Lot 26 was just up the road from Steve and Barb's. In his effort to show me every apple tree in The County, this orchard had slipped his mind. I was leaving next morning. Steve would be busy and wouldn't have time to check it out.

I went to sleep with visions of Dudley Winters dancing in my head. The site was only a minute or two away. We had to go, and of course, next morning Steve gave in. All right, one more trip. Then back to work. David, Steve and I drove up the Westmanland Road. The truck was packed with all our gear as well as four bushels filled with little brown bags of the apples we'd collected over the past few days, each one labeled in Sharpie. One last stop. This might be it.

A carload of bear hunters was parked at the entrance to the Johnson driveway. We didn't want to bother the hunters so we kept on going up to a neighbor's farm where we picked a bag or two of another Duchess lookalike, and then headed back to Steve's. It had been a rewarding trip. Although I hadn't found the Dudley tree, I did learn a lot more about the apples of the north. I felt as though I was beginning to sort out how Duchess and the rest of them found their way to The County.

Up ahead was the entrance to the long driveway up to Sophie Olson's orchard. The bear hunters were gone. I couldn't help myself. I convinced Steve to make one last stop. He pulled over and we walked up the long, rutted dirt road. I think I ran. I could see an apple tree up ahead, with others beyond.

Unlike Duchess, Dudley often gets scab (black spots)

Directly in front of us were three rows of old trees, six trees in each of the outside rows, with about half of the middle row now dead and gone. We waded though a ground cover of Redosier up one row and down the next. Here were more red-striped apples. By this time I was pretty familiar with Duchess. We'd been living and breathing and eating them for nearly a week. Then I noticed that the two rows of apples appeared to be different from one another.

We picked fruit from each tree. One row was distinctly different. Not only that, but the apples from the southerly row were soft and tart like all the other Duchess we'd tried. Those in the northerly row were hard and crisp. They were quite tasty as well. Steve and I looked at each other. One row had to be Duchess, and the other had to be North Star, a.k.a. Dudley Winter. We had found it! We said our goodbyes, dropped off Steve at the end of his driveway and drove four and a half hours south, back to the farm.

Waiting for me at home was a box of apples, not unusual for that time of year as the ID requests had begun to roll in, but this package was different. It came from Eugene Jackins of Houlton, an apple enthusiast who sends me letters every couple of years. I opened the box and pulled out the letter, which read, "...the smaller, striped apple, I've been told is a variety called the New Brunswicker." I opened the bag, and there it was. I set New Brunswicker on the dining room table in the sun next to the three others: Duchess of Oldenburg, Dudley Winter, and Nutting Bumpus. Perhaps someday I'll do that genetic test of the four of them to discover how related they all really are. For now, I'll just enjoy them.

Steve cut Dudley Winter scionwood for me the next winter. We have grafted and sold it many times in the Fedco Trees catalog. I've topworked it into our orchard. I've also planted it in the Maine Heritage Orchard over at MOFGA. It is a good apple.

How did we know it was Dudley Winter? Dudley and Duchess are similar but can be told apart. Dudley is somewhat later in season than Duchess, although this would be most apparent up North. Dudley is a more balanced dessert apple (less tart) than Duchess. Dudley is roundish-conic-oblate in shape. Duchess tends to be round and oblate but rarely conic. Dudley is slightly, though rarely deeply, ribbed. Duchess is usually not ribbed. Dudley's base ground color is usually a rich yellow while the Duchess ground color tends to be a lighter, washed-out yellow. Although both are striped and blushed, Dudley is more apt to include large areas of solid red. Duchess is more apt to be all or mostly striped. The biggest giveaway is Dudley Winter's susceptibility to scab—not a good thing, but a defining characteristic nonetheless. I've never seen Duchess with scab. The scab on Dudley Winter doesn't appear to get out of control, even without spraying, but it's usually there.

There are multiple watercolors of both apples in the USDA collection. The Duchess is listed as Oldenburg. One painting that epitomizes Duchess for me is the Oldenburg from South Haven, Michigan, painted on August 15, 1919. There are two good Dudley Winters. The first is from Ottawa, dated October 9, 1909. The second is from a specimen submitted by W. M. Munson of the University of Maine on September 18, 1901. Check them out.

I remain curious about the inclusion of "Winter" in the Dudley name. It's not a keeper for us down here in the tropics of central Maine, at least not like the classic winter apples, such as Baldwin and Black Oxford. That brings up the issue of growing apples in different locations. Some apple varieties will simply die when grown out of their range. You rarely, if ever, find Baldwin trees up in northern New England. They may have been planted up there from time to time, but they don't last more than a few years. This is another one of those useful bits of information when doing an ID. If you find a tree in northern New England and you think it's Baldwin, think again.

And it works in reverse. Many of the super hardy varieties do not do at all well farther south. You will rarely find a Duchess growing much south of the Maine/New Hampshire border, and with good reason. They don't do well down there. Duchess is a northern lady. She doesn't like warm weather. Some southerners can be downright nasty in their assessments of the apple. A southerner here means everyone south of about Portland. A few years ago my interest in Duchess provoked a number of exchanges via the internet, including this one from New York State:

> Here are some more demerits for Duchess of Oldenburg. In my experience (fortunately just one tree), the Duchess drops most of her fruit before they mature properly, while they are still inedible. The fruit also broke down quickly when ripe (like most other summer apples). After 7 years of exasperation as I was plotting to overthrow the useless Duchess, her final problem emerged—a very brittle graft union with M.9 rootstock. She abdicated at the union before I could get her to the guillotine, and now lies mouldering in my brush pile!

One evening I was relaxing on our couch by the fire. Something distracted me. It was the pillow on the chair. My mother had given it to me some years before. The pillow was printed with one of those old Western U.S. fruit box labels, the ones that make good pop art. I stared at the picture on the pillow.

The fruit label on the pillow depicts a large swan swimming in a placid pond in front of a yellow-orange sky. "Swan Brand Extra fancy APPLES Perham Fruit Company Sales Office, Yakima, Wash" is written on it. Was this the prelude to the next batch of chapters in the John Wesley Dudley story? John Wesley grew up three thousand miles east of the Perham Fruit Company in Yakima, Washington, but only eight miles from Perham, Maine. He died in 1921 at Big Four Apple Ranch near Ephrata, between Soap Lake and Stratford, about a hundred miles northeast of Yakima. He grew "Skookum" brand apples. Did the Perham Fruit Company actually get started a few miles from Perham, Maine, in a nineteenth century orchard on the Dudley Road in Castle Hill?

Over the years I've received some wonderful letters and emails from various members of the Dudley family. They've helped me piece together the story of the apple. On September 14, 2017, I traveled to Aroostook County and gave a talk at the Haystack Historical Society on Rte 163 in Mapleton. Before the evening began, I went out to the car one last time to get a couple of things and gather a few thoughts. My host came out to tell me it was time. We stood together and looked west. "That's Castle Hill," he said. "There's the beginning of the Dudley Road." After the talk a number of people came up to say hello. I'd been quoting Lena Dudley Kenney periodically during the evening, not knowing that she was sitting there in front of me with her son and daughter, Russell Kenney and Jane Kenney Hopkins. Also in attendance was Sophie Jacobson's granddaughter's husband, as well as Dana Allison, the owner of the property on which the Aroostook Sunset apple originated. That apple might well be the "sweet russet" Dudley described in 1889. "[A]n apple of very fine quality and a good keeper." I was among royalty on Castle Hill.

JPB, Lena Dudley Kenney, Russell Kenney and Jane Kenney Hopkins with Dudley Winter apples, 2017

One question, yet to be settled, is who planted that seed? I had thought it was the Dudleys' daughter, Grace, who grew up to manage the family apple business in Washington State and died in about 1980. But the apple was first planted a hundred years earlier. That would have meant that she would have had to have lived to be at least 105 or so. Perhaps it was Grace's mom who planted the seed that grew to become one of northern New England's favorite apples. Joe Dudley of Delaware left me with these words when we last corresponded some years ago, "It is really great to know that the 'Winter' apple is still around."

Farther north, Garfield King died unexpectedly on January 20, 2006, at his log camp in Eagle Lake, Maine. He was 77. Gloria let me know. Later, Venette told me that Garfield was sorting four cups of beans the evening he died. Earlier that day he had called his neighbors to tell them he was going to slice up one of his Blue Hubbards and wanted to know who would like a piece. He always had something to share.

Garfield King showing off a prized cherry in 2005

Rolfe is "shaded and striped with red; stalk short, inserted in a large cavity; calyx large, closed, in a rather large, regular basin; flesh white, fine-grained, tender, juicy, sub-acid; core small; " W.M. Munson, 1907; Stalk (stem) and cavity in photo above; calyx and basin below: photos by Abbey Verrier

## Thirty

# Just a Plate of Rolfe

"Not invisible but unnoticed, Watson."
Sherlock Holmes, *A Case of Identity*, p. 196

The apples that originated in Maine are spread out all over the state, though usually in predictable locations. When I search for a variety, I always begin as close as I can to the point of origin and then expand outward in concentric circles. Degrees of separation. If I'm going to locate an apple that originated in Vassalboro, why not start in Vassalboro? If possible, I visit the farm where the first tree grew. Then I expand my search to China or Sidney or Winslow. That would be the first degree of separation, the first concentric circle. With this strategy, I've had pretty good success. It was able to find Starkey in Vassalboro and Marlboro in Marlboro. Dudley was farther away, but still only three towns from Castle Hill.

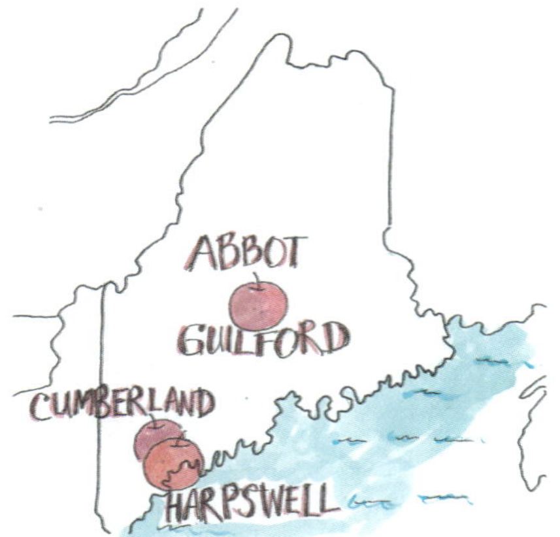


One year I decided to spend a day looking for the Stowe apple. I was on another one of those visits to see Steve and Barb Miller for a few days in Aroostook County, and while they were doing other things I had some time to go out on my own. I had a hunch that it should be possible to find Stowe, and that when I did, it would be in nearby Perham where the apple originated. The Millers live just two towns north. Stowe is a green-colored apple that ripens late. There are only a few green apples in Aroostook County, and as far as I know, only one late-ripening green apple. That would be Stowe. As you know, Aroostook County is the land of late-summer, red-striped Duchess lookalikes. So I spent a day in late September, after the Yellow Transparents had all dropped off the trees, meandering along in second and third gear up and down the back roads near Perham, occasionally pulling over to let the odd pickup truck zip by. I was looking for "greenings" still hanging on the trees.

Many of the older towns down in central Maine are what a geometrician might call polygons. Even those with four sides are far from square. I think you'd call Freedom a quadrilateral or maybe an irregular parallelogram. Whitefield has seven sides. Liberty has nine. China has ten. Somerville and Palermo both have eleven sides. They are all towns with interesting shapes and hardly a right angle in the bunch. Aroostook County was settled rather late compared to the rest of Maine. By then I suppose the surveyors were sick of weird angles. Most Aroostook County towns are perfect squares with the rare rectangle thrown in here and there. With Perham at the center of my search, concentric "circle" number two consisted of eight squares, four of which matched Perham's four borders and the other four that each touch Perham at a point.

Chris Drew with his "Playmate" filled with scionwood at the Scion Exchange; keep them cool!

After visiting a couple of orchards I had heard about, and driving down many quiet roads, I came to Washburn which is probably three miles from Perham. On the left, surrounded by potato fields, was a pretty little orchard next to a small farmhouse. I pulled into the U-shaped driveway and asked if I could wander through the orchard. The owner was pleased to have me do so. There were once fifteen trees. Now there were twelve. Six Yellow Transparents had dropped all their porcelain yellow fruit, but enough were still on the ground to recognize them. There were two Alexander trees and three trees with no fruit. The twelfth tree was the one I was looking for. The fruit was green and still on the tree—right season and right color. I pulled out my wanted poster and checked the description. It matched. I thought I had found Stowe. I had found it three miles from Perham and only one degree of separation. Might I have found it in Kennebec or York or Androscoggin Counties? Maybe. But I doubt it.

More recently Chris Drew of Mt. Chase, up near Mount Katahdin found a different yellow/green late ripening apple he wanted me to see. Chris has assembled a large collection of apples he's discovered, mostly since he retired as chief ranger at Baxter State Park. He thought he'd found a Stowe. I received the apple with this note: "Mystery Apple. Possible Stowes Winter from old Patten [Maine] orchard behind Richard Mcmanus home. Occasional Pink Blush. Picked in September. Very short stem. Fits closely the description of Stowes Winter. What thinkest thee? Keep smiling! Chris." Although a bit farther from Perham than Washburn, it also fits the description quite well. It's also beautiful and unusual looking. I know I've never seen it before. We'll include it in the Heritage Orchard. I suppose we should call it "Drew's Stowe of Patten."

Location, location, location! A couple of years ago, Robert McIntyre discovered an apple at the end of Harpswell Neck that he called "Lumpy Red." Whenever "Roberto" finds a new apple, I know it must be something interesting. He's endlessly on the prowl. If you live down that way, you may look out your living room window and see him wandering through your old trees. I love going exploring with him. Over the years he's introduced me to many old trees. Those include an ancient Baldwin, a beautiful Golden Ball, and an old seedling known to the family as "Norton Greening." Roberto's favorite apple is one he calls Firehouse. It's an ancient shell of a tree growing on the edge of the Harpswell fire department parking lot.

Robert McIntyre and Andrew Tufts at the Scion Exchange—note the two scions in Roberto's pocket!

So one day Roberto shows up with a fruit he calls Lumpy Red and immediately I know it's something I'd never seen before. The solid, brilliant red fruit was so deeply ribbed you might think it wasn't even an apple at all. It looked so foreign! Turns out it was. Evidently, right across the road from the tree, on the shore of Casco Bay, once stood the glorious Hotel Germania, back when the coast of Maine provided the summertime air conditioning for those who could afford to get away from New York, Boston and Philadelphia. Grand hotels lined the coast from the New Hampshire border to Canada. I pulled out my facsimile of *Pomologia: From Netherlands, Germany, France, England and other regions; described, represented and enlighted with their natural colors* by Johann Hermann Knoop

The Hotel Germania exists no more; can you see the apple tree?

Robert's nickname for Danziger Kantenapfel is "Lumpy Red"

(Nuremberg, 1766). Although I don't speak German, I love the dark, bold images of the fruit, all painted blossom-side-up and stem-side-down. I had a hunch that Roberto might have found a German escapee from the hotel across the street. If I was correct, I figured that Johann Knoop would know it. I didn't have to look far. There it was with seven other apples on plate 4, looking red, lumpy and very ribbed with an apex that resembles a pair of puckered lips: "Danziger Kantenapfel."

Most years I attend at least a couple of harvest celebrations and fall street festivals. I go for the day, set up a display with a few dozen historic apples, and then see what happens. The apples are the magnet. If there's enough of a crowd, the day becomes one continuous conversation. Sometimes I make a great new discovery. At a recent Maine Ag Trade Show I learned of a variety I'd never heard of before: "Bung Hole Russet" from Parkman. It's entirely russeted and the size of a Wolf River. I want it! I meet new and interesting people and hear fun stories about apples and farms and the history of Maine. I also attempt to answer questions. "Why doesn't my tree produce fruit?" At one festival a few years ago they set me up about thirty feet from the main street in town. I was on a shady side street where no one ever walked. There I stood all day watching a river of happy-go-lucky revelers flow by my little eddy. If only they would turn and look, they would see me and my apples. Those who walked anywhere near me were focused on getting to the action. I felt as though I was just inches away. I think I talked to four people all day. It was like one of those bad dreams when it's just out of reach. Fortunately, it's usually not like that.

In October 1997 I was invited to do a display at Maine Audubon's Apple Day. The event is held annually at their facility in Falmouth, which is plunked down in an old apple orchard a few miles north of Portland. A couple years later I made them a map of the orchard. It was primarily Northern Spy and Tolman Sweet trees with one Porter, one Golden Russet and a large branch of McIntosh that had been topworked long ago onto one of the Tolmans. On that day they put me just inside the main entrance. Everyone had to walk by me. I had my display, and I was chatting with people most of the day. Next to me was a second apple display, this one unattended. It was from Sweetsers, a popular commercial orchard in nearby Cumberland Center. Their display consisted of a dozen varieties set out on paper plates. During a quiet stretch that afternoon, I checked it out.

I have learned that commercial orchards are always worth visiting. It's one of the best places to learn about apples. It's also one of the best places to find the odd variety. Even those who grow nothing much more than Macs and Cortland and Honeycrisp usually have a gem or two tucked away. These are easy pickings, low hanging fruit and all that. The trees are there, and the apples are there, and so is the scionwood. Someone else has done all the work. They already know the names; no discovery is necessary. But despite all that, it is important to be on the lookout. Keep your eyes open. In the summer of 2015 someone in Beverly Hills tried to pawn four N.C. Wyeths that had been stolen two years earlier in

Portland, Maine. An FBI agent, whose job it was to look over pawnshop slips, noticed that something didn't seem right. She was correct. With apples, it's all about those names. Although I'll never know all of them, every so often I hear a little ding in my brain when a name appears that I haven't heard before. A thousand people could walk by that same table, and it would never dawn on them that they were looking at a rare, endangered, historic apple, maybe the last one, right there on this paper plate. It was immediately evident that Sweetsers has an interesting assortment of varieties. Red Gravenstein, Rhode Island Greening, Northern Spy, Nodhead, Wolf River. These were all apples I knew. And then there was one I'd never seen before: Rolfe.

Apple Day ended. The crowd went home. I packed up my display and drove directly to Cumberland Center. Less than a quarter mile from the blinking light at the junction of Rte 9 and the Blanchard Road sat a small, folksy, brick-red wooden farm stand with two overhead garage doors opened wide. Hanging between the two doors was a set of small square cubby holes. Above, a sign read "Today's Varieties." The "O" in Today's was a round, red apple. In each little square sat a different variety. There in one of the holes was a beautiful, glowing, reddish-pink, round apple with a label that read "Rolfe." I went in, bought a bushel, and headed home. That was easy.

How many other people noticed the plate of Rolfe at Apple Day that afternoon? Any? Why not? All sorts of things are invisible until we teach ourselves to notice. I still recall the experience of learning to recognize Forsythia one springtime long ago. The next day and for several weeks, I saw those yellow flowers everywhere. Was Forsythia invented the day it was introduced to me?

The lone Rolfe tree at Sweetsers Orchard in 1997; the tree is now gone

Was I looking for that plate of Rolfes? Can you be looking for something, not knowing what it is you're looking for? If that Rolfe had not been labeled, would I have known? Rolfe originated a hundred and twenty miles from Cumberland Center. Why was it way down there? Is it possible that such an obscure apple could travel so far? How many concentric circles and degrees of separation?

On the other hand, maybe Rolfe wasn't obscure way back when. Maybe some of the apple varieties were considerably more mobile than one might think. After all, most people today haven't even heard of Baldwin or Ben Davis, and yet, once upon a time, they were as well known as Honeycrisp or Red Delicious. Who was Rolfe, anyway?

Jeremiah Rolfe (or Rolf as it is sometimes spelled) was born in 1759 in the town of Buxton, in York County, when all of Maine was still part of Massachusetts. Little is known about his childhood. His father was most likely either Samuel or John Rolfe, two brothers who lived in Buxton at the time.

A typical Rolfe: red striped and blushed, with a cascade of tan russet; the blackish smudges (sooty blotch) and the tiny black dots (flyspeck) are harmless & tasteless summer "diseases"

Rolfe enlisted in the Continental Army at about the age of eighteen, serving for three years under the controversial General Horatio Gates, reportedly fighting in Saratoga in 1777 and South Carolina in 1780. That would have made him twenty-one when he was honorably discharged. He spent the next thirty-eight years winding his way north. All the way he farmed, from Rochester, New Hampshire, to Oxford County, Maine, to Foxcroft and Guilford in Piscataquis County. Eventually in 1818 at the age of fifty-nine, he moved a few miles west of Guilford to Abbot village which, at that time, had been settled for about ten years. There he established what was considered to be one of the finest farms in the county, a mile south of the village and a mile west of the Piscataquis River. Although details of his family are sketchy, we know that he married Fanny Huzzey, and together they raised a family.

He purchased the Abbot property from a family named Houston. The Houstons had cleared the land, and Betsy Houston had planted a number of seeds of Blue Pearmain. Planting seeds was still the most common method of apple tree propagation at that time. Blue Pearmain was a good choice for the seed. It was fairly common and quite popular.

By the time those Blue Pearmain seedlings were large enough to transplant, the Houstons had moved away, and the Rolfes had moved in. Sometime around 1820 Rolfe gave twelve of the seedlings to the Reverend Thomas Macomber in nearby Guilford. Macomber kept eleven for himself and gave the twelfth to his son who planted it on an adjacent farm. That twelfth tree was the good one. By the time it fruited, Rev. Macomber had taken over his son's farm. Perhaps his son had died or moved away. In any event, it was the reverend who owned the first "Rolfe" tree and presumably it was the good reverend who first discovered the rosy-pink, red-striped fruit and tasted its delicious, slightly tart, sub-acid flesh. For many years the apple was known as Macomber. Gradually it picked up the name Rolfe, or occasionally Coreless, due to its relatively small core size.

Rolfe attracted enough attention and admiration to be widely planted around the state, even up into Aroostook County, a testament to its hardiness. Unlike many other Maine varieties that were virtually unnoticed outside the state, Rolfe was even described in several of the most important fruit books of the day, including A. J. Downing's *Fruits and Fruit Trees of America*, John Thomas' *The American Fruit Culturist*, *The Apples of New York* and Hedrick's *Cyclopedia of Hardy Fruits*. In Downing and Thomas, it's called Macomber, but by the time of Beach and Hedrick, it's Rolfe.

100 years ago, Rolfe was more commercially popular in Maine than McIntosh

Beach wrote that Rolfe "is very hardy, vigorous and a reliable cropper. At the present time it is probably grown more extensively in Maine than in any other section of the country. It is regarded highly wherever it is known and is gaining in popularity among fruit growers. It is worthy of testing in those portions of the state [NY] where superior hardiness in a variety is a matter of prime importance."

In the Maine Agricultural Experiment Station *Orchard Notes* of 1902, W. M. Munson wrote that "a sprout from the original tree is still standing on the Macomber farm and produces annual crops of fruit. H. L. Leland of Sangerville, has more than a hundred trees of this variety in his orchard and says: 'The Rolfe in our local markets, sells better and at bigger prices than any other variety we grow. It sells well as a shipping apple, though not much known.' "

Like the vast majority of other historic apple varieties, Rolfe couldn't live up to the dictates of the twentieth century. Eventually it disappeared from all but the occasional orchard, like Sweetsers, that liked it and kept it around to satisfy a local fanbase. I've found one other source of Rolfe, that being another small commercial orchard, Olmstead's on Route 15 in Charleston Maine, much closer to Abbot.

Over the years I wormed my way into the lives of Connie and Dick Sweetser, and more recently their son Greg who has been taking over the business as Dick and Connie slow down. I collected scionwood from their big old tree and grafted a branch of Rolfe in our orchard. That branch has now been bearing fruit for several years. More recently I planted a tree at our place entirely grafted to Rolfe. It's that good. We've also planted it in the Maine Heritage Orchard in Unity. We offered it for sale in the Fedco Trees catalog, and in April 2000 I was able to give a talk in Abbot and bring a bunch of trees with me to help repopulate Piscataquis County with Rolfes. In 2002 I grafted five replacement Rolfe trees for the Sweetsers which they planted two years later. Those trees are also now bearing. Sadly, Sweetser's old tree began to crumble in recent years and was finally cut down in 2012.

Rolfe is medium-sized, roundish in shape, and red. It resembles McIntosh in size and shape and in its mid-fall season, but it's easy to tell the two apart. Rolfe is bright red striped and blushed. Sometimes when the apples are at peak ripeness, the ground color mellows to a rich yellow, blends with the red blush and the skin takes on an orange hue. Macs, on the other hand, are generally bluish-purply-red. Rolfe is more likely to be confused with Starkey. Starkey has a somewhat similar bright rosy-red color. Still, I would call Rolfe red and Starkey rosy-red. Both Rolfe and Starkey have prominent dots, but Rolfe's dots tend to blend in and get lost in the red. Rolfe's cavity is deep, and its basin is deep, wide and abrupt. Starkey's cavity and basin are both medium at best. Mac's basin is small, round and regular. Starkey also ripens somewhat later than Rolfe. Rolfe's cavity is usually filled with a russet splash that varies from small to large enough to spill out over the rim and drip down the sides.

Rolfe fits the same sub-acid, dessert-cooking niche as McIntosh. You could fool yourself into thinking it's a grocery store variety—it looks that good. In 2013 we did an apple sauce comparison of a couple dozen varieties we grow on the farm. Rolfe was the favorite of the fall.

So was it luck to find Rolfe so far from home? If it was, it was good luck. A year earlier that plate of apples might have been invisible. That day I happened to notice, and I scored. Jeremiah Rolfe became locally known as "Uncle Rolfe" as he grew older. He was described as having undefined "peculiarities" but also "many sterling virtues." He died on April 1, 1841, at the age of eighty-two and was buried in Abbot at the village cemetery. A hundred and seventy-five years later, Dick Sweetser died on October 11, 2016. But the Rolfe apple lives on.

JPB with Dick, Connie & Greg Sweetser; you can see the apple cubby holes behind Connie

# Thirty-One

# Russet Is a Skin Condition

There is nothing more deceptive than an obvious fact.
Sherlock Holmes, *The Boscombe Valley Mystery*, p. 204

Roxbury Russet

Some potatoes are russet potatoes. Nearly all Asian pears are russet pears. A surprising number of apple varieties are either partly or entirely covered with russet. For most contemporary commercial growers around the world, russet on an apple is considered a flaw. It can render the apple unsellable. But it wasn't always that way.

For two and half centuries, one of the most sought after commercial apples throughout the Northeast was the Roxbury Russet. Many thousands of trees were planted. While it was never considered to be a dessert apple of the best quality, many people did and still do love it fresh. It is also a good cooking apple. It makes fabulous sauce. Some cider makers like to include it in their cider. Long before electric refrigeration and supermarkets became the norm, everyone knew which apple varieties stored the best and longest. Roxbury Russets would keep in the root cellar well into the following spring, nearly until the new crop of Yellow Transparents, Early Harvests, Sweet Boughs and Astrachans were ready.

Russet skin looks like suede or unglazed pottery. It's not smooth, and it's rarely shiny. It is typically rough to the touch. The Jamaican pickers at Steve and Marilyn's orchard call them Guavas. Russet is typically colored some shade of light, tannish-brown, but it can also be yellow or rather dark, burnt umber. It can be greenish or reddish in tone, or tan on one side of the apple and reddish or orange on the other. On some varieties, there are patches of russet here and there, while other areas of the skin are as smooth as a McIntosh. Sometimes an apple will have a web of russet netting that decorates the surface of the fruit. Others have a splotch or splash of russet surrounding the stem and emerging like a star from the cavity. In some cases, as with Wolf River, Red Astrachan or Hurlbut, that russet splash is one of the apple's defining characteristics.

Silver Cup from England and Everett Cunningham from Maine showing off their almost identical russet cavities

I've had a difficult time finding a satisfactory definition of russet. I've asked many apple people over the decades, and I've read everything I can find. I even tried the internet. Years ago someone told me that it's a thin layer of dead skin cells. While that explanation may not be technically correct, it seemed reasonable enough. I was happy to accept it and move on until recently, when Cammy and I visited the USDA Agricultural Research Service apple collection in Geneva, New York. There we were able to spend an hour visiting with a graduate student named Ben Gutierrez, who is studying russets. We must have asked him a hundred questions. He was a good sport, however, and hung in there with us.

Later Ben and I communicated via email, and he attempted to correct some of the misconceptions I had about what he told us that day. It seems that we had the biology incorrect, but at least we were close on the concept —it is a skin condition.

Here's my updated take on russet with some of Ben's words quoted here and there: the outermost part of the apple, commonly referred to as the skin by most apple eaters, is technically called the epidermis. The job of the epidermis is to protect the inside of the apple. The epidermis is sometimes damaged during the first few weeks in the development of the fruit. Because the apples are growing fast at that time, microcracks in the epidermis can develop. When that damage occurs, the next layer, the periderm, steps in and takes over the job of protecting and "sealing off the fruit." Perhaps because of the "structural and chemical" differences between the epidermis and the periderm, the periderm takes on a very different look. It's corky, often tan and sometimes has superior flavor. Recent research appears to be showing that russet apples may also be more nutritious. Once the periderm has broken through the epidermis and made itself visible on the outside of the apple, it is called russet. Russeting may also result from epidermal irritation by the fruit's stem or by nearby twigs and foliage. This could explain the russet often found in and around the cavity. It's also worth mentioning that commercial growers are well aware that timing and variations in fungicidal spray materials can cause or trigger russeting on the fruit. One way or another, it's a skin condition.

At a glance, many russet apples resemble one another. One unglazed coffee mug looks like another, I suppose, regardless of the size and shape, until you jazz it up with colors and catchy slogans. But just like red, green and yellow apples, russets come in many varieties. Oddly enough, however, for as long as there have been apples in America, all the various russets for the most part have just been called "russets." Could be big, could be small, could be tasty, could be awful. Still, it's a russet. Even the most observant, knowledgable old-timers rarely bothered with any differentiation. "That's our russet."

Roxbury Russet is mottled green and russet; it's the best winter sauce apple

To add to the confusion, many of the traditional russets are of similar shape and size. They are often roundish, medium-small, and mostly or entirely covered with a light, yellowish-tan russet. When these varieties were not simply called Russet, they were called Golden Russet. By 1900, when there were a dozen or more russets being grown in New England, most or all of them were referred to interchangeably as Russet or Golden Russet. Even the best apple books of the time were confused when it came to russets.

In the fall of 1896, Ziba Alden Gilbert, President of the Maine Pomological Society, collected all the russets he could locate from around Maine, Massachusetts and the Canadian Maritimes. Gilbert ran a dairy and large orchards in the Androscoggin County town of Greene, in the heart of Maine apple country. He was one of the founders of the Maine Pomological Society in 1872. His goal, four years before the turn of the century, was to sort out the confusion around all the russets.

JPB in 1997, a hundred years after Z.A.Gilbert: will the real Golden Russet please stand up?

In his article "The Russets of Maine," which appeared in the 1897 report of the State Pomological Society, he wrote, "In the very start, however, of our effort to classify and name the different Russets grown, the difficulty is encountered that the authorities on fruit nomenclature are in a measure as puzzled and mystified over this matter, as our growers and exhibitors have been." Even with the Roxbury Russet, Gilbert lamented, it "would seem that everybody ought to know so common an apple as this long time favorite. Yet, they do not..." Gilbert makes a gallant effort to illuminate the differences between the russets of the day. With apologies to Z. A., here are a few tips.

Roxbury Russet was the most famous of the heirloom russets, and probably still is. It is also thought to be the first named American apple variety, originating in Roxbury, Massachusetts, within a few years of that city's incorporation in 1630. It keeps well, and it makes perfect sauce all winter. It's also unmistakable once you know what to look for; it's all in the details. There are two defining features of this medium-large sized apple. First, it's almost never entirely covered with russet. Instead, it's typically a patchwork of areas of smooth green skin alternating with brownish russet. The second is the oval shape of the fruit from a bird's-eye view. Looking from above, the apple is not round. While neither characteristic is unique in apples, there are no others that combine the two. Patchwork green and russet and an oval shape from above—that's Roxbury Russet.

Is your russet medium-small, roundish and mostly or entirely covered with light, yellowish-tan russet? It could be Golden Russet. The Golden Russet is the most commonly grown russet today. Ironically, it's also the most easily mis-identified russet. It's likely that as many as a dozen or more "Golden Russet" imposters are still being grown in the twenty-first century. The variety is fraught with multiple challenges for the identifier. All the Golden Russets are medium or small-medium in size. They are all round and a light, yellow-golden russet color. Even upon close examination they are difficult to tell apart.

And don't forget that a percentage of the ancient "golden" russets found in old orchards and along the roadside are not named varieties at all. Some might be grafted trees, but don't be fooled; often they were grafted from russet-fruited seedlings of undistinguished flavor, unestablished cider quality and questionable storage ability. They can look as though they should be Golden Russets, but they aren't. In fact, they're no variety at all.

Golden Russet of Western New York: photo (and hand) by Abbev Verrier

Fortunately for us, the different Golden Russets do vary in flavor, storage ability and tree form. If you get past the seedlings, and don't rely on the visuals, you can sort them out. Flavor sometimes works. If it tastes lousy, it's not Golden Russet. A few years ago I was going to do some topworking at a local orchard. The owner wanted to switch over a bunch of trees to Golden Russet. He already had lots of Golden Russets, so I collected the scionwood from his trees. When we chatted on the phone a week before grafting day, I mentioned where I had gotten the scionwood. "Oh no," he said, "I was hoping you'd get the wood off the trees up the hill. Those Golden Russets are better than the others." Sounds a bit like Duchess of Oldenburg.

What are you likely to receive when you purchase Golden Russet trees from a catalog? Most nurseries selling Golden Russet aren't even aware that there are multiple pretenders to the Golden Russet throne. Shouldn't all Golden Russets be Golden Russets? Chances are that you either get Golden Russet of Western New York, English Russet (sometimes called Poughkeepsie Russet) or American Golden Russet (sometimes called Bullock). They've all been sold as Golden Russet at some point. How are they different from one another? Is one better than another?

Golden Russet of Western New York may be the most common Golden Russet in the marketplace today. Unfortunately it has several misleading and confusing synonyms, including American Golden Russet and English Golden Russet. No one seems to know where the apple originated. Downing thinks England. Others don't bother guessing. Because of its tree form, Golden Russet of Western New York is probably the easiest Golden Russet to ID. Its branching habit is long and almost willowy. It practically droops and weeps. Its downside, which is another identifiable characteristic, is that it's not a great bearer. Its upside, which is another identifiable characteristic, is that the fruit is of excellent flavor. It ripens late and keeps extremely well. Many cider makers include this Golden Russet in their cider blends.

Is your Golden Russet's skin smooth enough to take a shine? Is it densely fruiting? English Russet, sometimes called Poughkeepsie Russet, is the second most common "Golden Russet." Keep it straight though: this is English Russet, not English Golden Russet. No one knows where English Russet originated, although almost certainly not in England. On close examination, English and Western New York Golden Russets are quite similar. Fortunately there are differences in tree form. Unlike the willowy Golden Russet of Western New York, the tree of the English Russet, according to Ziba Gilbert, can be distinguished because the "stout thick limbs and branches of the trees are filled with spurs which are stuck full of fruit clear to their junction with the trunk, which renders it an enormous producer. The peculiarity is not found in any other variety with which I am acquainted, and is enough of itself alone by which to identify the variety." There are also differences in the vegetation and the fruit itself, although you have to be willing to look closely. Writing ten years after Gilbert, S. A. Beach takes nearly two pages to describe the differences in fruit and twigs. English Russet can be "narrow towards the eye [the basin end]" and it also "takes a good shine." It has a tendency to be a bit more conic than Golden Russet of Western New York. That English Russet takes a good shine suggests that the surface is a bit less like unglazed pottery and more like old, buffed leather.

In the *Report of the Commissioner of Agriculture for the year 1875*, F. R. Elliott took his own crack at English Russet:

> [S]ize medium; form roundish, slightly conical, and also slightly angular; very regular; color dull greenish-yellow, mostly covered with russet, which is thickest near the stem; calyx small, closed, with pointed reflex segments; basin open, round, regular, of moderate depth; stem small, short, about even with the surface of the fruit; cavity narrow, pretty deep; flesh, yellowish-white, firm, crisp, yet tender, with a pleasant mild subacid flavor, classing it as "good" to "very good." Its eating period varies from October to April or May, according to climate. It is regarded as one of the best of keepers, and very productive. The tree grows very upright, forming what is termed an upright, rounded head. The young wood is of medium size, smooth, and of a reddish-brown.

That should make it easy to tell them apart. Or not. I did go back to Ira Proctor's orchard some years ago and collected some fruit, including his russet. I confess that I was not able to determine which one he had, although I think it might have been English Russet. The various "Goldens" look so similar. I reread my journal entry the other day from thirty-five years ago when I went to visit Ira and we cut scionwood together. Back then I hardly knew a Northern Spy from a Red Delicious, but still I knew there was something peculiar about Ira's russet. In my journal, I'd written, "The Russet tree looked very different from the Golden russets at The Apple Farm and Ira also mentioned the difference." What was the difference he mentioned? Too bad I didn't write it down in my journal that night. Too bad I didn't ask him if he knew the full name. Too bad I didn't ask him lots of things. Maybe the reason the russets looked so different back then, even to a novice like me, was that Ira didn't have the more typical Golden Russet. Maybe he had an English Russet.

Is your apple conic, partly russeted, partly green and keeps until just after New Year's? Bullock is another important russet that has also been called Golden Russet. S. W. Cole called it Hunt Russet. Beach calls it Bullock. Ziba Gilbert calls it American Golden Russet. They may be referring to the same apple. Like English Russet, American Golden Russet has a conic shape. It's not conic like Chenango Strawberry or Hudson's Golden Gem or even Yellow Bellflower, but conic nonetheless. We're talking subtleties here. Its skin is mottled—not solid russet. It does not keep nearly as long as the two previously described Golden Russets—only into January would be typical. It originated in New Jersey in the colonial days, and the flavor is excellent.

Hunt Russet is usually reddish blushed and is exceedingly rare

Does your russet have a reddish blush on the sunny side? This brings us to the Hunt Russet, originating on the Hunt farm in Concord, Massachusetts, in the mid 1700's. Hunt's synonyms also include—you guessed it—Golden Russet, American Golden Russet and Bullock, among others. While its shape and size resemble the other Golden Russets, Hunt Russet's distinguishing feature is the red-russet blush. None of the other Golden Russets have much real color, just a lot of muddy tans and yellows with some green peeking through. While the flavor of Hunt Russet is not as good as Golden Russet of Western New York, it keeps a long time. Eugene Carpovich, of Mt. Vernon, Maine, told me his Hunts were still useable after a year in his root cellar. Carpovich was an early proponent of what he called "natural orcharding." When I asked if I could get scionwood one year while he was away, he sent me a postcard with a small map and this note: "The Hunt Russet is at the south end of the orchard. A very big 100-years-old hollow-trunk tree. You can not miss it." He died a few years later in 2000 at the age of ninety-four.

Is your russet only about two inches in diameter, about the size of a large crab, with dark, grayish-russet skin? Pomme Grise is a flavorful, small, round russet. Beach suggests that it might be synonymous with the ancient French apple, Reinette Grise, dating it to about 1650. Presumably the apple would have then been brought over to the New World from Europe. Does that make Reinette and Pomme synonyms? Having read Hogg's description of the Reinette Grise, I'm skeptical. I'll vote Canadian. In any event, Pomme Grise is smaller than most other russets, slightly oblate, and tends to be darker in color, some would even say gray. Hence the name. It is still delicious past the first of the year but is not a long keeper. When Cammy was teaching agriculture to third graders in Roxbury, Massachusetts, many years ago, she helped the students stratify and plant their own apple seedlings. Because each seedling would be new and unique, the kids all got to name their new variety *in vitro*, so to speak. The naming might have been a bit premature since they were still seeds, but naming the apple was part of the fun. One little boy explained to Cammy the name he had selected: he calculated that he would be thirteen when his new apple tree fruited for the first time, and because the seed was from a Pomme Grise apple, he decided to name his new tree "PG-13."

Pomme Grise is smaller than the other russets

This brings us to the task of identifying six rare Maine russets, four of which Z. A. Gilbert mentions in his article but only two of which he thought he had ever seen. Fortunately all six have some defining characteristic that sets them apart from the rest. I've tried to straighten them out.

What if you find your russet in western Maine, and the fruit is medium-sized, resembling all the various Golden Russets but often includes faint-red radiating squiggles and white dots across the surface of the fruit? What if it keeps magnificently in the root cellar until well into spring? You may be looking at Winn Russet from Oxford County. Gilbert had heard of "Winn," but assumed that he had never seen it. Winn Russet is "only mentioned here as showing that possibly some of the russets now found in the State may be the native apples named by those fruit growers who have preceded us and have passed away, leaving only a few stray sprays of the records of their work."

JPB & Nat Peirce with a Winn Russet at the Common Ground Fair, 2017

I first heard about Winn Russet in the late nineties when a Fedco customer named Tara Peirce contacted me. Tara and others had a connection to the Winn farm, although I didn't learn the full story until many years later when I sat down with her husband, Nat, in the Hiram, Maine, offices of GrandyOats Granola one rainy April morning in 2018.

Hearing from Tara, however, twenty years earlier had piqued my interest. From Bradford, I learned that Winn was thought to be a seedling of Roxbury Russet, that John Winn and his family came up from Woburn, Massachusetts, and that they were early settlers of the Oxford County town of Sweden, not far from the New Hampshire border. I also had Bradford's rudimentary description of the fruit itself, although I knew by then that the russets are tricky to tell apart. I also knew that at some point I'd want to head to Sweden to see it myself. In the meanwhile, like so many other Maine apples, Winn floated away into a holding pattern in a far-off place in my brain.

A seemingly random road sign may hold the key

But not too far off. Winn Russet had a way of popping back into my consciousness anytime I drove down to Boston. I couldn't help but notice the bright green highway sign along that eight-lane stretch of Route 128 just north of the city. "Winn St. Woburn Burlington." It's unclear what town you're passing through as you zoom along, but with a name like Winn, I thought there must be some connection between the sign and the apple. I remembered that the Winns came from somewhere in Massachusetts. One day I checked the entry. Yes, they were from Woburn. I figured the sign must also be in Woburn and, yes, it is. So the Woburn Winns escaped from what would become suburbia, taking Roxbury Russet seeds with them, and headed up to Sweden in Oxford County just before 1800. Winn Street still connects Woburn Common with Cambridge Street in Burlington. Sweden's first settler, Samuel Nevers, came from Burlington. The pieces begin to fit together. Somewhere along the three miles of Winn Street must have been the original Winn farm whose orchard included the Roxbury Russet trees that bore the apples with the seed inside that John Winn and his family pocketed and took north and planted on their new farm. It was from one of those Roxbury Russet seeds that the first Winn Russet tree sprouted two hundred years ago.

The first Winn Russet crop from the tree at MOFGA, 2017

The Winns did well in Sweden. They built a large and prosperous farm, which became the focal point of a small vibrant community in the north part of town, not far from Lovell and Waterford and just below what is still known as Winn[s] Hill. They built a huge barn, a blacksmith shop, a cobbler's shop and even a store. They planted orchards and shipped much of the fruit to England. They dug the clay that became the bricks for the stately homes in Lovell. They thrived for generations. And there in the barnyard the Winn Russet tree grew to be large and productive.

But, as was the case with much of Maine, Sweden lost population after the Civil War. Some were buried down South. Others simply headed west. Over time, the web of busy roads below Winn Hill became unmaintained roads, then trails, and eventually paths in the woods. By the century's end, neither Bradford nor Gilbert had the time or inclination to venture off to Winn Hill to investigate. According to Bradford, the original tree was still standing in 1846. That was all he knew. The long beautiful Winn farm house fell into disrepair. By the 1940's, the last Winn descendant was gone. The massive barn collapsed and was burned. What little was left of the house became a porcupine hotel. But the tree remained.

Two hundred years after the first Winns migrated north, in one of life's inexplicable coincidences, another Winn found his way to Sweden. Nat Peirce grew up in Scarborough, Maine, but his family's roots were in Lexington, West Cambridge, and Woburn. Nat was a Winn. He and Christy Berry landed in Bridgton. Although Bridgton is in Cumberland County, it borders Sweden to the south. It's a busy tourist town sandwiched between Long Lake and Highland Lake, not far from Sebago. Nat and Christy started a small bakery and cafe called The Bountiful Berry. Christy attended the New Hampshire Institute of Therapeutic Arts next door. The director of the institute was a fellow named Dr. Robert Berube. When

Greg Marston and the Winn Russet tree, September 2005

the doctor learned that Nat and Christy were looking for a place out of town to spend the summer and have a garden, he suggested they check out an old abandoned farm off the Haskell Hill road. As Nat's cousin, Meredith Winn, said to me, "He felt like he was intuitively led to this piece of land."

Nat and Christy contacted the owner, who lived overseas, and made arrangements to farm-sit the place. They drove their blue school bus out to the site and parked it in the small clearing that remained, surrounded by the barn's foundation, only a dozen yards from a huge old apple tree. For the next few years, they spent the warmer months on the farm. They explored the old roads and cellar holes, dismantled the last of the various buildings, had a garden and ate the apples. Later Christy moved on; Tara entered Nat's life and she shared the world below Winn Hill.

One of the regular customers at The Bountiful Berry was a fellow named Greg Marston. Greg was a cheerful-looking guy with a toothy grin, a bristly beard and John Denver wire-rimmed glasses. Greg and Nat became friends and Greg also began to spend time out on the old Winn farm. According to Nat, it was Greg who ID'ed the apple. Perhaps he had found George Stilphen's reprint of Bradford's thesis and put it all together. He may have even met George who lived out that way. Greg told Nat and Tara. Then, Tara contacted me through Fedco. Nat and Tara eventually found their own place, started a family and only occasionally returned to the old farm. No one has lived there since.

And that's how it stood until one Friday afternoon in late September 2004 when I was invited to set up my display of apples at the Bridgton Farmers' Market. Greg Marston came to do some shopping at the market that day and at last we met. The apple display really is a magnet. The following year, Greg and I went to visit the huge, old tree in the small hillside clearing below Winn Hill in Sweden. I drove over with writer Michael Sanders and photographer Russ French. Michael was doing a story for *Downeast* magazine, and Russ was taking the photos. The three of us met up with Greg at the end of the long, rutted dirt driveway. We left our cars and walked in. Greg had been up since before dawn. "I couldn't sleep," he confessed. We were all excited. The farm looked as though it had been abandoned for many years. All the original buildings were gone. There were several old apple trees, overgrown and untended for decades. The largest and oldest had three apples in the upper branches. By climbing a few feet up into the tree, the

pole picker was just long enough. Two of the apples were in poor shape but the third was perfect. Medium-sized, roundish and russet, with faint red radiating stripes, sometimes a glowing orange-red blush and a sprinkling of small white spots, just as Bradford said. Was it the Winn Russet? It had to be. I labelled a branch with a small piece of ribbon. I returned on March 29, six months later to the day, snipped a couple of twigs of scionwood and that was it. It was saved.

If the russet you're attempting to identify "is dotted all over with rough, projecting dots," you might be looking at a Windham Russet. Sometimes I drive a hundred miles to find an old tree. It's almost exactly a hundred from our farm to Winn Hill. Fortunately, sometimes the old apple trees are right around the corner. I found one of my favorite russets just eight miles from home. Think globally, act locally. The Morse Road is one of the oldest roads in Palermo. The Richard Turner farm on the Morse Road is the first farm you come to when you turn off Level Hill Road. In the early 2000's the Turners still occupied the house, and the old barn directly across the road was in good shape. The barn is now gone, and the house is set to be pushed down and burned sometime soon. But the small orchard and the half dozen ancient trees around the house still remain. These include a Duchess, a Northern Spy and a very old russet. Like the various Golden Russets, this one is small-medium in size and roundish. Unlike any other russet I've found in Maine, the surface of the fruit is covered with dozens of small, raised, textured bumps.

The Windham Russet tree in Palermo

The dessert flavor is delicious. I collected scionwood and topworked it onto one of our trees. Assuming that I would never identify it, I simply called it "Morse Road Russet." Then I found Gilbert's article. He wrote about receiving a russet from D. J. Briggs of South Turner, "which is dotted all over with rough, projecting dots. Flesh firm, does not wither, crisp, juicy…of a best quality for their season." It's perfect on about November 1. We took fruit to the Franklin County Cider Days apple tasting in 2014 and, up against some rather heavy hitters, it won! Gilbert calls the apple Windham Russet and gives its origin as Massachusetts. But since there is no Windham, Massachusetts, maybe it's from Windham, Maine, Windham, New Hampshire, or maybe it's named after someone named Windham. We may never know.

Kennebec Russet is light yellow, netted and patched with russet

If your russet is slightly elongated, roundish-oval in shape, if the coloring is a patchwork of light-yellow and golden-tan russet, if some of that russet looks like fishnet hanging over the railing of a Cape Cod porch, if it has a scattering of small brown dots, and if the tree is located in central Maine, you might have found the elusive Kennebec Russet. Gilbert describes the apple in his article, though he didn't think he'd ever seen it.

I assumed I'd never see it either, until the fall of 2011 when Carol Godfrey contacted me. Carol is an old friend and fellow Fedco employee. She told me about an old apple tree not far from her house, on the outskirts of Waterville, deep in the heart of Kennebec County. Despite being near the Interstate, her neighborhood still has a rural feel. The nearest access was about a mile north, in the center of town. Rumors had it, however, that a new interchange was going in less than a quarter mile from the tree. New development was in the planning stage. Not only would traffic increase, the ancient tree looked as though it was destined to be buried under an off-ramp or maybe a Pizza Hut.

I didn't have time to visit the tree that fall, but I did head over in February and cut some scionwood. I had no idea what the variety was, but it was a beautiful, old tree. I grafted it at our place in the spring. At least now it was saved. I collected fruit early the following October and began to attempt an ID.

The fruit was round to slightly oval. The skin was yellow and mostly covered with a netting and patchwork of tan-golden russet. It had a long stem and practically no basin. The flesh had very little acid so it may be a good addition to cider. It cooked up quickly into an excellent sauce, skins and all. In fact the skins turned out to be delicious. The fruit was in good shape only until about mid-November. It was not a keeper. This was a nice apple and one that deserved to be saved.

I knew I hadn't seen it before. It was certainly not Roxbury or one of the various Goldens. Wrong shape, wrong color, and it didn't keep. The tree was very old. This all suggested something local. Waterville shares its southern boundary with Sidney. Sidney was a focal point of orcharding even before the Civil War. Maybe it would be one of those old Sidney apples. Could it be King Sweet? No. King Sweet is a synonym for Summer Sweeting, another of Earland Goodhue's favorites, a small dessert apple, one of the first of the season. Earland liked to tell me about collecting pocketfuls of Summer Sweetings as a boy, eating a few and having apple fights with the rest. They're about the size of a golf ball. Perfect for throwing at your friends. Summer Sweeting does occasionally have a bit of russet, but it is primarily a yellow fruit with a small, pinkish blush. Summer Sweeting is similar to the Massachusetts apple, Hightop Sweet. But don't be confused. They are two different apples. This new Waterville apple was too large, and its season too late. It wasn't Summer Sweeting or Hightop Sweet.

Summer Sweeting is small, conic and mostly yellow: photo by David Grima

I also considered Washington Sweet, another of Earland's Sidney discoveries. It's a large roundish, multi-colored, low-acid russet with a large deep basin. At a glance the two apples resemble one another, but only at a glance. This new Waterville russet has a small basin and no blush of any sort. It's not Washington Sweet.

It was about this time that I came across Ziba Gilbert's article for the nth time. I read it and reread it, and still re-read it every now and then. Here's what he wrote about the Kennebec Russet:

> [Kennebec Russet is] a name found in the Transactions of the old Pomological Society. None of our fruit growers now know anything about it, and we never find it at our exhibitions. I have no doubt that this apple is one of those stray varieties popping up at our exhibitions occasionally, usually labelled "Golden Russet" because it is tinted with yellow and nobody knows any other name for it. I have frequently noted a variety coming from the central part of the State of a beautiful yellow shade and covered all over with russet—a golden russet, in fact, though not true to the record. I have thought it probable that this is the apple referred to in those old records. Alas! who is there left today to set us aright!

So maybe this was Kennebec Russet. I pulled out my Downing, the 1886 edition. There it was: "Fruit medium, roundish conical, yellow, partly netted with russet, and sprinkled with brown dots. Stalk long, slender. Calyx small, closed. Flesh moderately tender, juicy, brisk subacid. Good. November, December." I think I've found it! If I'm lucky enough to go to orchard heaven, I'll tell Ziba.

I collected scionwood, and I was pretty sure we had the name. It was time to develop a plan to save the old tree itself. In May 2014 I heard from Carol again. Construction appeared to be imminent. A comment period would conclude in a few days. Could I write something ASAP? I sprang into action and submitted comments to the Maine Department of Transportation. I also wrote a letter to the editor of the *Waterville*

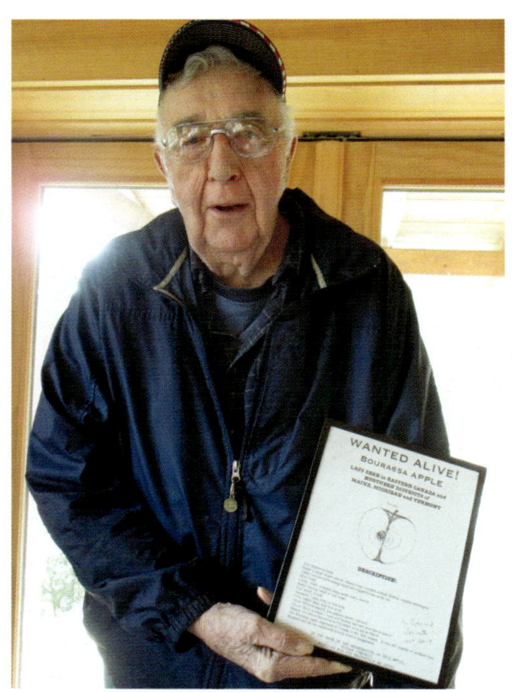

John Scates with the Bourassa wanted poster 2015

*Morning Sentinel.* A day or two after the letter appeared, I began to receive emails from the property owners and town officials. While it was unclear who actually owned the land on which the Kennebec Russet tree stood, all parties agreed that the tree should be saved. We were given permission to prune and care for the tree. Construction on the interchange went forward, and is now complete. We have done some initial pruning and clearing around the tree. It had a huge crop in 2015. The first young tree we grafted at our place died. Fortunately we have another small tree in the Maine Heritage Orchard. We also grafted more at home. We've disseminated scionwood through the Fedco Tree catalog; at some point we'll offer trees for sale. I'll go this winter, complete the pruning, collect more scionwood, and graft more trees this coming spring. We don't want to lose this apple.

What if your russet resembles Baldwin in nearly every way except that it is partly to largely covered with russet? If that's the case, you've found the Red Russet. As you may recall, I found one years ago in southern Maine. Many tens of thousands of Baldwin trees have been planted in the past two hundred years. Ten years before his Russet article, Ziba Gilbert wrote about the Red Russet. Evidently the first documented Red Russet appeared in an orchard in Hampton, New Hampshire. S. W. Cole believed that the apple was a cross between Baldwin and Roxbury Russet. Others assumed that it was a sport of Baldwin. Reports of Red Russets were sent to Gilbert over the years. Some growers loved the apples, while others did not. Why? Each russeted Baldwin sport is going to be different. Buds will mutate in their own unique ways. Each resulting sport will be its own individual. Some will have good flavor, while others will not. Is the Red Russet, as Cole believed, a cross between Baldwin and Roxbury Russet? I'd say no.

What if your russet is somewhat oval, a little bit conic, red and orange, excellent fresh eating and from somewhere across northern New York, New England or southern Canada? There are a number of russets that Gilbert missed altogether in his investigations, at least three of which are of interest. The first is Bourassa. One of our neighbors here in Palermo, John Scates, is a Bourassa. He pronounces it "boar-a-SAW." John and his wife Jeanette are local, old-timers. They've both been instrumental in teaching me about the apple culture in Palermo over the years. John's grandfather, Felix Bourassa, grew up in St. Nicholas, just south of Québec City and came down the Kennebec on one of the massive annual log drives sometime around 1900. The drive ended in Fairfield. Felix looked around and decided central Maine was a pretty good place. He met Addy King, got married and never went back north.

Although mentioned in several of the old fruit books, the origin of Bourassa remains a mystery. It was grown early on in northern New England and the Canadian lands that wrap around Maine. By 1750 the apple was popular in Montréal. Perhaps it was from somewhere near St. Nicholas. The apple's shape is oval and conic and slightly ribbed. The skin is orange, red and russet. It is unlike any of the other eastern russets. It is an excellent dessert fruit. Maybe Felix had a few of the apples tucked into his oil-skin coat as he danced across the tree-length logs, picking his way down the Kennebec a hundred and twenty years ago.

Claude Jolicoeur with two of his introductions: Ursa (L) and Banane Amere; in the background, an old Red Astrachan tree, November 2018

In 2010, friend, cider maker and author Claude Jolicoeur of Québec began a quest to locate the Bourassa apple. He was inspired by Paul-Louis Martin, author of *Les fruits du Québec, Histoire et traditions des douceurs de la table*. This was the same Martin with the beautiful plum orchards I visited long ago with Garfield King. That same fall Regina Grabrovac sent me three russets for identification from three trees in one yard in Dennysville, Maine, not far from the Canadian border in Washington County. It was Regina's husband Paul Molyneaux who introduced me to Beauties of Wellington. Although her timing was impeccable, Regina had stuck all three russets in one bag. After some examination, I became convinced that two of the apples were Bourassa. I contacted Claude and put him in touch with Regina. I also asked Regina to send me scionwood. Unfortunately, she wasn't sure which of the three trees belonged to which of the three russets.

Fruit identification is about patience. Like the fortune cookie says, "A handful of patience is worth a basketful of brains." So what's another year here or there? In 2011 Regina sent more fruit, this time labeled according to tree. It appeared as though two of the three trees were the same apple, probably Bourassa. The third apple was a different russet, one I have still not identified, possibly from seed. That winter Regina cut scionwood, and we grafted it for the Maine Heritage Orchard. On October 28, 2013, Todd, Abbey and I made a pilgrimage to Dennysville to see the fruit on the trees. They are old. Three years later, we offered trees in the Fedco Trees catalog.

What if your apple is quite large, roundish-oblate and entirely covered with grayish-tan russet? No green and no red and no orange.

Call me. You may have an apple we've been trying to ID for thirty-five years. Cynthia "Snee" Anthony has lived in Waldo County as long as I. We met Morris Dancing back in our twenties. A group of us danced to traditional English music throughout the summer, wearing bells and ribbons. It was part of the continuum of thousands of years of Anglo-culture, like Wassail and apples and cider. Snee caught the apple bug about the same time I did. She spent a great deal of time exploring the outskirts of Belfast, including Searsmont, Northport and Lincolnville. She has a keen eye. Just a couple of years ago she showed up at our house with an apple from Northport, hoping that I could identify it. She had found a Kilham Hill, a rare apple these days, but one that, according to Bradford, was "found in almost every orchard in that County [Waldo] and there considered a good fruit and profitable."

Snee started her heirloom apple collection before I planted a single tree. She introduced me to Wolf River and Northern Spy and Maiden's Blush. She brought me old books. She loaned me her copy of Cole's *The American Fruit Book* on several occasions as long as I promised not to fold the binding back as I read it. One day she arrived at our place with a brand-new pointed apple ladder she had made. I had never seen such a thing. I loved it. I had to have one. I was so excited, I could think of nothing else. I began to plan mine that same afternoon. Within the next week I had made my own.

Pruning the Kilham Hill tree in Northport; it laid down years ago but kept on going strong

Snee introduced me to what became one of my favorite summer apples. Coles Quince is not a quince (*Cydonia oblonga*), although it does visually resemble one. It's medium-sized, roundish and blocky in shape, and ribbed and furrowed, though not as deeply as Calville Blanc d'Hiver. The coloring is yellow with some russet in the cavity and small, scattered russet dots across the surface. There is rarely a faint blush of pink. The apple originated in Cornish, in southern Maine's York County, introduced by Captain Henry Cole. It is one of Maine's most widely disseminated apples, perhaps because Cole's son, S. W. Cole, featured the apple in 1849 in his new book just a few years after its introduction. It was a brand-new apple in a brand new book.

Commercial growers generally avoid summer apples. They have tender skin and therefore damage easily during picking. Most don't keep more than a few days which makes them difficult to transport or put on the shelf. Most ripen unevenly. Pickers can't go in and harvest them all at once; that means multiple pickings. Multiple pickings get expensive. Customers are also not thinking about apples in August, so there's not much of a market for them.

Cole's Quince is ribbed and yellow and resembles a quince; it also resembles Calville Blanc d'Hiver but ripens much earlier

Everything that the commercial growers hate about summer apples makes them great for the home orchardist. Coles Quince ripens over a two to three week period in late August. It doesn't rot on the tree. When it drops, it sits on the grass for three or four days before becoming mushy, giving you time to harvest them. We let our Coles Quince ripen on the tree and drop. I go out barefooted in the morning and check beneath the tree in the wet grass for three or four freshly fallen fruit. No need to pull any off the tree. Coles are best when they're perfectly ripe, and when they're perfectly ripe, they drop.

I collect the fruit like you'd collect eggs in the chicken coop or a few tomatoes from the garden, then head back to the house, cut them up on the spot and make them into sauce. The timing couldn't be better. The late August morning temperature is just cool enough to warrant a quick fire in the cookstove. It's the time of year to start eating oatmeal again. Nothing is better with oatmeal than fresh Cole's Quince apple sauce. You can also save them up for a few days and use them in a pie. The pies are excellent. They're also quite good fresh.

Interestingly enough, I've never located an old Cole's tree. I asked Snee recently if she remembered where she found the scionwood she gave me. She didn't. Perhaps the variety is not long-lived, and the old trees are gone. Or maybe, because it's a summer apple, the fruit has already melted into the ground or been cooked up in the saucepan by the time I'm out there searching the old orchards in the fall.

Snee also discovered that large roundish-oblate, grayish-tan russet I've now been trying to identify for thirty-five years. She called it "Route 52" because the ancient tree was right on Route 52 in Northport, not far from the Lincolnville line. She also called it "Haunted House" because the tree was near a locally notorious, picturesque and haunted old farm house. She didn't know the apple's real name and never spent much time trying to identify it. She did get me scionwood, and I grafted it onto one of our trees. Meanwhile the haunted house was finally burned at the stake. The tree also died and returned to dust. For

The "haunted house" in Northport, site of the Lincolnville Russet tree; house and tree now both gone: hand colored photograph by Liv Kristen Robinson

many years our one branch was probably the only bit left in the world. I've spent a lot of time staring at the Route 52 apple. I decided to give it my own provisional name. It deserved something better than Route 52. "Haunted House" is fun but never sounded like an apple to me. "Snee's Russet" would have been a good one, but I chose "Lincolnville Russet" despite the fact that the apple isn't in Lincolnville at all. It's over the line in Northport. The apple is distinctive. It is larger than any of the other old russets here in Maine. It has no blush. It does have slightly raised bumps, although nothing like a Windham. It is rather oblate. If only Audubon had been painting apples in 1830 instead of all those birds.

At the time I was researching Fletcher Sweet, I made contact with a number of Lincolnville and Northport residents, one of whom was Isabel Morse Maresh. Isabel was a wealth of knowledge about the old days in Waldo County. She attempted to guide me towards those who might have known the apple, one of whom was Carl Dean. I never was able to meet Mr. Dean. Apparently he had a russet in Lincolnville that he called "Banana Russet." According to Lincolnville historian Diane O'Brien, a Banana Russet was entered in the local Tranquility Grange Agricultural Fair in the 1920's. The grower was Allen Miller, who lived not far from the Haunted House, at the corner of the Slab City Road and what is now Route 52. No Banana Russet appears in any of the literature that I've been able to find. There is the famous Winter Banana, but that apple is bright yellow with a rosy-pink blush and no russet.

Lincolnville Russet is large & entirely russeted; might it be the Bung Hole Russet?

Although I believe that the Lincolnville Russet is most likely a

Cynthia "Snee" Anthony, at the 2014 Maine Heritage Orchard planting day

Waldo County selection, I have scoured Bailey and Bradford many times, looking for its true identity. Zabergau Reinette and Pumpkin Russet would be two possibilities. Both are large russets. Zabergau, however, doesn't appear in either Beach or Bradford. As far as I can tell, it probably didn't get over to the states from Germany where it originated until well into the twentieth century. Pumpkin Russet originated in New England and is large and oblate. That all fits. But the descriptions all say it is a patchy, netted russet, not solid russet like our Lincolnville apple. We have an old russetted variety in central and southwestern Maine known locally as Pumpkin Sweet, almost certainly a synonym for Pumpkin Russet. It bears no resemblance to Lincolnville Russet. So we'll count out Zabergau and Pumpkin Russet.

Bradford mentions two varieties that have some promise. Monstrous Russet is one candidate. It was offered for sale by Taber's nursery in Vassalboro, fifty miles or so from the Haunted House, but there is no description. The other could be the Winthrop apple called Jerry Brown, described by Bradford as a "fall apple, large, oval, russet brown." We know that apple varieties traveled, but Winthrop is even farther from Lincolnville than Vassalboro. Also, oval and oblate are two different shapes. For now we'll say it might be Allen Miller's old Banana Russet from Lincolnville or Northport, near the rocky shores of Penobscot Bay in the southeast corner of Waldo County, Maine.

One final thought: maybe it's actually Gloria and Tony Poulin's "Bung Hole Russet" all the way down from Parkman up in Piscataquis county. Wouldn't that be cool. Crank up the press!

Called Pumpkin Sweet in central and southern Maine and known farther south as Pumpkin Russet; partly russeted and largely blushed brilliant orange and red

Roxbury Russet tree spared long ago when the farm became a motel lawn near the beach in Ogunquit

Regina Grabrovac in an unidentified Russet tree—probably a seedling—on a quiet road in Addison

# Thirty-Two

# Kavanagh

It is one of those cases where the art of the reasoner should be used rather for the sifting of details than for the acquiring of fresh evidence.
Sherlock Holmes, *Silver Blaze*, p. 335

What if the surface of the fruit before you is a patchwork of shiny green and tan russet resembling the skin of a Roxbury Russet? But unlike Roxbury, your apple is rather large, bulbous and conic. What if you found the tree in one of the coastal Maine communities of Lincoln or Knox Counties? What if you looked at the side of the apple with the stem pointing up and the shape reminded you of a cat's head? You've found a Kavanagh.

The coastal area of Maine that later became the Lincoln County towns of Newcastle, Nobleboro, and Damariscotta was all part of the Pemaquid Patent, granted by the Plymouth Council in 1631 to Robert Aldsworth and Gyles Elbridge, both merchants from Bristol, England. The area was perfectly suited to become one of the most important in Maine.

The Pemaquid peninsula borders the Damariscotta River (pronounced dam-rah-SCOT-tah). The river begins nineteen miles upstream from the ocean at the foot of Vaughan's Pond, which is now known as Damariscotta Lake. At the intersection of the lake and the river there are multiple waterfalls. The falls drop over forty feet, empty into the Great Salt Bay and then into the river itself. This was the perfect location for water-powered mills—lots of them. The first grist mill was built in 1730. Many mills followed, including additional grist mills, fulling mills, lumber mills, an iron foundry, a match factory, a leather board factory, and electricity generators all powered by the water from the huge lake and the immense watershed beyond. Along the river below the mills, above what is now Damariscotta's and Newcastle's downtowns, were ten shipyards and a brick factory that produced the bricks for Boston's Back Bay.

The long river downstream protected the mills and the town from severe coastal weather, while also providing easy access to Portland, Boston, and beyond. Over time, Damariscotta Mills became one of the most important ship producing locations in Maine and thus in the world. The coastal waters were the Maine Turnpike of the time. Every minute of every day, ships would pass by, transporting hay, lumber, granite, potash and much more. Boys went to sea when they were ten. Teenagers became captains when they were only nineteen.

The Kavanagh House in Damariscotta Mills

Fifty years after the opening of the first Damariscotta grist mill, events in Ireland and the U.S. were destined to have a profound impact on the future of the apples of Lincoln County, Maine. The American revolution was desperate for allies. For a brief few decades, pragmatism overcame hatred and discrimination. Catholics and native people in New England were pitched a bit of tolerance. "Papists" were allowed to partake in the American dream. Meanwhile, across the Atlantic, the younger sons of prosperous Catholics were looking for opportunities abroad. Ireland had run out of room. In 1784, two young Irish Catholic friends left for Boston. James Kavanagh (1756-1828) and Matthew Cottrill (1764-1828) immigrated that year from Inistioge in County Kilkenny. Kavanagh was the son of a farming family in nearby Cluen. Cottrill was the son of a merchant in Inistioge. The two of them were in search of opportunity.

The Cottrill House in Damariscotta

After four years in Boston learning the ropes from the small local Irish Catholic merchant community, they headed north for Newcastle where there was already a small Catholic presence. There they opened a "general mercantile store" in 1788. The lumber trade was booming. Nobleboro had a large, double saw mill at the falls and was already a busy shipping port. Kavanagh and Cottrill's services were immediately in demand. They became wealthy.

In 1795 they formed a partnership and purchased Vaughan's Mills at the falls. They bought two thousand acres of prime timberland surrounding Damariscotta Lake. They built a fulling mill and then a shipyard. They began to build big ships, completing twenty-five by 1800, eight of which they kept for themselves. Their ships took vast amounts of lumber primarily to Ireland, and returned home each excursion with a few more countrymen. They both built large, expensive, Federal-style homes and surrounded themselves with one of the largest Catholic communities in New England.

Kavanagh and Cottrill made a great match. They were like the country mouse and city mouse. Kavanagh, with his farming background, lived at the base of the lake where he farmed and managed the mills. Cottrill managed their overseas trading and the shipyard while living in town.

Over time they built the first bridge to join Newcastle and Damariscotta. That bridge is still the defining feature of the downtown. Without it, there would be no Damariscotta. They built the road connecting the bridge to the mills north of town. They even built the first alewife fish ladder. (Damariscotta is an Americanization of the Algonquian word Madamescontee which means something like "place of an abundance of alewives.")

They built what is now the oldest Catholic church in Maine. St. Patrick's opened in 1808, in Newcastle, just a short walk from Kavanagh's estate. Constructed with Damariscotta bricks and Irish limestone, it features one of the last bells cast by Paul Revere. At one time, the French claimed to own everything east of the Kennebec River, more than two-thirds of the state. I always assumed that the oldest Catholic church in Maine would obviously have been French. But I was wrong. Even today many in the area remain grateful for all of James Kavanagh's and Matthew Cottrill's accomplishments. But greater still than the mills and the bridges and all those ships, there was an apple.

In the early nineties I met an artist named Natalie Wriggins out on Whitehead Island, off Spruce Head. She became a friend over the next few summers. Natalie lived in Damariscotta. A year or two after we met, she brought her sister and brother-in-law, Norman Hochella, over to visit our farm. Norman was interested in apples, and Natalie thought we'd enjoy one another. During our conversation, Norman told me about an apple variety I might be interested in, one that originated in Nobleboro. The variety was called Kavanagh. He also told me about a retired University of Maine professor named George Dow who was Director Emeritus of the Maine Agricultural Research Station and a local historian. George had researched and written several articles about the Kavanagh apple. Norman thought that George and I should meet.

George Dow "opening" his annual Christmas apple basket: photo courtesy of the Dow family

George and I met on a spring-like day in the winter of 1995 when he took me to what he thought at the time was the last living Kavanagh apple tree, a gigantic hulk of a stump on the West Neck Road in Nobleboro. The top two-thirds of the tree had long ago broken off and rotted away, but the base was still peppered here and there with a few living sprouts. We collected scionwood, grafted fifty trees that spring, and then sold them two years later in the 1997 Fedco Trees catalog. That may have been the first time Kavanagh ever appeared for sale in any catalog anywhere.

The Kavanagh fruit is highly unusual in appearance. Sometimes it's called a Cathead. Once you see it a couple of times, you should be able to identify it. It is quite large and distinctively conic, wide at the stem end and tapering to a small blossom end. The tough skin is lime green, turning somewhat yellow as it ripens. It's about half to two-thirds covered with brownish russet. The stem is medium in length and thickness. The flesh is white, firm and juicy. It is tart, pretty tasty and ripens in October. It grows to be a huge, long-lived tree.

Kavanagh is also called "Cathead"

Some like Kavanagh for fresh eating. It is one of the best apples for sauce. It cooks up into a foamy, frothy, thick, smooth, tasty sauce in barely a few minutes. No need to remove the skins; they dissolve. It's also a treat fried: core and slice thickly with the skin on. Fry in pork fat. Add water as necessary till soft, then add "a few dollops" of molasses. Serve with biscuits.

During the next few years I collected Kavanaghs from the old Nobleboro tree and began to get to know the fruit. I found a second tree on the West Neck Road. Then I found another, near the James Kavanagh farm in Newcastle. All three were very old. The one in Newcastle was in the best shape. It bore a modest crop in 1996. At chest height, it was eight feet in circumference. The two trees in Nobleboro are only about forty feet apart by the side of the road. It's odd that we missed the second tree that day when George and I first went out together. The second one was in much better shape than the first, although I haven't seen either now in over fifteen years. The second tree also bore a medium crop in 1996, and I was able to collect and eat fruit from both.

Kavanagh is mostly conic but not always. These are are all from the same tree: photo by Abbey Verrier

From my journal September 28, 1996: "I'm eating one now as I write. The skin is tough. The flesh is white, firm and juicy. It is tart and is probably still not quite ripe, but it's pretty tasty and I think would make a good cooking apple." It does.

How the Kavanagh apple originated will likely remain a mystery forever. There are stories, like the one that appeared in a 1950 *Boston Post Magazine* article written by Grant Lee. An old Kavanagh tree was found to be bearing that year behind the home of Edward J. "Josh" Lincoln, Sr. in Newcastle. Everyone had thought the tree was dead. As far as anyone knew, no one had grafted new trees and it was assumed that the apple was gone forever.

Lee provides the following account of the apple's early history:

> At that time [c 1800] Maine had no apples, and one of the settlers brought in some apple seeds and planted them in a field. They were not very good apples, but they were all there were in the area, and the farmers were beginning to grow them extensively here.
>
> But James Kavanaugh [sic] remembered the fine big apples of his native land, which were far better than these scrubs, and so he sent one of his ships to Ireland for some seeds—or seedlings, and records are obscure on this point.
>
> The Irish apple flourished on the estate, and soon the farmers were coming to his door and asking the favor of a few seeds. In the years that followed, the seeds from Kavanaugh [sic]—the name of his estate—were planted in the fields for miles around, and soon supplanted all other kinds of apples. The scrubs were forgotten.

Edward J. "Josh" Lincoln Sr. cradling four beautiful Kavanaghs, Newcastle, 1950: courtesy of the family

Our writer could have used a lesson in apple pollination and genetics, but we'll let that pass. Early on, apples were planted from seed everywhere in Maine, including Lincoln County. Most of the early seedling apples were pressed into cider, so the "scrubs" were valued for the cider barrel. Occasionally one of these scrubs would prove to be of excellent quality. Those became the named American varieties, like the Kavanagh.

Over time, Lincoln County found itself populated with its own unique assortment of apple varieties. Each county of Maine—in some cases each town—had its own favorites. Some varieties were pretty common throughout the state, some were unique to one town or another, and others were regional. Every so often, someone locates a list from an orchard or a local fair. I'm always on the lookout for these lists. You can use them to predict which varieties you're likely to find in which general areas. That's particularly useful when doing ID's. You shouldn't be surprised to find a Baldwin from about the middle of the state and south, but even though it's a classic heirloom variety, you'll never find it north of Bangor.

One such Lincoln County list comes from William B. Hall's 1870 orchard on the East Neck Road in Nobleboro. The orchard contained sixty trees of eighteen varieties. When someone brings me an apple to identify from the Nobleboro area, before I do anything else, I always refresh my memory with his list. Knowing that these varieties were all growing in the same orchard at the same time is useful information. I have learned over the years, for example, that when I see Baldwin or Northern Spy, I should be on the lookout for Granite Beauty, Hurlbut and King of Tompkins County.

On a rather dark and stormy night, October 14, 2006, I packed up my Corolla and drove down to the Nobleboro Historical Society. With me I had as many of the William Hall apples as I could locate. George Dow would not be joining me. He had died four months earlier, just shy of age 101. I wish he could have been there. I hope he knew I was attempting to return the favor. Giving talks where people have shared their varieties and stories is a way of expressing gratitude. It's also fun to see the mix of people who want to hear about their apples. There was a good crowd that evening and I didn't mind driving home late that night in the pouring rain.

Could Kavanagh be a local synonym for the Irish apple called "Cathead" or "Catshead?" Catshead is mentioned in 1629 by John Parkinson as an English baking apple. It also appears in various other British texts. In a 1989 letter to George Dow, Richard Bradford quotes an unnamed source, probably apple explorer Wendall Mosher:

> The account of the Kavanagh is very interesting and so is the history of it. This Kavanagh apple is probably the old English variety "Catshead." I have English books that go back to 1790 but no "Kavanagh" is mentioned. James Cavanagh [sic] probably knew the apple as "Catshead" but probably local people did not know what it was and named it "Kavanagh." It originated in Great Britain in Elizabethan times and was the most common cooking apple in the British Isles in 1790—the description of it is identical down to the last detail.

It's possible that James Kavanagh imported the apple from Ireland. It could have easily traveled over on one of his ships. Between 1808 and the English blockade of Maine in 1812 he made several personal crossings to hire and bring back workers. On one of those trips, he could have easily stashed away a few grafted trees, a pocketful of seeds or a bundle of scionwood. In the case of seeds, the Maine apple would be its own unique variety. If he brought young trees or scionwood, they would be identical to the original Catshead. Some day we'll head back to England and Ireland. We'll attempt to track down the fruit. Apparently it's still being grown in the English West Midlands. Maybe we'll find out if our Kavanagh is a seedling or a grafted tree.

On this side of the Atlantic, I assumed that I'd only find Kavanagh trees close to Damariscotta. That assumption proved to be incorrect. Within a few years I began to hear of Kavanagh sightings elsewhere in the mid-coast area. The first was in Arrowsic at the home of Lisa Holley and Jim Arsenault; the second was in Blue Hill at a location I've yet to visit.

Jim Arsenault and Lisa Holley and their Kavanagh tree, 2018

Then, in the fall of 2005, Jack Kertesz appeared one day with an apple to show me. He was excited. He's forever prying another gem out of the landscape. This one was going to be a gift to me. I was flattered. He'd found a huge russeted apple next to the parking lot at R & D Automotive repair shop in Freeport. I took one look: "Jack, this is a Kavanagh," I said. "Darn," he said, or words to that effect. I think he was both impressed that I knew the apple instantly and disappointed that he hadn't found something new. Either way, it was an awesome find, and extended the range of the apple another twenty or thirty miles down the coast. This would be a total range of close to eighty miles by water, thirty west to Freeport from the mouth of the Damariscotta River and fifty east to Blue Hill.

The old Kavanagh tree shading the cars at R&D Automotive Repair

The locations of Kavanagh trees suggest that the apple found its way to new orchard sites by traveling up and down the coast, not overland. A schooner landed. Sailors and merchants got to chatting, one thing led to another, and next spring scions were dropped off.

Aside from needing pruning, the Freeport tree was in good shape. The DeGrandpre family, who own the repair shop, knew that they had a great tree and are not about to chop it down.

Undoubtedly this Kavanagh is the last remnant of an old farm orchard that might have numbered a dozen trees or so. The farm is gone, replaced by Main Street Freeport, Casco Bay Yacht Sales, Antonia's Pizzeria, Lincoln Canoe and Kayak, R & D Automotive, a few parking lots and a lot of cars. Fortunately, a guardian angel opted not to cut down the tree. It would have been so easy to do it. For years the Kavanaghs grew and ripened and dropped down on to the hoods of the Toyotas, Subarus, Fords and Chevys. Occasionally one of the mechanics would eat one. In more recent years an unofficial fan club has materialized. Most of the members don't know one another and don't even know the name of the apple. Still they drop by each fall to pick a few apples or collect them off the ground. When I'm in the area, I always stop in to see the tree. I take my quota of ten or twelve apples, say hello, and off I go.

I'd rather have a Kavanagh in the backyard than a Lexus any day!

In 2014 I was contacted by Mary Pols, a staff writer for the *Maine Sunday Telegram.* She was planning to write an article about the Kavanagh tree at R & D Automotive. I was pleased to help. Her long article came out in the October 12, 2014, *Sunday Telegram's* "Source" section. It included new information about Kavanagh and Cotrill, an extensive history of the property on which the Freeport tree stands and the people who planted and tended the tree over the generations.

There were no additional developments in the case of Kavanagh until the fall of 2015 when it seemed as though every apple tree in central Maine was covered with fruit. Calls and emails were pouring in. We were racing around every day visiting one tree after another. For many of the most senior trees, this could be their last hurrah. This could be our best—and maybe even our last—chance to see fruit before they moved on to orchard heaven. No time to waste!

Kavanagh tree by the side of the road in Union— note graft just above the snow

In early October we received an email from Nancy Stevick: "While bicycling, my husband discovered an ancient apple tree on Wotton's Mill Rd. in Union. The apples are delicious, a little tart, but great in apple crisp and sauce. The tree is not being tended and does not look very healthy. We thought you might be interested in preserving this variety. My husband Peter is going to bring you a couple of apples today when he picks up our CSA share in Belfast." Peter Stevick dropped off two apples, and, yes, they were Kavanaghs.

Although I've been through Union many times, I'd never been down the Wotton's Mill Road. Next day, Cammy and I had to go to Rockland, and so we planned a detour to check out the tree. Before we did we went to the famous Olson home in Cushing where Andrew Wyeth painted "Christina's World" in 1948. We mapped the orchard (sadly not in the painting) and did a recreation photo shoot we call Bunk's World. It was a rainy day. On the way home we found the Wotton's Mill Road and headed south. There was the tree, only a few feet from the side of the road, about fifteen miles inland from Nobleboro and Damariscotta. It was beautiful and very old.

A week or two later I was heading back towards Union to look at several other old trees when I found another inland Kavanagh, this one in Washington. I'd passed that ancient tree many times but had never stopped. In fact I didn't remember seeing apples on it before. It was also very old. I noticed the fruit and decided to stop. It was another Kavanagh, this one also about fifteen miles inland. Across the road is a dairy farm that had been vacant for a number of years. It was newly occupied by Conor MacDonald and his family, another of the thirty-somethings who are re-populating the farms of central Maine. It is a great sight. I introduced myself and told him about his apple tree. I wrote the name in Sharpie for him on the door frame of the milk room. It shouldn't fade for another few years.

The Kavanagh tree in Somerville; why are so many old trees only ten feet from the paved road? The roads were all dirt wagon trails when these trees were planted

These two new discoveries have meant a slight revision of the Kavanagh migration theory. It did come inland, after all. Looking at a map suggests that the apple was planted along the road from the coast up to Augusta. That would make sense. Many of the roads in that part of the state converge on Cushnoc, the early trading post on the Kennebec that later became Augusta, the state capitol. From there, it was a luxurious ride down the river to the world beyond.

Distinctive apples like Kavanagh or Black Oxford are relatively easy to find. This makes them great for putting together patterns of plantings. Plotting their locations allows us to hypothesize about travel routes. Isabella dalla Ragione has been doing a more sophisticated version of this work in Italy. She has been studying frescoes in old estates and monasteries. Long before wallpaper, wet plaster walls were painted by traveling artists. The subject matter was often the vegetables and fruit that were growing in the local gardens or hanging on the trees. As a result, the walls became records of what was growing on the estate. Where the fruit shows up can be plotted like a connect-the-dots drawing. The fruit varieties traveled with the people. By inventorying the subject matter of the walls, Isabella has been able to reconstruct patterns of long lost roads. Despite the fact that most of the trees are long gone, she has been able to rewrite portions of Italy's history through art and the locations of trees.

Two classic Kavanagh apples: green ground color; largely covered with golden russet; apex and basin on the left; base and cavity on the right

Unfortunately for us in New England, not many farmers painted frescoes on their "best room" walls. Some did make maps, however, like the one on that barn wall in South China. Too bad everyone didn't get out the paint brush back then and document their orchards. Unlike Italy, we're looking back a mere two hundred years. Fortunately, some of our oldest grafted apples are still here. Where will the next Kavanagh appear? What will the tree have to say?

The days of the russet may be about to come back. During our visit to Geneva last fall, we learned something else about russets that could be of interest one of these days. There appears to be mounting evidence that russet apples are more nutritional than the smooth-skinned varieties preferred in the grocery stores.

Gladys and Dick Johnson with their Fall Jenneting tree

Damariscotta returned to my life again when I received a call from a woman named Gladys Johnson. I got a good feeling about Gladys, right off the bat. She got right to the point: "Have you ever heard of Fall Jenneting? We have a Fall Jenneting and a Porter and a Sweet Mac over at the farm. We used to have a commercial orchard years ago."

Later that winter I was in the truck heading to Bunker Hill in Jefferson, high above Damariscotta Lake not far from the land of the Kavanagh. My apprentice Abbey Verrier was with me. We had duct tape, pruners, a pole pruner, Sharpies and a big plastic bag for the scionwood. Always put apples in paper bags and scionwood in plastic. We met Gladys and her husband, Dick, at their new house, a mile or so past the farm. We sat and chatted about apples and the orchard they ran for many years.

"We have something to show you," they said. Dick brought out a framed magazine photograph of an apple orchard. The two-page photo showed five adults and two children surrounded by bushels and barrels of apples on a beautiful fall day with a lake off in the distance. The photo had come from the October 4, 1958, Saturday Evening Post. There was Gladys sitting in the grass with two of their three children. Dick and his brother and their parents were all gathered around a makeshift table. In the distance was Damariscotta Lake. In the boxes and barrels: Northern Spys. Autumn's Bounty is the title of the article. "...But now, during an afternoon break, all hands are sampling their own sweet cider and apple pie before the harvest starts its way to lunch pails, dinner tables and the tops of teachers' desks."

An hour later Abbey and I headed back up the road, pulling into the old farm to cut scionwood. Dick was with us. There was the Fall Jenneting. Next to it was the Porter. We grafted Fall Jenneting the following April. Another variety saved.

Later on that spring, Cammy and I spent an evening with the Johnsons and many of their neighbors at the Bunker Hill Church. We ate dinner, and then I gave a talk on apple history in Maine. It was a good evening. In September we returned again, this time to collect fruit. I had never seen or tasted Fall Jenneting before. I didn't know what to expect. It's medium-large and oblate. Surrounding the stem and

Fall Jenneting of Maine is yellow with netted russet on top: photo by Abbey Verrier

spilling over the rim of the cavity is a large splash of russet. Although Porter is already growing in the Maine Heritage Orchard and on our farm, it's good to know of another tree out there too. I still have to go back to check out the Sweet Mac. That will be this coming fall. What a gift. That's the kind of phone call I like.

I topworked Kavanagh at our place onto a tree along with Cole's Quince and Winthrop Greening. The graft took and we got fruit. But then that one branch began to suffer from an undiagnosed malady. A year later it died and I pruned it off. We still had one other Kavanagh tree but it was young and wouldn't fruit for several years. It looked as though it would be slim Kavanagh pickings for some time to come. Maybe a little scavenging from one of the ancient trees on my travels to Freeport or Union. But I was wrong.

Up at the nursery we have a scattering of trees we've grafted over the years. They're mostly apples but also a few plums and pears. Some of them are seedlings that appeared uninvited. Rather than dig them up or cut them down, we put them to use. We grafted them in place. A few were nursery trees that should have been for sale but just never got dug. Most have fruited and we now have an impromptu orchard around the perimeter and in between the nursery rows and vegetables. They do get in the way but they also provide us with quite a bit of fruit.

One of those random trees has been particularly stubborn. It didn't want to produce. It was an apple but I didn't know what variety. It had been grafted by one of our apprentices years ago. Nathan Dorpalen had learned about the farm from our daughter, Phoebe. He was a student at Bates College in Lewiston. Phoebe was going to Colby. They met through Nate's twin sister Erica, one of Phoebe's dorm-mates and good friend. Nate was looking for farmers to interview for a film he was making on chickens and the chicken industry in Maine. Phoebe suggested he talk to me. Nate and a friend came over and we spent a few hours together. It was a fun visit. The film had a limited release later on that winter. They called it *Fowl Play.*

The following spring Nate returned and did a three-week internship on the farm. We planted peas, we rebuilt the grape arbor, we cleaned up the raspberry patch and, it being topworking season, I taught him to graft. That first month turned into many days together over six years. Sometimes we wouldn't see Nate for long periods of time. He'd be working on a farm in North Carolina or Alaska, or traveling in Europe or Russia or who knows where. Then he would resurface and stay with us on the farm for a weekend or a week or even a couple of months until some other new adventure would lure him away. He and I grafted many trees together, collected scionwood, went on fruit exploring adventures, worked in our gardens and talked endlessly about agriculture and life.

One spring I was heading out of town for a few days during mid-May. Nate stayed behind to watch the place. He wanted to do some grafting. We looked at several candidates to topwork and I told him to go for it. While I was away he grafted an assortment of varieties onto an assortment of trees around the farm. He was skilled by then and I knew they'd all take. I also knew that he could be a bit on the casual side when it came to documentation. "Sorry, John, I forgot to bring along any tags."

Then, tragically, Nate died one October while hiking in Scandinavia where he had entered a program in agricultural policy. He was twenty-six. I assumed that Nate would deliver the eulogy at my memorial service, not the other way around. But such is life. And the fall turned into winter and winter into spring and then it was time to graft again, but no Nate.

A few years later a funny thing happened. Nate reappeared on the farm. The anonymous grafts he had done that one weekend when I was away began to fruit. Not all at once. There would be one this year, a couple more the next year and so forth. They were gifts from Nate. I could hear him speaking to me. He had a unique way of saying John. He would say, "Jawn." I could hear him saying, "Here's a Starkey, Jawn. Like that Wickson, Jawn?" They were presents and puzzles. The branch would fruit and I'd think, "Oh here's another one. What is it?" Most of them were fairly easy to identify. I had a good sense of what scionwood he had access to. After all, we'd cut it all together. But some were surprises. He must have really dug through all the bags to pull them out.

At some point they had all fruited and I had tagged them all. All but one. He had topworked a small seedling on the edge of the nursery. For whatever reason that particular tree was stubborn and would not fruit. Every year I'd watch. No fruit. Then in the spring of 2017 it flowered profusely. All right! The flowers faded. The fruitlets formed. Spring turned in to summer. And one day the fruit was large enough to ID. Several weeks later I was eating the best applesauce central Maine has to offer. It was a Kavanagh.

Nate Dorpalen at the farm during the plum bloom in 2006

The multiple stems of this tree, combined with its location a few feet from the road on a narrow slope above Sedgwick Harbor, with no houses nearby, are a giveaway: it's a seedling. The apple itself is a bittersweet that Cammy named Galatea

This group of seedings in Palermo has survived by locating itself next to a mailbox and newspaper tube. The mailbox deters snowplows and mowing machines from getting too close: another example of the partnership between people and apple trees!

# Thirty-Three

# Blake Is Willing

"It sounds high-flown and absurd, but consider the facts!"
Sherlock Holmes, *The Naval Treaty*, p. 464

In January 1849 Charles Dickens made a quick trip to the coastal fishing town of Great Yarmouth, a hundred and fifty miles northeast of London. By then Dickens was a famous man. He stayed for two days at the Royal Hotel on Waterloo Road. He was looking for inspiration for his next novel. He soaked up the local surroundings, met the local folk, smelled the local smells, and then went home and wrote his famous semi-autobiographical novel. He wanted local color and he found it. Great Yarmouth became the setting, and a number of the people he met became the models for assorted members of the cast of characters. It all went exactly as he hoped.

One of the locals he met, most likely servicing the hotel, was a truck driver named Blake. Blake owned a business called "The Old Forge" in Blundeston in Suffolk, a few miles south of Great Yarmouth. Some would call Blake a cart driver. He was the mid-nineteenth century horse and wagon version of a short-and long-haul trucker. He made a big impression on the author. Whether or not Dickens made it down to Blake's Blundeston Forge we don't know, but he liked Blake, and he liked the name Blundeston. The full title of his new serial novel became *The Personal History, Adventures, Experiences and Observations of David Copperfield the Younger of Blunderstone Rookery (Which He Never Meant to Publish on Any Account)*. Or *David Copperfield*, as we know it.

Altering the spelling in a Freudian slip sort of a way, Dickens had his hero born in *Blunderstone*. "I am David Copperfield, of Blunderstone, in Suffolk…" Blake himself became the inspiration for one of the more famous characters in the novel, again with a slight change in spelling. Barkis is the painfully shy, and rather glum, cart-driver. He made his fame by uttering three words, famous enough to merit his own entry in Bartlett's *Familiar Quotations*. Barkis is in love with David's childhood nurse, Clara Peggotty. He asks David to give Clara a message for him. Although David doesn't understand what it means, he cheerfully complies. The famous message is simply, "Barkis is willin'. " "That Barkis is willin' I repeated, innocently. 'Is that all the message?' "

For the Maine fruit explorer, the name Blake is noteworthy for another far more important reason. The cart-driver may have played a key role in the preservation and eventual rediscovery of one of Maine's most elusive apples. That would be the one that coincidently bore his name and originated in Westbrook, Cumberland County, not far from Portland, thirty-two hundred miles from the Forge at Blundeston.

The yellow-fruiting tree on the Jones Road turned out to be Grimes Golden; note the two cleft-grafted scions that grew to be side-by-side trunks

In November 1995, nearly a hundred and fifty years after Dickens' visit to Great Yarmouth, George Stilphen received a letter from a man named Gerald G. Fayers. The letter came from Great Yarmouth, England. George had only two years earlier self-published his revised version of F. C. Bradford's 1911 University of Maine thesis, *Apple Varieties in Maine*. George called his book *Apples of Maine*. Mr. Fayers had been told by Rex Welland of the British Columbia Fruit Testers Association of Stilphen's work. Gerald had obtained Stilphen's address, although he had not yet seen a copy of the book. He was writing to George in reference to an apple he had recently discovered called Blake. "The three trees are growing at Blundeston near to the prison. The Waveney Valley is a triangle of land between

Great Yarmouth, Lowestoft and Norwich which years ago was an important fruit growing area but regretfully not so today. In fact the large fruit store is on the point of becoming derelict. Mainly due to large imports of French Golden Delicious apples which the Supermarkets like and the public buy."

Gerald Fayers had made a hobby of tracking down rare apples. He was a true fruit explorer and collector. He was born in 1926 and lived most of his life in the Great Yarmouth area. "In my garden I have 600+ known named varieties of apples and more than 200 awaiting identification/evaluation." Over the years, he had traveled throughout eastern and western Europe, Kazakhstan and North America in search of apples.

Gerald first heard of the Blake apple in 1990 at the Gressenhall Farm and Workhouse, a local rural life museum. He did some cursory research and concluded that Blake may have originated somewhere nearby. He knew hundreds of apples but not Blake. He was relatively sure he'd never seen the fruit. This must be one of those countless local varieties.

About the time he first learned about Blake, Gerald met a traveling hardware salesman named Neville Turner. Turner caught the fruit exploration bug from Fayers and took up the Blake torch. "Some of my enthusiasm has rubbed off onto him with the result that on his journeys about the countryside he had looked for interesting old apple trees, frequently stopping to talk to the old boys on his rounds, then bringing apples to me to see what I can find out about them." Then one day in 1995, Turner showed up with the apples from the three trees in Blundeston.

Gerald decided to dig deeper into the history of the Blake. It was then that he learned of the Westbrook, Maine, connection. Was Blake an English apple? American? Were there two Blakes, one American and the other British? Had it originated in Blundeston and been taken over to Maine? Had it originated in Westbrook and been taken to England? How would it get from Westbrook to Blundeston? These were the questions he was attempting to answer and the reason for the two letters to Stilphen in the fall of 1995. The first was dated November 13 and the second, December 1.

George Stilphen and I first connected in 1984. I contacted him first. He had written a brief article in *Pomona*, the quarterly journal of The North American Fruit Explorers. He was in search of rare Maine apples. So was I. We wrote back and forth a few times a year. It was the era of letters, and I kept them all. He had just discovered Bradford's thesis but had not yet decided to update and reprint it. His letters are filled with a joy and zest for life and apples.

JPB and George Stilphen in 2001; Stilphen made Bradford's 1911 thesis available to hundreds of enthusiasts in 1993

"Haute tige" means grafted high in Normandy

George's book is copyrighted June 1993. As I recall, the first printed copies were available sometime in '94. He sent me a copy in March 1995. In a letter dated six months later, he wrote to me about a new discovery: "P.S. I've just opened this envelope and I have the most strange news! The old Maine BLAKE has just been discovered in (are you ready for this?) the Waveney Valley, Great Yarmouth, Norfolk, England!…Truth is stranger than fiction!…In fact, this whole episode is so mind-blowing that I'm enclosing herewith copies of both letters I recently received from Mr. Fayers!"

Although George was clearly still filled with plenty of passion and fire, I was younger and probably had more energy on the ground. I could also tell that he was beginning to wind down. He had put a great deal of effort into creating *The Apples of Maine*, and he was probably not about to charge off to find the Blake apple in Westbrook. I read Fayers' letters over and over again. This was a search I was ready for. In one of the letters, Fayers wrote, "If you are not able to progress the information any further do you know of others who may be interested?" A few days later I sent off a letter to Great Yarmouth.

Gerald Fayers and I then began our own correspondence. Gerald had put together a theory of how the apple got to England. It's commonly known that many plants have been brought to America over the past few hundred years. Early classic apple varieties, such as Blenheim Orange, Ribston Pippin, and Sops of Wine, all arrived here with no questions asked and no serious negative consequences. Most of our iconic pear varieties did the same. (Other plants, however, have wreaked havoc. We wish they'd never come.) But plant material was also going the other direction. North America was the exotic New World, and North American plants

Other grafts are very low like this one, just above the spreading rootstock. The very small green spouts are root suckers: Gramercy Park, New York City

were in hot demand in England and on the continent. There were no import/export regulations back then, and the plant material flowed back and forth with ease. The famous Kew Gardens just outside London has some of the largest specimens of American trees in the world, all imported long, long ago. Some are enormous. Some may be larger than anything back here in their native land. Check out the Black Locust if you're ever in Richmond, England. Then when you're back here, compare it to one in Richmond, Maine, or Richmond, Virginia, or anywhere else in the United States. You'll be amazed.

Collections of American apples became popular in Europe; one of the best known was that of John Scott in Somerset, England. Scott was a contemporary of Dickens and Blake. He was also a collector. Gerald had obtained a copy of his 1873 catalog, *Scott's Orchardist; or Catalogue of Fruits, Cultivated at Marriott, Somerset,* which includes five hundred varieties from England and the continent and "200 sorts from America." One of those seven hundred apples was "Blake." Seeing it in Scott, Gerald began to suspect that Blake might not be a local Blundeston variety. Maybe it was from America. Unfortunately, the American varieties in the catalog are not labeled as such. Rather, they are all blended into a combined list. It's not possible to tell if Scott knew of Blake's origin. There also could have been two Blake apples, one from Blundeston and another from Maine. There were two Collins and two North Stars. Why not two Blakes?

A lot of American apples did make their way to England, however, and Blake could have been one of them. Westbrook is next door to Maine's largest city. It's not out of the way like Cherryfield, Castle Hill or Temple. Maybe someone from Thomas Jackson's Forest City Nurseries in Portland cut the wood and wrapped it in wax paper and canvas. It would have been an easy trip from Portland onto one of the dozens of ships destined for Liverpool. From there it was just a cart ride to a nursery in Merriott in Somerset. But then, how would it get to Blundeston?

Evidently, when Barkis/Blake wasn't doing short hauls to Great Yarmouth, he was doing UPS-type long haul treks across England's mid-section. One of these trips might have been two hundred and seventy-five miles to Somerset, where he could have stopped by to check out the famous Scott's Nursery when he had a little free time. While there, Blake would have learned that an apple in the collection had his name. Being as vain as the rest of us, he had to have a tree. (A few years ago a collector in New York sent me scionwood for a Bunker Hill apple. I guarded the wood and grafted it onto a prominent branch not far from the house. It still hasn't fruited, but I'm sure it will be great!) We can assume that Blake would have purchased a couple of trees, packed them into his wagon and headed home to the northeast coast.

Over time, the large, yellowish-green cooking and dessert apple was propagated and passed around Suffolk and Norfolk. In February 28,1998, Gerald wrote to me that "Old people in the village of Lound [Blundeston] are quite adamant that this is an old local apple originating from many generations of Blacksmiths at the local Forge... most certainly the BLAKE is a very historical one in that area." According to a handout he sent me, "An elderly former resident of 12 The Street [sic] always knew 'Old Blake' as a local variety. This orchard was planted in the 1870's." Thanks to the modernization of the apple industry, along with hundreds of other varieties, Blake eventually slipped into obscurity. Then came the day in 1995 when Neville Turner just happened to be in the right place at the right moment, and with the help of Gerald Fayers, the Blake lived again.

A brief but useful description of Blake can be found in A. J. Downing's 1872 edition of *Fruits and Fruit Trees of America*, "Originated in Westbrook, Cumberland Co., Maine. Fruit medium to large, roundish, greenish yellow, quite yellow at maturity. Flesh firm, fine, crisp, juicy, subacid. Good. October to January." In *The American Fruit Culturist,* John Thomas also describes Blake, but his description appears to be copied out of Downing. Bradford describes Blake, but his description does not add much either, except that the apple was "more or less widely distributed fifty years ago [1865]."

Meanwhile I began to look throughout central Maine for a Blake tree. I familiarized myself with the descriptions. I knew I was looking for a fall-early winter, medium-large yellow apple. That's not a lot to go on. I knew I'd have to use other strategies if I was going to succeed. For one thing, I needed to find the Blakes in Westbrook.

The next clue may be painted on the side of a brick building

I contacted the Westbrook Historical Society. They knew nothing about the apple or the Blakes. I made a wanted poster and added it to the stack. I got my first break when I was buying a loaf of bread in Portland. It came from an old sign I admired over the years but never thought much about. I saw it that day and I knew I was on the right track.

I love commercial signs painted on brick. Could it be that the bricks hold the paint in some unique way that helps to make the lettering so visually interesting? It probably has something to do with the old fonts of the letters as well. To me those old billboard-like signs are beautiful.

At the end of one of the blocks of brick buildings on Commercial Street in Portland, above the Standard Bakery parking lot, is another one of those old gigantic signs. This one reads, "W. L. Blake & Co. Mill & Industrial Supplies." It dates from sometime after 1924. There's an earlier version from a 1924 City of Portland tax-revaluation photograph with slightly different wording. Under the same W. L. Blake & Co, it reads, "Steamship Supplies Pipe Fittings Valves Etc." The W. L. Blake & Co. warehouses and machine shops were down on the Portland waterfront for many years. Now all that's left is the beautiful lettering on the brick wall.

Did W. L. have something to do with the Blake apple? Could he have planted the first seed? Maybe his daughter planted it, or his sister or his cousins or his aunt. Maybe one of them discovered an apple seedling out in the pasture at one of the Blake homes in Westbrook. Maybe I'm on the trail to the Blake.

Early on, Westbrook was known as Saccarappa. Next it became part of Falmouth, then it became Stroudwater, and finally Westbrook in honor of Colonel Thomas Westbrook (1675-1744) of military fame. (Some would say infamy for his role in the horrible murder of the Jesuit Sebastian Rale at Norridgewock on August 23, 1724.) Westbrook has been a mill town since its beginnings, powered by the Saccarappa and Congin Falls. Its best known mill is what we always called S. D. Warren, the paper company started by Samuel Dennis Warren in 1854. Like most large American mills, it has changed hands multiple times; S. D. Warren is now owned by a South African company. The other Westbrook mills are mostly shut down although some of the long, brick, riverside buildings avoided the bulldozers and have been refurbished into an assortment of businesses. A great young company is growing high quality mushrooms in one of them. These days you might call Westbrook a bedroom for Portland.

About the time I was gazing up at the old brick signage in Portland, email was suddenly taking over the world of communication. The letter was doomed. It was a defining moment for everyone over about forty. Most of our friends and family embraced email but a bunch of them did not. Cammy's mom embraced it. My father did too, but my mom never did. Neither did George Stilphen. I don't believe he ever used it. He also moved and left behind his orchard of Maine apples. He never replanted. Another collection lost. In 1998 he and I discussed the possibility of my writing a revised edition of his book with the inclusion of my own discoveries and other additional information. That never happened.

Grimes Golden originated in West Virginia but became popular long ago in central and southern Maine

He had plans to have his book printed in paperback but that also never came to be. He did do two more printings of *Apples of Maine* in hardback. The book had enough popularity for St. Lawrence Nurseries in Potsdam, New York, and Fedco Trees to subsidize what turned out to be the final printing. That was in 2001. I don't think it will ever be printed again.

Meanwhile, my own fruit exploring life was taking off in many directions, though rarely out of New England. I was traveling a lot from late August to Thanksgiving. Cammy was living near Boston, and I was making many trips back and forth. Those trips took me away from the farm but did provide me with a quiet place to work. I would spread the apples out all over her dining room table and leave them there for days at a time. I had Blake on my mind, but it was clearly on the back burner. The Blake engine had left the station and was picking up steam, but only ever so slowly. I had many other apples to fry.

At one point I thought I'd found a Blake tree in Palermo. That was about 2004. There was a tree a mile up the road from the end of our driveway that I had been curious about for decades. It was old and grafted. I liked the fruit very much. It was one of my favorites, and still is. Could it be Blake? Truth is stranger than fiction, but finding the last Blake on earth a mile from the farm was a little over the top. The fruit is yellow, and it ripens in mid-late fall, and it keeps until January, just like Downing said. It seemed unlikely, but there it was. Was I really going to find Blake in Palermo? Shouldn't it be down near Portland. Was this too good to be true?

It *was* too good to be true. After convincing myself that I'd found it, I then had to disappoint myself with the truth. I was incorrect. It was a false alarm. It was not a Blake. I confess I was skeptical, especially after it occurred to me that it might be a Grimes Golden, the classic heirloom from West Virginia that wound its way north, possibly with returning veterans after the Civil War. Grimes Golden was grown throughout New England. I received a letter from a neighbor about fifteen years ago listing all the apples she remembered on her parents farm growing up in Palermo in the early twentieth century. One of them was Grimes Golden. I sent fruit to Tom Burford. Tom is the grand master of American apple history and varieties. I knew that Tom would know, and he did. It was Grimes Golden. It's a great apple and a wonderful find. By then, I had already topworked most of one of our trees with it. We had all the Grimes we could ever want. But, alas, still no Blake.

Whenever you go to a commercial orchard, you can be fairly certain all the trees are grafted

# Thirty-Four

# Saving the Apples of the World

But we hold several threads in our hands, and the odds are that one or other of them guides us to the truth.

Sherlock Holmes, *The Hound of the Baskervilles,* p. 693

Gerald Fayer's Old Blake

My correspondence with Gerald Fayers continued along. Like Stilphen, he was not an email guy but, unlike George, he did keep writing. Once or twice a year, we'd trade letters. He was in his late seventies, and he and his wife had decided to leave the orchards and move to a smaller place. He sent me a description and photographs of his Blake, but nothing beats the real thing. I wanted to meet him. I wanted to see the fruit up close and personal. I wanted to photograph it, key it out, taste it, hold it. In the fall of 2006 I finally had my chance.

Cammy and I were invited to attend the Slow Food conference in Turino, Italy, in late October, 2006. My Dad was living in Twickenham, just outside London. I rarely had the opportunity to see him. I decided to go to London and spend a few days with him and then meet up with Cammy in Milan. It would be great to see my father, who was getting on in years. I was also thinking about Great Yarmouth. The timing would be perfect for a side-trip to the land of Barkis and Blake.

Off I went on October 24, 2006. I took the train into London from Twickenham, then picked up the 9:30 AM at Liverpool Station bound for Great Yarmouth. At Norwich I had to switch to a bus for the last half hour —something about cleaning the tracks. It was a long six hours. Gerald met me at the bus in his tiny car. He had turned eighty-two days earlier. He had a mop of silver hair and a beaming smile. He was wearing a dress coat and a tie. This was a big occasion. It was the world conference of the Society for the Resurrection of the Blake apple. Just we two.

Gerald Fayers with the Old Blake (Cooper), 2006

We decided to skip going to a restaurant and head for his place instead. We only had a few hours before I would have to head back to London. The Fayers had moved to a small house close to Great Yarmouth. Their modest yard was already packed with fruit trees. One of them was the Blake. We both had a lot to say. I filled him in on my stateside efforts to locate Blake. He filled me in as well. I blurted out whatever else I could think of in the small amount of time we had. He had Blake fruit from two different locations for me to photograph and take with me to document and eat. He called one of them Old Blake Cooper; he called the other one Old Blake Brogdale.

The Cooper Blake appeared to be the one from Blake's Old Forge in Blundeston. Cooper was the name of the current owner. It was unclear where the Brogdale Blake had been found. Might it be an entirely different variety? Might one of them be the Maine Blake? The apples looked different from one another, but maybe one was more ripe. There wasn't time to straighten it all out.

Then all of a sudden it was over, and I was back on the train and heading south in the dark for the long trip back to London and Twickenham. I had met Gerald. I had Blake apples with me. I pulled one out and looked it over. It was beautiful. I opened up my notebook and took out a pen. I wrote:

> Blake as grown by Gerald Fayers Oct 2006: Shape roundish somewhat blocky somewhat ribbed Overall color is yellow-green, not russet; very slight orange-ish blush, almost non-existent, more like a glow, flecked and dotted with russet. Cavity russeted, wide and medium deep. Stem medium to long and medium thick. Basin shallow and wide and somewhat crowned.

Gerald's Old Blake (Cooper)—base, cavity, and stem

When I returned to the states, the search for the Blake apple continued, and the momentum built. Every year I would sift through the ID and orchard visit requests, always keeping an eye out for clues. While I focused my efforts on Westbrook and Portland, I also expanded my search to the towns a second or third degree of separation away. When I found myself in the area west of Portland in mid-late October, I would take some time to scout around the oldest roads looking for ancient trees with yellow fruit. I stopped at many places. I followed the occasional lead. I met many interesting people, but there was not a Blake to be had.

I was also putting out the occasional Wanted Poster when I had the opportunity. People were seeing them and the word was getting out. David Buchanan got involved. He found a possible Blake in downtown Portland, not far from the Blake brick wall painting. I tracked it down. It was a seedling.

Gerald's Old Blake (Cooper)—apex, calyx and furrowed basin

After discovering the trees in Blundeston, Gerald sent scionwood to the Brogdale Heritage Trust in Kent. Brogdale is where the UK protects and preserves its fruit varieties for researchers, breeders and future generations. I was anxious to import Blake scionwood and get it growing in our nursery. I contacted Gerald, and he directed me to Brogdale. Then I contacted the Maine Department of Agriculture who connected me with the USDA in Beltsville, Maryland. My contact at Brogdale was a woman named Mary Pennell. It was an arduous process, but by 2009 it was underway. In the Brogdale nursery the Blake tree was taking forever to grow. I assumed that it was grafted onto a dwarfing rootstock with no vigor. The tree was too small, I was told; I would have to wait until it put on enough growth before Brogdale could spare any scionwood. In addition Mary said that she might not be able to release the wood until she could verify that it really was Blake. Wasn't

Old Blake (Brogdale): did Gerald have two Blakes or one?

that last piece my job? At this rate I might not live long enough to get that Blake grafted in our nursery.

I had read that Brogdale was one of the many victims of the privatization of the nineteen eighties. While Reagan was saving us lots of money by privatizing America, Margaret Thatcher was busy doing the same thing over there. Budget cuts took their toll on England's public institutions. The fruit and cider research program at Long Ashton Research Station in western England was ended in 1981, as the Prime Minister famously declared, "There is no such thing as public money…" Brogdale was not killed outright but also lost funding. It was subsequently privatized and then descended into decades of bickering, factionalism and outright chaos. By 2008 its management had been taken over by the University of Reading and something called the Farm Advisory Services Team. "FAST" had a lot on its plate, and taking care of heirloom apples was not high on the priority scale. The Brogdale Trust was finding it increasingly difficult to take care of the collection. In February 2009 Mary Pennell wrote to me, "There have been big changes at Brogdale in the last year which has meant that the staff who were working on this observation plot are no longer employed at Brogdale and the work as a result has been put on hold for the moment."

The U.S. has its own version of Brogdale. The USDA Agricultural Research Service (ARS) at Geneva, New York, is charged with preserving and protecting our apple heritage. Although Geneva will never have the space or funding to preserve thousands of rare local varieties, it is nevertheless one of the premier collections in the world. It is a national treasure. It's an essential resource for fruit explorers, collectors, breeders and all those attempting to create a secure food system for the future. Unfortunately, it too is in danger of being bled to death by well-meaning politicians in Washington whose one goal in life appears to be to eliminate burdensome public services. While we may have entrusted our government with doing too much, we might want to think twice before following the British model. Maybe plant protection shouldn't be the job of government. Maybe it should be up to us. If that's the case, we better hop to it. Meanwhile, write to your congressperson! Save Geneva!

While I never intended to take over the job of the United States government, by this time I'd been collecting apples for about twenty-five years. When I'd find a new one, I'd topwork it onto a tree in our orchard. I was replacing Washington one tree at a time. Our farm was becoming a mini-version of Geneva and Brogdale. There were, however, severe space limitations on our two acres of reclaimed woods. I would never be able to protect even a tiny fraction of the apples out there. I didn't have to worry about budget cuts because I had no budget. Land was my issue. I made the decision to strip down our collection to the rarest of the rare along with a small number of personal favorites.

Private collections, be they apples or bean seeds or books, may be the best last resort for battling off extinction, but they too are fraught with risk. Someone must pay to keep them going. It's not going to be the public. And what happens when the collector dies? There have been many excellent private fruit collections in the United States and Canada, not to mention elsewhere around the world, that have gone to the grave with their creators. The apple pharaoh dies, and all his apple trees get buried in the pyramid with him. That's not a good plan.

The orchard at Tower Hill Botanic Garden, September 2018

One apple collection that has continued to inspire me over the years is that of the Worcester County Horticultural Society at Tower Hill Botanic Gardens in Boylston, Massachusetts. The S. Lothrup Davenport Collection at the The Frank L. Harrington Sr. Orchard consists of a hundred and nineteen historic New England varieties. Each variety is grafted onto two different semi-dwarf rootstocks—probably M7 and M26. Having them on semi-dwarfs greatly reduces the total space required for the orchard. The trees are only ten feet apart. Having two each is also a hedge in case a tree dies. The orchard is located along both sides of the driveway coming into the Botanic Gardens so everyone who visits drives right through it. Tower Hill hosts an apple festival every fall and occasional orchard classes throughout the year. They also sell scionwood. The orchard has been endowed by the Frank Harrington family, so presumably it's safe from our elected officials.

The Tower Hill collection was largely assembled by Stearns Lothrop Davenport, who was born in 1887 in the Worcester County town of North Grafton. He was known as Stearny by some and Davey by others. (I use Davey, which rhymes with savvy, not gravy.) Aside from brief stints in other parts of the state, Davey lived all of his life in North Grafton. During the Depression he was put in charge of a Works Progress Administration (WPA) funded project to cut down thousands of old apple trees and distribute the remains as firewood. The idea was to clean up feral trees that might host disease or insect pests. Like most of the rest of the New England, Massachusetts was transitioning from a three hundred year orchard culture of vast diversity to a commodity form of commercial orcharding based around one apple: McIntosh. The thousands of ancient trees in the small orchards of Worcester County had become a nuisance and had to go. It was Davey's job to get rid of them.

Fortunately for the rest of us, they had hired the wrong guy. It dawned on Davey that because of his work, many classic American varieties might be lost forever. Generations of plant heritage would be gone. There was danger and folly in what he had been asked to do. He was the executioner. Though no one ever doubted the good intentions of the program, the result would be agricultural genocide. As he was overseeing the removal of old orchards, he began to collect scionwood and bring it home in his lunchbox. He then topworked it into his own orchard on Creeper Hill.

In the 1940s Davey became secretary of the Worcester County Horticultural Society. He continued to grow apples in North Grafton and dabble with apple breeding; he also remained troubled by the loss of American pomological heritage. He felt guilty. In 1953 with the assistance of WCHS president Myron Converse, and after consultations with the University of Massachusetts and the New York State Agricultural Experiment Station, he and a Fruit Committee from WCHS devised a program to create an official collection of sixty heirloom apple varieties at his "Creeper Hill Experimental Orchards" in North Grafton.

A 1969 *Food Marketing in New England* article, describes what he did for much of his life. Davey "has brought back many an old (and loved) variety of apples into the picture, has been a veritable Sherlock Holmes tracking them down, and then has raised them in an orchard devoted to this project and has made scions available for planting throughout the country and abroad."

The article continues that Davey "is well qualified to judge whether an apple is true to variety or not. Apples are often sent to him from universities, growers, homeowners and breeders for identification. He can confirm their identify by his knowledge of variety characteristics, apple shape, color and flavor. His is a rare knowledge, for most horticultural students today know only a few varieties left in commerce, whereas Davenport's knowledge goes back eight decades in fruit growing."

Who knows where you'll find that long lost variety?

Like many other privately owned plant collections, Creeper Hill Experimental Orchards is long gone, sold in 1967 and destroyed a few years later. Fortunately, Davey's collection did not die. With the guidance of J. E. Auchmoody, a Vice President of Old Sturbridge Village as well as a Trustee of the Worcester County Horticultural Society, an arrangement was made between the two organizations: the trees would be planted and maintained in Sturbridge, and scionwood would be made available through the WCHS. The collection had by then expanded from sixty varieties to a hundred and nineteen. In April 1973, young trees were planted at Old Sturbridge Village in Sturbridge, Massachusetts. Davey died later that year, but the collection had been saved.

In 1986 the Worcester Horticultural Society moved to its new Tower Hill facility in Boylston, and four years later the S. Lothrop Davenport collection was planted along the roadside coming up the hill into the site. In the intervening years, thousands of scions have been sent throughout the country and countless apples have been tasted by visitors to Tower Hill. Some varieties in the collection, like Baldwin and Ben Davis, are not in danger of extinction. Others, such as Sutton Beauty and Newtown Spitzenburg, are rare. Neither Tower Hill nor Sturbridge Village has been able to continue to search for apples. Although a hundred and nineteen is an admirable number, there were hundreds of varieties being grown in New England generations ago. There's also a downside to being grafted onto dwarf and semi-dwarf rootstocks: they have a limited lifespan. By 2016 many of the trees were beginning to show their age. They will all have to be re-grafted and replanted within the next few years. That will be a huge job.

David Buchanan planting a tree at the Maine Heritage Orchard, 2014

Sturbridge, on the other hand, had its own challenges. Over the course of several decades, it appears that labels were lost on some of the trees and other trees were incorrectly identified. Some varieties may even be gone. I was able to visit Sturbridge in the spring of 2016. This is no early Americana theme park. It's a major institution with a huge facility, including dozens of historic buildings. It's definitely worth a visit. Unfortunately the orchards may not have received optimal attention at several points along the way. Still, Sturbridge and Tower Hill remain inspirations for the fruit preservationist. Without them, Davenport's work would have been lost. Next time central Massachusetts has a good apple crop, I'll head down to Sturbridge with my pole picker, some bags and some Sharpies to do what I can to correct any errors, create some new maps, and re-label every tree.

Not only was Russell Libby Executive Director of MOFGA for seventeen years, he was also a passionate fruit explorer. No matter where he was heading, he loved to pull over by the side of the road and check out old apple trees. "I brake for old apples." In October 2009 Russell, David Buchanan and I spent most of a day together driving every road in Westbrook, looking for the elusive, medium-large blocky-yellow Blake. We drove slowly, we pulled over, we irritated impatient drivers who were used to going sixty on the back roads of Cumberland County, not twenty-five. We scoured the landscape. We knocked on strangers' doors. We had a wonderful day together, but found no Blake.

About that time, Russell and I began to concoct a new preservation plan. We had talked about a large orchard at MOFGA many times but always in vague terms. When MOFGA purchased its permanent home in 1998, I was asked to develop two small orchards on the site. I was surprised and flattered by the offer. There had been an attempt in 1995 to put together a similar collection of Maine apples at what was then called The Pine State Arboretum in Augusta. George Stilphen provided much of the scionwood. But for one reason or another, the timing was not right. The Pine State collection was not able to receive all the attention it required. Trees had died or lost labels. Maps disappeared, and although some trees remain, the effort was largely abandoned.

There were, and still are, some good private collections in the state but nothing that is organized to preserve and educate. On May 4, 2000, the Fedco Trees crew descended on MOFGA and planted the first fifty trees. It was the only time the staff had ever worked together out in an orchard. At Fedco we're always tagging, sorting, assembling and shipping orders inside a cold, wet warehouse. It was fun to be outdoors together and see the crew wielding shovels in the dirt.

Robert "Roberto" McIntire collecting scionwood from an ancient Baldwin tree on Orr's Island, 2018

The idea was that one of the two orchards would consist of varieties originating in Maine. The North Orchard, which is an acre, would accommodate about forty standard-sized trees. I would select the varieties, graft them and plant them over the course of several years. The South Orchard would be about two acres and would include a mix of tree fruit species grafted onto an assortment of rootstocks, including standards, semi-dwarfs and dwarfs. It would include popular modern varieties as well as a few heirlooms. Both orchards would be interspersed with vegetable gardens and nurseries. They would become demonstration orchards.

The plan went well, though it was fraught with challenges. MOFGA had never owned land before. Despite being all about farming, MOFGA staff didn't get to spend much time outside in the sunshine, the exception being the Common Ground Country Fair. Even the Fair was always held on someone else's rented fairgrounds. After the Fair, the staff went back inside their small cramped office in Augusta for another eleven months. At the new two-hundred-acre site in Unity, there was a real landscape and unlimited potential for visitors year-round.

The North and South orchards were located in good spots, both right on the fairgrounds. People could wander through them for those three days in September. The North Orchard, with its collection of Maine heirlooms, filled up fast. I was finding a couple of new Maine varieties every year. We were packing them in, but running out of room. The North Orchard was squeezed by the fairgrounds and expansion was out of the question. We were finding more than we could fit. Others, notably Robert McIntyre, Pete Jenkins and Todd Little-Siebold, were leading their own explorations. Todd and Pete were scouring coastal Washington, Hancock and Waldo counties. Roberto was doing the same down in the Brunswick area closer to Portland. They were all making excellent finds.

Yellow Transparent tree in Pittston; fifty feet across and thirty-six feet high: that's a lot of apple sauce: photo courtesy of Steve and Kris Oliveri

Another challenge we were facing was what to do with the unnamed varieties we were discovering. We knew they were grafted trees. We knew they were exceedingly old and rare. I had enough experience to know which ones I'd never seen before. Do we exclude them from our collection as long as they remain unidentified? Some of these were excellent apples with wonderful stories. Some had even become personal favorites. We just didn't yet know their names. To exclude them would mean the end for some varieties. To include them would be opening the flood gates. We needed more land.

I was also meeting other Mainers who were bitten by the fruit exploration bug. Pete and Todd and Roberto weren't the only ones. By the early 2000's, I was giving talks here and there around the state each fall. I would talk for about an hour, then take questions from the audience and then spend additional time identifying apples and listening to stories. Sometimes I would speak to half a dozen people, other times to a hundred or more. Everyone seems to love apples and old apple trees. Apples are non-sectarian, non-denominational, asexual, apolitical and multi-racial. The apple talks turned out to be a great way to gather up humanity and entice them into one room. Give an apple talk, and suddenly it seems as if the world's greatest problem is to come up with a name for every apple tree. An apple talk is a lovefest at a time when we need more lovefests. No more quibbling and bickering. A little less hatred, please. Let's ID some apples.

Sometimes I carefully prepare a talk and even read it. As Mark Twain said, "On average it takes me two weeks to prepare for an impromptu speech." Other times, especially when the group is small, I don't prepare much at all. I just show up, and we sit and talk and share stories and strategies. Sometimes, of course, I can't predict. I think the group will be large, but it turns out to be small. In that case I set the notes aside and sit down and see what happens. In other cases, I expect it to be small and the room is packed.

There are a few set things that I expect to have to explain. Most people don't know about seedlings and grafted trees. I usually talk about the migration of the apple from Kazakhstan. Beyond that I have no set speech. I may follow the history of the apples in Maine through the past three centuries, but I usually allow myself to drift or even veer off course as the spirit moves me. I have a theory that, even if someone has never heard me talk before, they know instantly if I'm repeating the same old thing. So I deviate. I learn about their town or county or special varieties. It keeps the talks fresh and fun. It keeps me on my toes. I'm glad that my audience knows that I'm forever traveling in uncharted territory. I try to do two things at once. It's like chewing gum and rubbing your head at the same time. One thing they hear: that's your voice. The second thing is your brain, which is a split second ahead of your voice, planning what to say next and guiding you towards that next word. Maybe they hear that too. I get lost sometimes. I stop sometimes. I take notes while I'm speaking so I don't forget the thing that I just remembered I want to say in a few minutes.

I avoid using PowerPoint. I used slides twice. The first time was okay; the second time was not so good. I have a hunch that your audience wants *you*. They want a real person and real human contact. They can watch videos on their computers all day at work.

Unless I'm doing a detailed history of an apple or a region, I avoid reading except a quote or two. I also avoid using too many notes. I like to take one three-by-five card with me and allow myself to write on one side. That's my roadmap. I put eight or ten words or brief phrases in tiny writing on my scrap of paper. Call it my set list. I introduce myself and then improvise on each of those ten words for the next hour. I had the opportunity to hear the eighty-year-old tenor saxophone player, Pharaoh Sanders, with his quartet at the Village Vanguard one winter not too long ago. The set began at 11:00 PM and ended at 1:30 AM. The streets of New York were still bustling when the concert ended. The group played four or five tunes. Each one lasted twenty minutes. It didn't seem as though they memorized a thing. It was all internalized; there was nothing to forget. You just open up the gate and direct the flow as it pours out.

On October 23, 2013, I gave a talk about the Roxbury Russet at the First Church in Roxbury at the corner of Putnam and Dudley streets. It was a full house. The church was built about the time the Roxbury Russet apple was discovered. One of the apple's synonyms was Putnam Russet. We were on Putnam Street and Dudley Street. Who could ask for more? Maybe I was standing on the site of the first Roxbury Russet tree.

I spent a lot of time preparing for that talk. I wanted to have the places, the dates and people correct. Still, although I relied heavily on several pages of notes, I didn't read a lot. I looked and then improvised. I knew I wanted all that information close at hand, but I also knew that I wanted to keep it fresh and inject a big dose of immediacy into the evening. The most magical thing about the night was the response of the audience. They were extremely knowledgeable about their community. As I talked and asked them questions, they filled in the details around my apple story. It felt as if we were a team of detectives unraveling the mystery of the Roxbury Russet together.

As fate would have it, it was also the first night of the 2013 World Series, and the game was in Boston. Fortunately the talk was at 7:00, and the game wouldn't start until later. Everyone was out of there in time to get home and turn on their TVs. We zipped over to a nearby restaurant and watched the Sox crush the Cardinals 8-1.

Laura Sieger *inside* the Tolman Sweet tree in New Sharon; the tree is still producing after 200 years—or might it be 300 ?

These talks have been one of the best of places to find rare apples. Out of the woodwork they appear: people bringing apples, people bringing stories, people bringing you the old orchards. I could have driven down a road a thousand times, and I never would have known what was behind that barn or over that rise. I go home with future fruit exploring adventures sketched out on scraps of paper and orchard maps drawn on paper bags.

So Russell Libby and I talked about the orchards at MOFGA and how we'd run out of room. We needed to create a preservation and educational orchard in Unity. It would need to be large and never be subject to the whims and winds of Washington. It would not die with me or with Russell or with anyone else. It would include the apples that originated in Maine. It would also include the apples that originated out of state but were traditionally grown in Maine. It would include those that we had identified as well as those unsolved mysteries. Russell told me about a piece of nice farmland MOFGA had purchased with a ten-acre gravel pit tacked onto it. It had been a pit for decades, and thousands of yards of sand and gravel had been removed. No one else had plans for the pit other than to close it and renovate the land. Why not turn it into a terraced orchard? There would be room to preserve hundreds of varieties while demonstrating how to renovate depleted land. It was exactly what I'd been hoping for. It would be a dream come true.

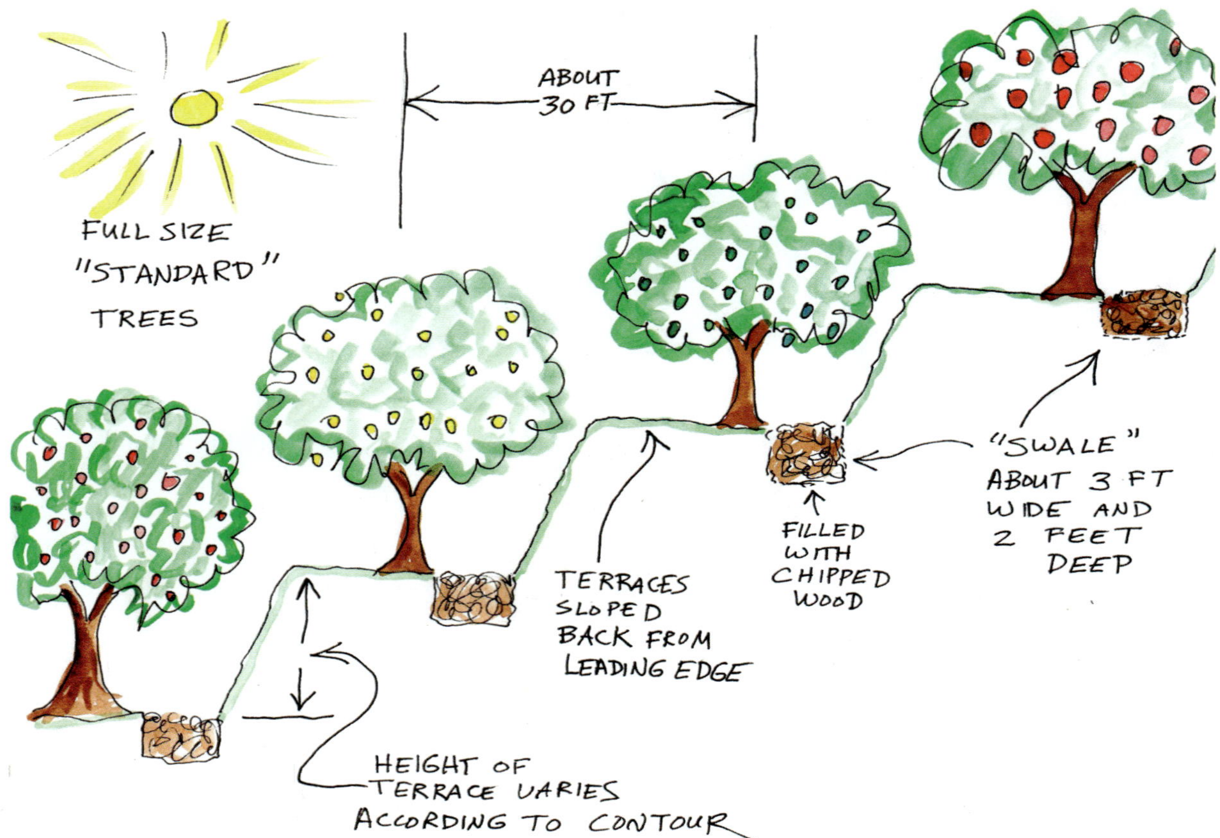


First we terraced the gravel pit with heavy machinery. Then we dug a swale along the back edge of each terrace, filling each swale with wood chips. We planted the apple trees in rows along the terraces. The terraces catch the rain and direct it into the swales. The chips soak up the water like a sponge and slow-release it to the trees and other plantings.

# Thirty-Five

# The Maine Heritage Orchard

"We are coming now rather into the region of guesswork," said Dr. Mortimer. "Say, rather, into the region where we balance probabilities and choose the most likely. It is the scientific use of the imagination, but we have always some material basis on which to start our speculations."
Sherlock Holmes, *The Hound of the Baskervilles,* p. 687

Russell's Russet

My next Barkis/Blake break happened at the 2011 Fair when I met a fellow named Peter Christiansen who thought he might have the apple. In my September 23 journal entry I wrote, "May have located the Blake apple!" Mr. Christiansen's tree was located on what was once his family's farm on the Flaggy Meadow Road in Gorham, one town west of Westbrook. His great-great-grandmother was named Mary Blake. He thought I might like to visit. He no longer owned the place but he would arrange to meet us there. Two weeks later, on a rainy October 4, Cammy and I stopped by to see the gigantic old tree. The fruit was yellow. It was medium-sized. It was mid-October. It was only a few miles from Westbrook. This could really be it. Sadly, however, I decided that it was another seedling. The fruit was bitter. It might make an excellent cider apple. It might be "a" Blake apple, but it wasn't "the" Blake.

Peter Christiansen & Cammy on the Flaggy Meadow Road in Gorham, 2011

Six weeks later I returned to England on a three-week orchard tour with a dozen students from College of the Atlantic. It had been five years since I visited Gerald Fayers. Todd Little-Siebold was leading the trip and convinced me to go along. I hemmed and hawed and avoided making a decision. I had enough to do in Maine. Why would I want to go to England? Then it was too late. The decision was made. A ticket was purchased, and I had to go. It turned out to be great. The students, the orchards, the cider makers, the apple museums—everything was perfect. I was able to fit in wonderful visits with my father at the beginning and end of the trip. By then he was ninety-one. It was our last time together.

Todd rented a large van and we all headed west, spending the first two weeks driving through Herefordshire, Gloucestershire and Somerset, visiting cider makers and cider orchards. Every day was a new adventure. I thought I was too old to stay at youth hostels, but I survived.

James Marsden reaching into one of his cider apple trees; Gregg's Pit, Much Marcle, England, 2011; note the graft!

After leaving Somerset on December 13, 2011, we headed for Kent and the Brogdale Trust. Spirits were high. Everyone in the van was excited to see one of the largest collections of apples in the world. My efforts to import Blake scionwood were well underway by then. This would be my opportunity to see "my" Blake tree and to meet my English contacts.

We rolled into Brogdale about mid-day. We met with a nice man who enthusiastically talked about modern orcharding in the U.K., but didn't seem to know about the preservation work. Some of us got the feeling that all those rare apples had become an albatross for "FAST" to deal with. There didn't appear to be much room left for the past. FAST had contracted with the U.K. government to take care of the trees, but the collection didn't appear to be high on the priority list.

My journal from the trip includes this note: "Collection is on the way out?"

I asked our hosts if I might meet Mary Pennell, my contact at Brogdale. We had emailed several times about importing Blake scionwood by then, and I was sure she would know my name. "Yes, we'll see if she is available."

She appeared and the two of us stepped aside and chatted for several minutes while the rest of our group continued to learn more about FAST and the future of apples in England. We had a good conversation. She referred to the apple as Old Blake not simply Blake as I knew it in the U.S. As we stood there, the thought crossed my mind: What if Blake and Old Blake are two different apples? Maybe there was a Blake of England and a Blake of Maine. Maybe one Blake originated in Westbrook and a totally different Blake originated in the Norwich-Great Yarmouth area of England.

Maybe Gerald was showing me two entirely different Blakes that day back in 2006. Maybe the Blake apple at Scott's Nursery was the American. When Blake went to Scott's, and saw this American Blake, did he have to have both? Did he bring a tree back to Blundeston and plant it? Now there were two Blakes, one from the U.K. and one from the U.S. Will I ever sort this out? Will Brogdale ever send Blake or Old Blake or new Blake or any Blake scionwood to the U. S.? I could only hope.

2012 featured another Blake lookalike. On October 2, I visited a farm tucked along one of the main streets in Gardiner. I'd driven past it a hundred times but never noticed it was there. I was contacted by the new owners, a young couple named Alex and Veronique, about an old tree on the farm. They had been in touch with the past owners who called the apple "The Ghost." It had the Blake coloring, the size and the season. The tree was exceedingly old. But the tree was in Gardiner, about fifty miles from Westbrook. I was skeptical. I resisted the temptation to declare it a Blake. I returned that winter and collected scionwood. I topworked it at our place and grafted one for the Heritage Orchard. It's an excellent apple, whatever it is. I don't think it's Blake.

Russell Libby & the Russell's Russet tree at Locke's Corner in Mt. Vernon, March 3, 1999; the tree is now gone

2012 also saw big changes at MOFGA. With the help of Russell Libby, we were able to move along the Maine Heritage Orchard project. On October 14 the Board of Directors unanimously approved the orchard. We had our orchard at last! But the joy of approving the orchard was greatly overshadowed by the realization that Russell was too ill to continue as MOFGA Executive Director. October 14 was his last Board meeting. For months his health had been deteriorating. He was very ill. Russell was the face of

MOFGA for close to twenty years. He devoted a huge part of his life to promoting agriculture, community, and traditional ways. He was a poet and a consensus builder. How fortunate I was that we became friends and fruit exploring buddies. Without him, much of the work I'd done never would have happened.

Russell's Russet is, more than likely, a chance seedling

Russell was the one who, twenty-five years earlier, said to me, "You need to vary the apple varieties you offer in the Fedco Trees catalog every year." It was Russell who called me late at night to tell me about some old tree he just discovered, like the Judy tree or the Munson Sweet or one I still haven't identified but just call "Russell's Russet." He was the one who squeezed in the time to go fruit exploring. He was always on the lookout, despite the fact that he always had a hundred far more pressing and important matters on his plate. He was the one who suggested that we renovate the gravel pit. Though he never got to see its transformation, his inspiration was the catalyst. Often I felt as though he was the true visionary. I was just the guy with the time to make it happen.

On December 6 I visited him for the last time. Beforehand I went through the apples in our root cellar and picked out the best, most perfect, most beautiful Black Oxford apple in the world. His wife, Maryanne, met me at the door and led me to his side. Russell was resting comfortably in a bed in their living room. He was not talking by then. He appeared to be entirely at peace. I filled him in on my latest adventure. I showed him the Black Oxford and set it next to him. He smiled. I told him I loved him. Three days later he was gone.

Fall turned into winter, and winter means time to cut scionwood. Anticipating the Board's approval of the orchard, we grafted the first hundred varieties in the spring of 2012. Those were set to be planted in 2014, after two years in the nursery. It was relatively easy to come up with that first batch of varieties. I'd been on the trail long enough to generate a big list. Now it was time to create a grafting plan for spring 2013. I decided I'd like to go to Oregon that winter to visit one of the largest private apple collections in the world and cut scionwood there.

I had heard of Nick Botner for many years. In the Seed Savers Exchange Yearbooks, he annually listed scionwood for many dozens of fruit varieties. I was always curious about who he was. In 2009 I was invited to an apple exploring meeting in Madison, Wisconsin. There I met Dan Bussey as well as Nick. At the Madison meeting Nick described for us the astonishing fruit collection he'd been assembling for twenty-five years, with over forty-five hundred apple varieties.

Nick's orchard is an immense, densely planted scionwood bank for every apple he can find. After moving south from Alaska to Yoncalla, he started collecting and never stopped; he also had a couple thousand pears and several hundred grapes. If you sent him scionwood for ten or twenty or a hundred apples, he would use whatever name you gave him and bench graft or topwork them into his orchards. He did not attempt to verify varietal names or even to vet out potential duplicates. Identification was never his priority. This is a big, inclusive collection where any and all are welcome. Every winter he sold scionwood, primarily through the mail. There was always some risk when you obtained a variety from his collection, as there was no guarantee that it was true to type, only that Nick had received it as such.

Just one of Nick Botner's note books: document everything!

I obtained his list and went through it multiple times, name by name—all four and a half thousand. There were many varieties I knew were historically grown in Maine but that I had not yet found in the state. My hope was to locate a number of them in Nick's collection. I'd purchase scions, bring the wood back home, graft them and eventually plant them in the new Maine Heritage Orchard at MOFGA.

Then I began to hear alarming rumors. It sounded as though Nick's collection was in serious jeopardy. Nick was in his eighties, and wanted to retire. He hoped that someone a third his age would take over the farm and steward the trees. But, as it is for many Americans, the farm was also his retirement. He couldn't just give it away. He'd have to sell it. A young person dedicated enough to want to take over an orchard of several thousand different apples would be hard to find. The odds of such a person having the resources to purchase the farm were even smaller. It looked as though one of the largest fruit collections in the world would be lost.

In August 2012 Cammy and I attended a wedding in Oregon. While we were there we made a detour to visit Nick and his wife, Carla. We hadn't intend to stay long, but we did. Nick took us on a tour through the orchards and then they insisted we stay for dinner. We left late and drove back to the Portland area, nearly three hours away. It was a magical visit.

Later that fall I made the decision to return to Yoncalla and cut scionwood of "Maine" varieties for the Heritage Orchard. In doing so, I would also help preserve some of Nick's rarest apples. I also thought it would be fun to do something with Dan Bussey. Dan was working for Seed Savers Exchange in Decorah, Iowa, where he was also assembling a preservation orchard. I thought he might like to come along. He loved the idea. We contacted Nick, and he was delighted as well.

On February 12, 2013, Dan and I met in Portland and drove down to Spearheart Farm in Yoncalla. We intended to stay at a motel. We didn't want to bother Nick and Carla, but they would hear nothing of it; they insisted we stay with them. They were generous and gracious hosts. We ate and drank like royalty. We had long leisurely, wonderful conversations. One day there were eight of us for lunch. Next day there were

Shaun Shepherd, Nick Botner, JPB and Dan Bussey, 2013

nine. We spent all day cutting scionwood in the orchards. I had my lists, and Dan had his. Sometimes I wouldn't see him for hours. Then we'd pass one another with our hand pruners and our clipboards, mumble a quick hello, and disappear again down another long row of trees.

One of those also cutting wood was Shaun Shepherd of the Home Orchard Society and the Temperate Orchard Conservancy. Shaun and others were cutting scions of everything Nick had. Their goal was to re-establish Nick's collection a hundred and fifty miles due north in Portland. The scale of the project was immense, and the work would take years. It was great to meet Shaun, another fanatic fruit explorer and apple identifier. It was also heartening to know that Nick's collection might not be lost after all. In early April Cammy and I grafted another hundred varieties for the Maine Heritage Orchard, forty of which I brought back from Nick's, carefully labeled, wrapped in plastic bags, and tucked into my suitcase.

That summer, eight months after Russell's death, an excavator and a large bulldozer descended upon the gravel pit, not to extract more sand and gravel this time, but to carve terraces from the eroded hillsides—terraces that we would plant with the first hundred and two historic apple trees the following April. Let the renovation begin.

The arrangement with MOFGA's Board of Directors gave the orchard operational autonomy and financial independence. Not wanting to be a drain on the MOFGA staff and budget, we decided that the orchard would operate under the guidance of a volunteer committee supervised by a staff member, create its own budget, and raise its own money. Do what you can, and pay as you go. Our goal was to get the varieties planted and protected while continuing to renovate the pit. The bells and whistles, the staff and the multi-million dollar visitor center can wait.

Nick and Dan looking through one of Nick's notebooks, 2013

Back in the early planning stages of the orchard, Russell and I visited a husband and wife who expressed interest in the project. We arrived at their house hoping we'd endow the entire orchard before we dug the first hole. That turned out to be a shovel-full of wishful thinking. The couple listened patiently to our pitch. "How much are you asking for?" We gave them a figure, and they said no.

Newly created terraces in the old gravel pit, summer 2013

Then they asked us if we'd like some advice. Of course we said yes. They told us to begin the project with whatever money we could scrape together. They said not to worry about how much we raised. Just begin. As you progress, people will see the goodness in what you're doing, and they will want to contribute. Someday you should endow the orchard. You will want to keep it going for generations to come. But not yet. Just get going. Dig and plant. Despite having been turned down on our big ask, we drove home that day feeling great.

Russell was no longer there to cheer me on or run interference on my behalf, but, with a new and enthusiastic committee in place, we raised enough money to begin the project. With the assistance of Lisa Fernandes, Jesse Watson, Haas Tobey and others of the Maine permaculture community, we staked out and then created about twenty contoured terraces of various lengths sculpted into the hillside of the first three acres of the pit. For several days we had some big machines on site doing the work. When it was done, it reminded me of Peru or Vietnam, though we weren't planting potatoes or rice. This would be a poly-culture preservation orchard. We marked out a hundred and two spots for the first trees, then fertilized each location with a pile of rock powders and compost. It was October 2013. On November 10, a large group of volunteers dug the two-year-old trees from our nursery and stored them for the winter in the Fedco Trees warehouse. Planting day was six months away.

Planting the first hundred apple varieties on the terraces at the Heritage Orchard, April 2014

Although many thousands of Baldwin trees perished in the terrible winter of 1933-34, you can still find the occasional Baldwin thriving on the edge of the yard near an old house, like this one in Harpswell

# Thirty-Six

# Blake: the Movie

"There go two of my threads, Watson. There is nothing more stimulating than a case where everything goes against you. We must cast round for another scent."
Sherlock Holmes, *The Hound of the Baskervilles,* p. 696

The fall of 2013 featured another episode in the adventure of the missing Blake. In the spring I was contacted by a French magazine writer named Christelle Gerand, who asked if she could come to Maine to meet me, in the hopes of writing an article. A month or two later I was contacted by a filmmaker from the West Coast named William Kirkley. He asked if he could come visit and film for three weeks in the fall. Both William and Christelle had read an article by Rowan Jacobson about my work in the April 2013 issue of Mother Jones magazine. I said yes to both of them.

William arrived in early September. He stayed in our friend's empty cabin down the road. We were together every day. Abbey Verrier was usually with us as well. Abbey was a student at College of the Atlantic, doing a fall internship with me. I had met her on the trip to England two years earlier. She had learned a lot and we were a good team. Most days, it was the three of us, with all of William's gear, packed into my vehicle, racing around central Maine, visiting orchards, identifying apples, and giving talks. Although William was happy to go along with whatever we were up to, he also "needed" certain shots, like the one where Abbey and I tried to look normal as we put up wanted posters in downtown Ellsworth, while amused bystanders looked on.

Abbey Verrier up in a tree on a perfect fall day, 2013: photo by Emily Skrobis

In many ways, however, it was a typical fall, filled with an assortment of overlapping exploration projects. Some were in their final stages, others were just beginning, and the rest were somewhere in the middle. They can take years. Sometimes an apple I'm looking for, or attempting to ID, will slip into the background for long periods of time, only to reappear at some random moment. As we drove around, I entertained Abbey and William with apple stories. One was about the Blake. Perhaps because of the Mother Jones article, it kept coming back to the Blake. It became the obsession of the month. We had to find the Blake. Of course, the next break in the Blake case could be tomorrow, and it could be in twenty years. Maybe it'll never come at all. I had plenty of other things to do. Lots of other apples were waiting to be rediscovered, each with its own unique story. But for William, it seemed as though only the Blake would do.

The beautiful September days were gliding by, and the Common Ground Country Fair was approaching. William had only a few days left. I had totally forgotten about Christelle when she called a day or two before the Fair. The Fair would be over on Sunday. William would be leaving on Monday. Christelle would be arriving Saturday and leaving Monday as well. I had told them they could both come. Christelle was driving up from New York City. William had come all the way from LA. Both of them wanted my attention. The Fair also demanded a great deal of me. I had a lot to do.

Christelle arrived. She was looking for a great hook for her story too. She also wanted me to find the Blake. But Blake really isn't that big a deal. It's just one apple I'm trying to find. There are lots of other apples out there. Let's go find *them*. But to William and Christelle it was the Blake. The blankety-blank Blake.

By this time, I was pretty sure poor old Mr. Blake was rolling over in his grave wondering why on earth he was getting all this attention after a hundred and fifty years of peace and quiet. No one had given a whit about his yellow, medium-sized, fall cooking apple for generations, and now this? Maybe the apple's no good anyway. Maybe it's a spitter. Maybe that bitter apple on the Flaggy Meadow Road in Gorham was correct. Can't we just forget the Blake and go find a Givens or a Naked Limbed Greening?

The Fair went well. Sunday night we dismantled the booth and I went home and collapsed. Christelle returned to her motel. William returned to the cabin down the road. Both would be leaving the next afternoon. I might actually find some time to relax for a day or two before the plunge into serious apple season. After all, October was still ahead and October is the best apple month of all. Before I went to bed I decided to check my email. I flipped through this and that, and then saw a curious one from CJ Walke, the orchard manager at MOFGA. It had a phone message attached to it. I'd never received one of those before and never have since. I clicked on it. It was long and urgent.

> I need to talk with John Bunker. He wants to look at a Blake…An apple tree that I'm pretty certain is the Blake that he's been looking high and low for that was first originated in Westbrook, Maine. My tree is in Hiram, Maine…It's got apples on it and I would like him to see it…If I could talk with him that would be great. I would like his phone number or him to contact me. I tried through Fedco and got no response. I left a message two weeks ago. My name is Foote. Thomas Foote.

Usually the day after the Fair is a quiet day. At Fedco it's an unofficial holiday. Everyone is spent. No one comes to work. When William showed up in the morning I played the phone message to him. Then I played it again. Then he filmed me listening to it. I acted surprised. I was surprised. Then Christelle arrived. I knew I would have to go to Hiram. I knew they would both be coming too. I called Tom Foote and reached him. The tree was on an old farm he owned, but where he didn't live. He would not be able to join us but was happy for us to go without him. There was an old, abandoned house. He gave me directions.

William Kirkley and JPB with possible Blake tree, Hiram 2013: photo by Christelle Gerand

This could be the real deal. Was it the end of the rainbow? Who would get the scoop? William? Christelle? What about me? Actually I was the least of my worries. I just wanted to get to Hiram and back without falling asleep at the wheel. Christelle had called me first, but William had just spent nearly a month with me. Christelle had showed up right at the last minute. I was feeling like a

Supreme Court Justice—it's five to four, you get to see the apple tree first. Somehow the three of us were able to come to an agreement though I can't remember exactly what it was. At least we were able to make a plan: all three of us would go.

At 4:00 in the afternoon we got into three cars and headed for western Maine. It was about a hundred mile trip and would take us two hours to get there. That would give us an hour of daylight if all went well. William was packed. He would be heading for Logan Airport and Los Angeles. Christelle would drive back to New York City. I would head home and go to bed.

Tom Foote's directions were good. We found the road and the old abandoned house, and our three-car parade pulled in. There was the tree, exactly where Tom had said it would be. It was all according to plan. The fruit was medium-sized and greenish-yellow. William filmed. Christelle wrote. I wondered if I could somehow make this into the Blake in order to end the search at this perfect moment. Was truth once again stranger than fiction? Could I just declare it to be the Blake and that would be that?

Time to scan the mental check list. We were in Hiram. That's about thirty miles from Westbrook. Not a huge distance for a variety to travel, but at least a few degrees of separation. Was it a seedling or a grafted tree? Had someone topworked it many years ago? The tree was not in an orchard. There were a few other apple trees scattered about, but there was no pattern to the group. It was on a stone wall. Most of the trees on stone walls are seedlings, but many thousands of those seedlings were topworked in past generations. Grafted trees on stone walls are not rare. The tree was old but not ancient. The tree was large, but maybe it was too young to be a Blake. Blake disappeared from cultivation over a hundred years ago, or at least we think it did. I couldn't find any evidence of a graft. There was no obvious bulge, indentation, ripple in the bark, or color change that would indicate a graft. Instead,

Christelle Gerand, Hiram 2013: photo by William Kirkley

the trunk looked as though it was made up of multiple stems that grew up close to one another and fused together over many decades. There was no evidence the tree had ever been pruned. The fruit was too green in color. Perhaps it just wasn't ripe. It was still September.

Each one of these factors was by no means a deal-breaker. But, taken in total, they probably were. I had a feeling this fruit was not going to turn yellow. I didn't want to make any great pronouncements right then, but I was pretty sure that we'd found another seedling. This was not a Blake.

I circled the tree and looked and thought. I think I was hoping I'd find a tag on the tree that said "Blake tree 1885 just for you." Meanwhile I babbled out loud, trying to sound at least vaguely intelligent. We did impromptu photo-ops. Christelle took photos of William and me and the tree. William took photos of Christelle and me and the tree. I could tell they were both happy. It was okay that I had brought them both along. I was happy too. The tree was beautiful. No doubt it was a fitting conclusion to our adventure.

By now it was getting too dark to see. The sun had gone down. William packed up his cameras. We said our goodbyes. William headed for the airport. Christelle for New York City. I headed home. I never saw either of them again.

Though the Hiram "Blake" tree is old, it wasn't Blake; the three trunks grown together and the uncultivated/unpruned look are giveaways: it's a seedling!

The Grasslings tree in an overgrown field in Castine with no sign of ever being pruned: clearly a seedling

## Thirty-Seven

# The Key to the Tree

"Oh, he rates my assistance too highly," said Sherlock Holmes lightly. "He possesses two out of the three qualities necessary for the ideal detective. He has the power of observation and that of deduction. He is only wanting knowledge, and that will come in time."

Sherlock Holmes, *The Sign of Four*, p. 9

Ribston Pippin

With desire, intention and a bit of coaching along the way, observation and deduction are learnable skills. The first is about the acquisition of data. The second is about the use of that acquired data to answer a question. For the CIA, the FBI, Watson, Holmes, and the apple identifier, both are essential daily activities. They are the keys that unlock the door. We seek to train ourselves in such a way that the two become internalized. They become second nature. They become automatic.

But although observation and deduction are powerful tools, it's the accumulated experience that holds the answer. And that third piece takes time. You can't know what it is if you don't know what it could be. You look, you process, you remember. That's knowledge, the accumulation of the experience stashed away over time. And that takes patience. Someone tells you it's a Forsythia, and you file that bit of information away in your brain somewhere. Then when you zoom by a large yellow-flowering shrub on the berm by the overpass on the interstate, your brain sifts through and eliminates roses and lilacs and oaks and crabapples. It eliminates hundreds of plants you don't even know. That process of elimination, that process of deduction, might take a minute the first time. "What was the name of that plant she told me about?" The next time it might take thirty seconds. The third or forth time, it's immediate. You see, you know. You carry this knowledge base, this data base with you. You observed, you deducted, you knew. That's Starkey. That's a Pound Sweet. That's a Somerset of Maine. It seems as though it's instant, but it's not. It's time slowed down.

The knowledge database consists of a lot of information you obtain and store over many years. Hopefully it's stuff you love so you enjoy the data entry. Some of it you need to have in your head or at your finger tips. Or maybe both. Some of it you don't. You always need to know your McIntosh. And don't forget Cortland. Then if you get inspired, learn twenty more. Don't worry about the other ten thousand. Once you've got the first twenty down, the next twenty take half the time. And the following twenty, even less.

While you fill up your brain with the essentials, keep notes in file folders in a box or a drawer or on your device. It's okay to store some of it off-site. It won't all fit in your head anyway. Buy a few books, and flip through them frequently. Get yourself a copy of *The Illustrated History of Apples in the United States and Canada*.

During the past few years, as a supplement to the drawings, photos, file folders, and books sprinkled around the house, I've been assembling a searchable computer database of apple varieties. It's roughly analogous to accumulated information technicians obtain when they extract the DNA from crushed apple leaves. Both are databases with which to compare the next mystery apple. This "genetic fingerprinting" work is being done at a few research labs around the country, including Pullman, Fort Collins, and Geneva, New York. They have the equipment and the trained staff. The information they gather can be incredibly useful to breeders, in particular. They can identify traits that the apple parents, grandparents and even more distant ancestors are apt to contribute to their progeny. The process, however, has its challenges for the variety identifier. As the University of California at Davis Plant Services Foundation explains, "To identify the variety, the DNA profile [a.k.a. fingerprint] of the client's sample is matched to a reference database." In other words, you have to have a correctly identified specimen with which to compare. Unfortunately, the last person alive in the U.S. who could identify a Blake may have died a hundred years ago. With an incomplete "reference database," what do you use for comparison?

The same is true for the thousands of other mystery apples. What is needed is an army of fruit explorers swarming over the landscape, studying old diaries and deeds, identifying the ancient apple trees before they're all gone. That's the most important information. Meanwhile, the technology continues to race along. Someday…..?

With the help of computer gurus at Fedco, as well as Cammy and the apprentices, I've been slowly assembling the searchable database. It's a gigantic taxonomic key of material assembled from old books, letters, notes, conversations and firsthand observation. It's turning out to be a good place to store a lot of information that doesn't need to be with me all the time. It's information that becomes useful when bags of apples arrive in the mail or when I hit the fruit exploring trail and need to bring myself up to speed again. It's not foolproof and it's got a long way to go, but I think of it as a computerized version of Watson, assisting us as we attempt to eliminate the impossible.

When it comes to doing ID's, I'm at the same disadvantage with our key as the USDA genetic testers are with their genetic fingerprints. I need positively identified specimens of all these rare apples. Then I can enter detailed descriptions into the key and use that information to help with future ID's. Where is that army of fruit explorers?

There are several classic apple books with identification keys. I'm familiar with three of them. *The American Fruit Culturist* by John J. Thomas (1875) and *Systematic Pomology* by U. P. Hedrick (1925) have much to offer to the identifier. Neither one, however, is complete. Hedrick includes a hundred and fourteen apples "which students of systematic pomology are likely to find in modern orchards the country over." Thomas includes about three hundred varieties, useful for mid-to-late-nineteenth century plantings. The third key I've spent time with is in Robert Hogg's 1884 *Fruit Manual*. His is the most detailed and extensive. His key is based on the relative position in the calyx tube of the stamen threads, as well as the shape of the calyx tube, the shape of the seed carpels and the configuration of the sepals. He gives no attention at all to the size, shape or color of the fruit. Having determined if the apple's stamens are marginal, median or basal, the identifier is directed down various steps towards a final answer. This is pretty detailed stuff!

Unfortunately (or fortunately) for us, Hogg was over in England, and most of his varieties have never been grown here so we're spared from having to deal with those stamens.

In Thomas and Hedrick, you're presented with a set of questions in a prescribed sequence. Both begin by asking if the apple is sweet or tart. Depending on your answer, you're directed to your next question, skipping over questions as the key dictates, all the while narrowing the pool. Eventually you arrive at an answer. For more common varieties, Thomas and Hedrick are good. But do you have a common variety?

Before beginning my key, I studied Thomas and Hedrick. I also consulted other keys I'd seen. Maybe they use techniques that could be helpful. My grandmother was a classical pianist. She had a concert grand in her living room and taught lessons for many years. She had a book by Harold Barlow and Sam Morgenstern called *A Dictionary of Musical Themes*. She kept the book on a shelf next to her other reference books. The *Dictionary of Musical Themes* contains ten thousand tunes along with an alphabetized "notation index." In my late teens and early twenties I regularly visited her and my grandfather in Boston. She and I would sit together on the couch at night and listen to classical music on her record player, following along with the open score, which lay on her lap.

From time to time I would remember a piece of music from Ermano Comparetti's music appreciation class but couldn't remember the title. I would whistle what I could remember to her. She would write down the tune in the key of C, then turn to the notation index where simplified versions of thousands of tunes were written out using the alphabet. Once found there, she would then flip to a corresponding page where the full theme was written out with the composer's name and the name of the piece. "Oh, that's Mozart's D major K. 451, 2nd movement, dear....are you ready for dessert?"

The great feature of the musical key was the section of simplified versions of ten thousand tunes. That would be like having a thumbnail sketch of every variety grown in Maine. Each thumbnail sketch would lead to a more detailed description. You'd read that description, and you'd know if you had that apple. Unfortunately, we don't have more than a name for many of the old apples. Hard to do IDs without decent descriptions. But we find more each year, and into the key they go. For each description we include our apple version of the musical thumbnail.

My grandmother, my mother, and many of their friends owned Roger Tory Peterson's *A Field Guide to the Birds*, which first appeared in 1934. *A Field Guide to the Birds*, and all the field guides that followed, are basically modified keys. About the time I turned ten, I was given my own copy. I think Peterson knew Sherlock Holmes well: "Identification by elimination plays an important part in field work...It is often quite as helpful to know what a bird could not be as what it might be...A quick field observer who does not temper his snap judgement with a bit of caution is like a fast car without brakes."

One thing I particularly like about Peterson is that each description ends with a few words about the "range" of the bird. You aren't going to see black-capped chickadees in Florida. In 1966 *A Guide to Field Identification: Birds of North America*, illustrated by Arthur Singer, went a step beyond Peterson by including a small map of the range of each bird in the corner of the description. The recent *Guide to Birds* by David Sibley follows that same protocol. These small maps are great. Knowing the range of an apple can save the identifier many hours, or even days.

Gene Frey and Clayton Carter at the Fedco Warehouse, 2018

Gene Frey and Clayton Carter have been two of the computer guys at Fedco for many years. Over the course of two or three winters, when they had some time, they created the innards of my searchable apple database. The key. I provided Clayton and Gene with the apple characteristics, and they took it from there. They got it to the point where I could use it in my own work, though not yet to where we could put it up on our website. We still have a lot of information to gather and enter.

Once Gene and Clayton were able to hand off the key, my job became entering data for hundreds of varieties. Each variety receives its own page. To supplement the key, Cammy and I also created a glossary of all the terms used in the form; if you're going to use the lingo, you've got to know what it means.

The database is only useful to the degree to which the information is complete and correct. This is the knowledge that Sherlock says "will come in time." I've limited the database to apples thought to be grown in the New York-New England area during the past four hundred years. Someone else will have do the rest of the U.S. I know there will be unintentional omissions and errors. Fortunately, the database is flexible; it allows me to continually edit and add new varieties. My decision to focus on the Northeast was out of necessity. The data entry piece is so huge that, even with a restricted geographical focus, the key already includes over sixteen hundred varieties. Once it's entered and double-checked, it will be formatted for the web, and made public.

The key is also only as good as the identifier is clever. As it is with Thomas and Hedrick, the method behind the key is to ask the database a series of questions, each one related to an apple characteristic. One advantage of a computerized key is that the questions can be asked in any order. With each question the pool of potential candidates is reduced, sometimes by only a few and other times by a lot. By eliminating the impossibilities, the key reduces the number of suspects. Once that number is down to a manageable number, say twenty or thirty, I abandon the search function and read the complete description of each of the remaining apples. Hopefully, one of them is it!

My strategy has been to play a game akin to Twenty Questions with the key. Because of the lack of information about many varieties, I ask most of the questions in the negative. If I ask for an apple with a long stem, I'll miss all the varieties for which the stem length is unknown. And there will be many. So instead, I'll search for all the apples that do not have a short stem. By approaching it in the negative, I reduce the pool by eliminating the short-stemmers while retaining all the long-stemmers and the unknown-stemmers. As I continue to search in the negative, I reduce the pool one search at a time until I get it down to that manageable number of possible matches. At that point I read those full descriptions and keep my fingers crossed.

Each page in the database also includes a space to enter varieties that resemble the apple in question. When someone gives me a bag of apples that resemble Baldwin, if I've entered my data well, I can select for all the varieties that resemble Baldwin and instantly eliminate thousands of impossibilities. It would likely lead me to Stark, Milden, Red Canada, and all the other Baldwin lookalikes I've become familiar with over the years. Then I would pull out the full descriptions and read. I'd look at photos. I'd look at the watercolors.

Over time I've assembled a short list of twelve distinctive and well-known apples to use for comparison. I've trained myself to use one these varieties in my "resembles check-off box" whenever I enter a new description. Each of them has become its own category for aiding in IDs and for describing varieties to friends and colleagues. Saying an apple resembles Fletcher Sweet may mean nothing to you, but telling you it resembles Tolman Sweet might be very useful. I want the twelve to be well-known apples, small in number but comprehensive in scope. I want this to be helpful but not misleading.

Recently I've had some inspiring conversations with a few computer geeks. Ultimately we should be able to fill out an apple description questionnaire, press a button and stand back as the computer instantly compares the completed form with everything we have on file, spitting out a list of suggested matches. If only I live so long. For now I still rely on the bag of apples on the dining-room table method. It works pretty well some of the time.

Unlocking the mystery: photo by Emily Skrobis

Maybe one of those bags will be from you. I'll gently pour the apples out and spread them around on the table. I'll look at the basin and the cavity. Abrupt or obtuse? Russet? Dots? Furrowed? I'll cut them up both ways. Axile or abaxile? Core lines meeting or clasping? Stamens, calyx tube? I'll send you an email and ask you a few questions. We'll go back and forth. You'll send me a few photos. I'd like to see the trunk. Where it is located? What state, what town? When was it planted? How old is the tree? Is it in an orchard? Is it a seedling or a grafted tree? When is it ripe? I'll jot down a bunch of notes. I'll flip through my books and photographs. I'll fiddle with the key. I think it resembles Golden Delicious or Grimes Golden. I'll press a button. I'll eliminate this and eliminate that. Wouldn't it be fun if I find myself staring at the page on the screen: "Blake."

# Thirty-Eight

# The Discovery

"I've told you all I know myself now, for the rest is mere surmise and conjecture."
Sherlock Holmes, *A Study in Scarlet*, p. 34

The Windham Blake

In February 2014 I received an email from David Buchanan about an old apple tree on Rte 202 in Windham. Windham borders Westbrook. This would only be a few miles from where Blake originated. David hadn't been there, but it sounded promising:

> I met someone who bought an old farmhouse in Windham recently, and was asking me about a tree in his yard we may want to look at. He took a grafting workshop at MOFGA last year and approached me to ask about saving it. According to him, it's a golden apple that hangs very late into the year, well into winter. He believes it's not a seedling, that it has a nicely maintained shape, and it's quite old, twenty-five feet tall, several feet around and hollow. Interesting…

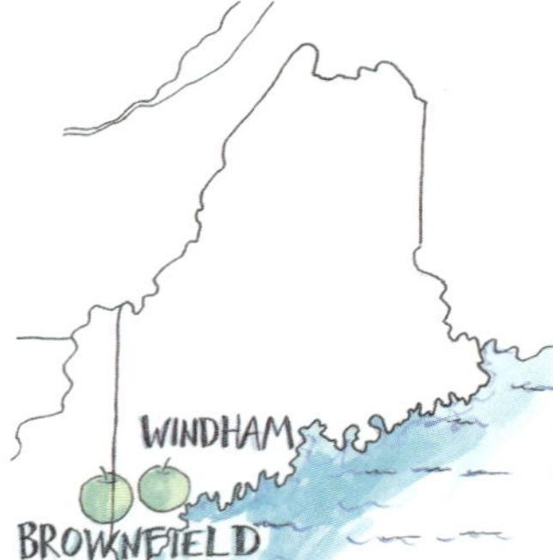


On October 27, 2014, a year and a month after the trip to Hiram with William and Christelle, I drove to Windham with Natalie Beaugard and Laura Sieger, our two new fall college interns. After some poking around, we were able to locate the home of Don and Shannon Johnson. Shannon was home. "Of course you can go check out the tree." There it was on the lawn, less than fifty feet from the pavement. Why are so many old apple trees so close to busy roads? Not so long ago those big roads were gravel and not so long before that, they were dirt. The roads were lined with apple trees back then. This was one of the lucky ones. The others are gone.

The tree was old enough to be the right vintage. It was grafted. It was hollow. The fruit was yellow. It fit the description. We took lots of photos. Unfortunately, the fruit was mostly gone by, but we collected all we could. We took it home and keyed it out. We compared it to my description of Gerald Feyer's Blake. According to research Todd had done, the Johnson site is only seven miles from the Blakes who lived near the Duck Pond at Pride's Corner in Westbrook. That sounded pretty good, maybe even convincing. Come back, William and Christelle!

The hollow Windham "Blake" tree in 2014; this one might be it

There were still other leads to follow. One was from a fellow named Norman Blake. I don't recall how Norman and I met; it may have been at the Fair. Norman had connections with old Blake farms in the western Maine town of Brownfield. Long ago the Westbrook Blakes swapped farms with folks in Brownfield. There's a good chance they took scions of their apple tree with them. It might still be there. According to Norman, the Blakes came across the Atlantic between 1630 and 1640, made a pitstop in Massachusetts, then headed north about the time that Massachusetts annexed Maine in 1691. They moved to Brownfield around 1840, as the Westbrook area became too crowded. In those days, when cash was not readily available, farm-swapping was not uncommon. Norman wrote to me, "It shouldn't be hard to find a round green palatable apple if one is still up there. If I find anything that looks suspicious, I'll tie some flag tape on the tree and let you know."

The Windham Blake with apples inside

In the fall of 2016 I decided to contact Norman Blake and see if he was up for a trip to Brownfield. He was game. On October 20, four of us headed west. Our first stop was the "Old Blake Farm." There was no one home but we had permission to wander around. The buildings were in excellent condition. The grass was recently mowed, but, unfortunately, the orchard was gone. One lonely fruit-less apple tree remained. A Blake? It's not unusual to find well-maintained sets of old farm buildings with no orchards. Apparently someone thinks they're doing a good thing by cutting them down and cleaning up the mess. It's okay to leave them. Gnarly old apple trees are beautiful. Please let them live out their days. It felt good to stand on the spot where bushels of Blakes were once picked and eaten, but we wanted the real thing.

The night before we headed to Brownfield, I went through my file folders and pulled out the phone number of someone named Dean Robinson. Dean also lived on a Blake farm in Brownfield. He had called us a year earlier, and later sent an email with photos of six trees on his old farm. He was looking for assistance in identifying and saving the trees.

> "... I acquired the property about 15 yrs ago and it is one of the Blakes houses (1860's) in what is called here the Blake neighborhood. The original Blake "starting" house is about 1 mile down the road and there were quite a few Blake family houses around here in the day I hear? This part of Brownfield was about the only spot that wasn't burnt in the fire of '47..."

Norman Blake and JPB in Brownfield, 2016: photo by Laura Sieger

I called Dean in the morning before we headed out, but there was no answer. I left a message. Norman was unaware of the other farms or anyone named Dean Robinson. Still he was up for an adventure. None of our cell phones got reception but we stopped at a neighbors and borrowed their land line. I called Dean Robinson, and he answered. "Yes, come on by."

Dean gave us the deluxe tour. It was wonderful. The buildings looked a bit tired and the lawn a bit shaggy, but not one apple tree had been removed in a hundred years. They were exquisite in their ancient decrepitude. These are the trees I love. Out came the paper bags, the Sharpies, the pole picker and the camera. I made a map. We took samples. We said our goodbyes.

I've given much thought to Dean Robinson's "Blakes." His farm belonged to one of the siblings or cousins of the manicured home down the road. Being on a Blake farm makes them all Blake apples, but are any of them *the* Blake apple? Several are yellow. They all ripen late. One in particular has promise. I call it #5 for now. We'll go back again for another look before too long.

Possible Blake from Brownfield

I've seen a lot of Blakes over the years. I've lined them up and studied them. I've sliced and diced them every which way. I've compared them. They're all in the key. They're all similar. Maybe one of these days I'll take leaf specimens out to Cameron Peace and his crew at the RosBREED project in Pullman, Washington. I'll watch as they grind up the foliage, extract the DNA and compare the genetic fingerprints. They won't be able to tell us which one is Blake; it's too obscure. They will be able to determine who's the same as who. That will be very interesting. Will the real Blake *please* stand up? I could make a guess. I like the "Windham Blake" the best. Sherlock might not approve, but he too came up with the occasional "provisional theory."

Meanwhile, in the spring of 2014, Brodgale finally sent scionwood of their "Blake" to the Plant Germplasm Quarantine Program in Beltsville, Maryland. It's now been checked for viruses. It's undergoing a process designed to clean it up. Someday the USDA will release it. I hope I'm still around to topwork the wood. If I'm lucky, I'll get to see the fruit, cut it up and taste it. I'm looking forward to making lots of applesauce.

And when that Brogdale/Norwich/Charles Dickens/Barkis/Blake apple finally fruits, by then I don't think we'll really care if it's correct. We'll just be glad to close the case. We'll shut the book. We'll send out a mass email and throw a huge party in Westbrook at the Westbrook Historical Society. We'll invite Peter Christiansen and Christelle and William and Abbey and Laura and Natalie and Don and Sharon Johnson and David Buchanan and Thomas Foote and Norman Blake and Dean Robinson and every remaining Blake descendant in the state of Maine, and we'll fly in Gerald Fayers and everyone from Brogdale. You can come too. Then we'll set up a big tent under the W. L. Blake sign in the Standard Bakery parking lot on Commercial Street in Portland and hand out Blake apples from Norwich and Windham and Hiram and Brownfield and everywhere else. We'll cook up a hundred Blake pies and stew up a gigantic vat of Blake sauce and tap a barrel of beautiful golden Blake cider. We'll sing and dance under a Blake moon and shout "wassail!" We'll graft a thousand Blake trees and pass them out on the streets of Cumberland County. Blake is back!

CONICAL CALYX TUBE... STAMENS MEDIAN ...
CORE LINES CLASPING... CARPELS...

# Apple Facts
# Getting To Know an Apple

Let us see that we have our facts correct before we start.
Sherlock Holmes, *The Case Book of Sherlock Holmes*, p. 1095

It's nearly impossible to describe an apple without knowing the correct terminology. Before attempting to identify one, it's a good idea to look at the fruit and learn all the parts. Most of us know a wing from a beak and a bat from a ball, but do you know a basin from a cavity? We're going to dissect an apple together. The goal is to get to know the words we use everyday.

Though it's not absolutely necessary, the exercise may be more useful if you do it too. All you need is a sharp knife and a few apples. I'm going to use McIntosh since it's usually available in this part of the world. Any apple will do, but if we both use Macs, we can compare details as we go along. You should be able to purchase a half peck at your local orchard, farm stand or grocery store. A half peck is a small bagful, about five pounds. A bushel of apples is about forty pounds. Two pecks make a kenning. Four pecks make a bushel. Three bushels fill one of those old-fashioned wooden apple barrels.

A classic Mac: Marshall McIntosh a.k.a. Worcester Strain (sport), Fitchburg MA, 1975

***Name***: We start with the name. In this case, it's pronounced like the computer, but not spelled that way. It's McIntosh. What do you do when you don't know the name? Let's say that you're keying out an apple you hope to ID. What now? Make up a provisional name that somehow makes sense. I typically use the name of the tree-owner or the tree's location. Over the years I've assembled half a dozen candidates for the Blake. I've got them labeled "Blake Gorham," "Blake Portland," "Blake Hiram," "Blake Windham," etc. I've got another apple, one I call Green Monster. It's big and green. I know it has a name, and I plan to find it someday, but for now it's Green Monster.

***Parentage and origin***: This information can be very helpful in doing an ID. Usually you'll have to look it up, although sometimes even the source you consult will have it wrong. Better check two or three and see if they agree. Beware of the web; internet sites often cut and paste from each other. Be on the lookout for the moment when someone presents you with an entirely brand-new story. Then you won't know who to believe. Be prepared for surprises.

***Synonyms***: Most historic varieties have multiple synonyms. Some have a dozen or more. List them all. Synonyms can often lead you to an identification. If X=Y and Y=Z, then…

***Resembles***: **What varieties does this apple resemble?** I love meeting my friends' siblings for the first time. I may not know it's the nose or the mouth or the eyes. But somehow my brain knows instantly. Of course, they're your brothers. Could that be one of the ways those FBI agents track you down? They study photos of your brothers and sisters?

A new variety or a synonym for an old favorite?

When you locate a new apple and it looks like a McIntosh but something tells you it's not, you want to know what it could be. The usual suspects in this case would be Cortland, Fameuse, Macoun, and Paulared. Maybe it's one of them. Of the more obscure varieties, consider Brock, McClellan, Noyes, Parlin, Pomme de Fer, Scarlet Pippin, Rolfe or even St. Lawrence. The old apple books rarely tell you who resembles who. If only they did, our work would be so much easier. Ben Davis resembles Jonathan. Stark resembles Baldwin. Ortley resembles Yellow Bellflower. Whenever I key out an apple, whether I know its name or not, I list all the varieties it resembles, even if only obscurely. Sometimes I'll note just a color similarity, or the size or the shape. Having a record of the varieties an apple resembles provides a starting point when it's time to dive back in and give it another try.

***Twelve General categories*** (by appearance): I pigeon-hole every apple I describe into one of twelve general categories, each being a particularly iconic variety. When I do an ID, I categorize the mystery apple, and then consider all the other varieties that fit that category. It's another way to eliminate the impossible!

1. Baldwin: large round-conic & red 2. Calville Blanc d'Hiver: ribbed 3. Duchess: medium & striped 4. Fameuse: med-small & red

5. Golden Delicious: conic & yellow 6. Golden Russet: russet 7. Northern Spy: large, irregular, red striped 8. Delicious: conic & red

9. Rhode Island Greening: large & green 10.Summer Sweet: small round 11.Tolman Sweet: med & yellow 12.Wolf River: huge & red

***Season*: An apple's season is the time of year when it's ripe**. Unlike many of the varieties in the grocery store that are available year-round, heirlooms have a definable period when they're ripe. Learning the season of the apples you're likely to see in your area is good information whether you do IDs or not. Every chef should know when the Red Astrachans or the Gravensteins or the Ashmead's Kernals are ready. You will be happiest if you eat them at their best. For the identifier this information alone can dramatically reduce the pool of possibilities. Ripe in late October? It's not Wealthy or Duchess. They would have gone by. It could be Starkey or Black Oxford. Is it ripe in August? It could be Red Astrachan, Yellow Transparent or Early Harvest. How about the first half of September in Maine? Could be Cole's Quince. The season in Maine for Macs is generally October. That's when they're at their best. People want them in early September but they're not ready. You can get them in the winter and spring, but they're never as good as they are in mid-fall. And they don't grow Macs in the Southern Hemisphere so, unlike Gala and Braeburn, you aren't going to get a new Mac in May.

***Vintage***: Is your apple an heirloom, a modern introduction, a wild seedling, a European import or a Russian import? These are groups we've identified to further reduce the potential pool of candidates as you ID. Every apple will fit into one or another. **An heirloom is an apple grown before 1934**. This would make Mac an heirloom. 1934 is just a convenient date. It was the last "test winter" when millions of apple trees died in the Northeast. Although by 1934 the traditional orchard culture had been faltering for a few decades, that winter was like the final nail in the coffin. **An apple introduced since 1934 is a modern introduction**. If you're unsure when your tree was planted, make a guess.

**A wild seedling is an unnamed seedling** that is yet to be grafted. Keeping track of European and Russian imports is also useful. While some of them were brought over early enough to be considered American heirlooms, many didn't immigrate to the States until recently. Some areas of the U.S. were swamped with Russian introductions in the late nineteenth century. Being aware of them can be helpful.

***Was it grown in Maine***: Is there any evidence that the apple you think you've found was ever grown in Maine (or your state)? There are thousands of varieties, and we're looking for ways to reduce the possibility pool. Any time we do an ID up here in northern New England, we immediately eliminate thousands of southern apples from contention. It's not them! They weren't grown up here. It is true there's a few exceptions. We do have a Grimes Golden right up the road and Ben Davis trees are all over central Maine. Where did Ben Davis originate, anyway? But, for the most part, most southern, mid-Atlantic and mid-western varieties weren't traditionally grown up here.

***Found growing in these towns***: This is a more focused version of the last question. When we find a Black Oxford or a Starkey growing in one town or another, we take note. When we find an old list of apples historically growing in some specific community, we copy it down or commit it to memory. When I find a mystery apple, the first question I ask myself is, "what varieties were growing in this town at this time?" While Macs can be found in most Maine communities, many varieties are only found in certain spots.

***In which states was the apple grown***: This is like an apple's alibi. If you're looking for an old Smokehouse tree, forget Maine. Try Pennsylvania. You'll find McIntosh throughout nearly all of New York and New England as well across a belt of the northern states and provinces from Nova Scotia to Michigan.

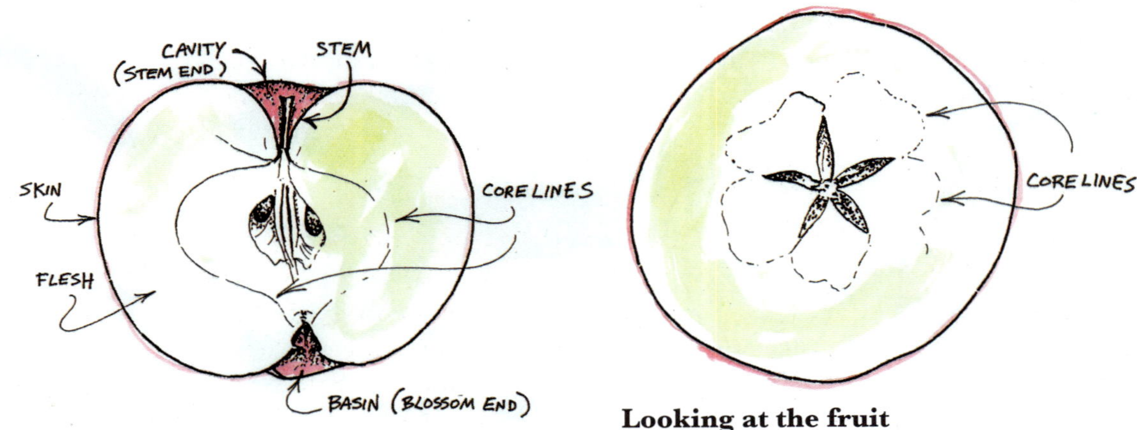


**Looking at the fruit**

Set the McIntosh apple on the table before you. The stem is likely pointing up. Major point number one is that you're looking at the apple upside down. The top is the bottom, and the bottom is the top when it comes to apples. Your apple is resting on the **apex**, while the stem is protruding from the **base**. Find a book of European apples and look at the images. Books from England and the mainland depict hundreds of apples all "upside down." Why is that? Pinch the stem between your fingers, and hold the apple in front of you with the stem down and the other end up. This is way the apple grows on the tree. **What we think of as the bottom of the apple is actually the top**. When it comes to apples, the Europeans got it right and the New World got turned upside down.

***Fruit size***: What size is your apple? Take out a ruler. I have fourteen randomly selected Macs here on the table. The smallest is 2¼" in diameter, and the largest is 3¼". Most are somewhere in between. The smallest one looks too small to be a Mac. The largest one looks too large. One apple specimen can be deceiving. "This couldn't be a Mac. It's too small…". Or, "This is too large to be a Mac…" I total them all up and divide by fourteen and get 2⅞". That's solidly "medium."

very small: under 1½"
small: 1" - 2"
small-medium: 2" - 2½"
medium: 2½" - 3"

medium-large: 3" - 3½"
large: 3½" - 4"
very large: 4"+

What makes a **crabapple** a crabapple? A better question would be, "what makes an *apple* a crabapple?" John Fiala, deceased but still the reigning world authority on crabapples, defined a crabapple as any apple under 2" in diameter. **A crabapple is just a small apple**.

***Shapes***: Now set the apple on the table stem side up and look at the shape. It can be round, conic, oblate (flattened) and so forth. More often it will be some combination of shapes. Beach calls McIntosh "roundish to somewhat oblate." He calls Northern Spy "roundish conical, sometimes inclined to oblong, often noticeably flattened at the base, nearly symmetrical, sometimes regular but often noticeably ribbed." It is rare to find a variety that's strictly one shape or another. Ben Davis is another good example. Some Ben Davis fruit can be practically round, while others can be as conic as a Red Delicious. Our fourteen Macs are mostly round, or "roundish." Some are truncate, that is, with a flattened apex or base. Some are slightly conic or slightly oblate. All, however, are roundish. We'll give a nod to the variations, but give special emphasis to round. When I stand back and look at my Macs, they are round. We'll call the Mac's shape roundish.

***Regular/irregular***: We also describe the apple's shape when turned so that the stem is pointing directly at you. In other words, you are looking at the base of the apple head on. From this view you can see if the apple is round, oval, elliptical or angular. All the common Russets are round from this point of view, with one exception. Roxbury Russet is distinctly elliptical, or oval.

Regular is also used to differentiate ribbed apples from the non-ribbed. Continue to look down at your Mac as though you were a bird in the sky. From this perspective you can also see if your apple is ribbed. Some varieties, such as Northern Spy and Yellow Bellflower, are slightly ribbed. Others are deeply so. Calville Blanc is one of the most famous of the ribbed varieties. **When looking down at the base from above, if the shape is round, the apple is regular. If oval, elliptical or angular (ribbed), it's irregular.** Our Mac is regular.

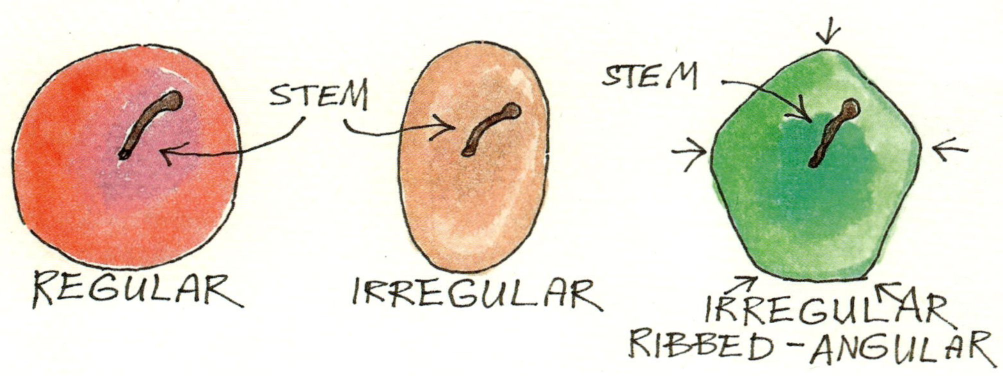

***The Base and cavity***: At this point in time, your apple should be sitting on its apex, stem-side up. **The stem end of the apple is the base**. **The depression surrounding the stem is the cavity**. The stem emerges from the apple at the deepest point in the cavity. The cavity can be any gradation of widths and depths. The angle of the slope descending into the cavity can be any degree of steepness. Some are really steep while others are rather gentle. When assessing the steepness factor, we're looking at the angle created by the stem and the side of the cavity. If that angle is really steep, we say it's **acuminate**. If it's only medium steep, we say it's **acute**. If it's pretty shallow, we say it's **obtuse**. In leaf morphology, acuminate describes a leaf that comes to a sharp point. When you look into an acuminate cavity, the angle formed by the stem and the side of the depression will be long and pointy. An acute cavity will be not so sharp, and an obtuse cavity will be shallow and wide. Mac's cavity is large, wide and acuminate.

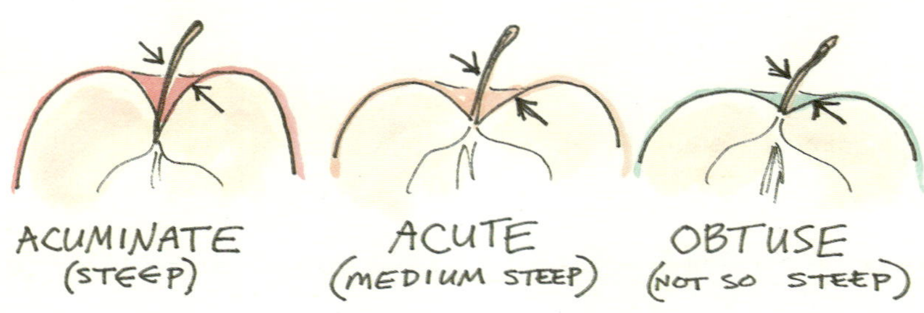


***The Stem***: The stem can be thick or thin and practically any length. Looking at our McIntosh stems, they are all short to medium in length and not particularly thin or thick. A few rise up slightly higher than the base of the apple. Several do not. Some apples have distinctive stems. Of the common modern varieties, Golden Delicious will nearly always have a long stem. Our McIntosh stem is short to medium in length and medium in thickness.

***Stem area coloring***: Take a look at the coloring in the cavity and around its perimeter. Sometimes that color is the same as the rest of the apple. Other times it's quite different. It can be a greenish or brownish russet. Sometimes the russet coloring emerges from the cavity and cascades over the surface of the entire apple in patches or even as a sort of netting. With some apples, this stem splash is so distinctive, it can be an instant giveaway. Wolf River nearly always has a huge patch of russet around the stem that's impossible to miss. Most of our Macs have a small contained patch of russet around the stem.

***Lipped***: Some varieties have a weird lump or protuberance of flesh that fills up part or even all of the cavity. The stem emerges from under this blob. In belly-button terminology, this would be an outie, not an innie. Yellow Bellflowers have it now and then. Others, like Pewaukee, frequently have it. Pomologists called these apples **lipped**. Beach says of Pewaukee that the stem is "often inserted under a lip." This condition is uncommon, but does show up from time to time. Macs are not lipped.

***The apex and basin*: The apex is opposite the stem-end.** Flip the apple over, stem side down. If it doesn't sit well, twist off the stem between your fingers. Then place the apple back on the table with the stem end down. You are now looking at the top of the apple or the apex. Note the depression with the small shriveled-up grayish tuft in the center. **The depression in the apex is the basin,** sometimes called the blossom end or the eye. We can assume it's called the basin because it looks like a miniature wash basin.

**The grayish tuft in the center of the basin is the calyx.** The calyx is what remains of the apple's flower after it's dried up. The calyx can be **open**, **partly open** or **closed**. A closed calyx looks like a small solid mass of brownish-gray leafy material. An open calyx has a small hole in the center through which you can see into the core of the apple. Sometimes you'll find open and closed calyxes in different specimens of the same variety. The calyx will also vary in size from one variety to another. This is about relativity. Note whether the calyx is small, medium or large. A couple of the calyxes on our fourteen Macs are partly open, but all the rest are closed. I think we can safely call the Mac calyx small and closed.

Sometimes the skin immediately around the calyx is pinched into tiny pleats. This is called **corrugated** or **wrinkled** depending on which source you consult. The McIntosh skin is smooth around the calyx.

**The miniature shriveled-up dead-looking bits that make up the calyx are called calyx lobes,** sometimes called sepals or segments. They look like tiny leaves. If the lobes stand erect and the tips come together forming a cone, they are **connivent**. If they stand up tall, meet one another but then arch or bend slightly away from each other at the peak, they are **erect convergent**. When they lay flat on the apple's surface and the tips point towards the center, they are **flat convergent**. If they lay flat and the tips point away from the center, they are **reflexed** or **divergent**. Sometimes you get more than one configuration per variety. Mac calyx lobes are, for the most part, erect convergent. A few are reflexed, but we'll call them erect convergent. Hedrick avoids the details and simply calls them erect.

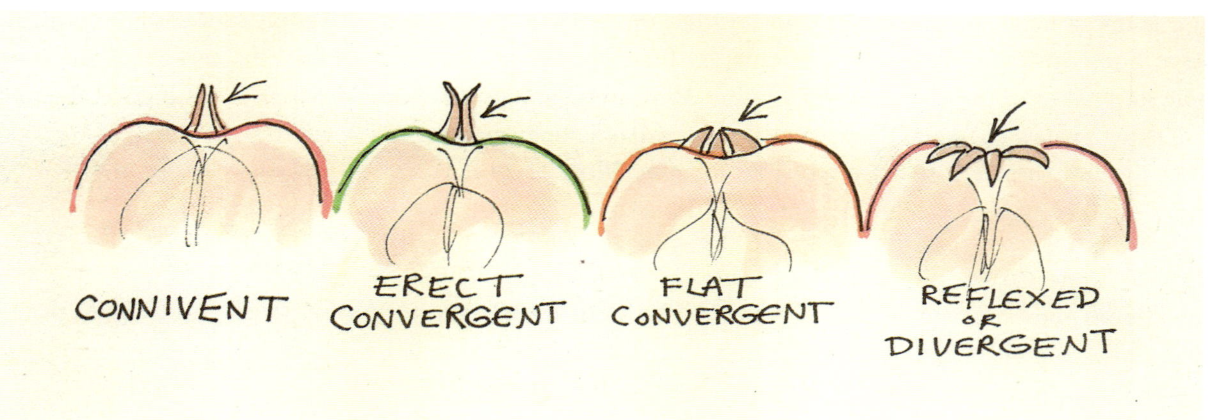

The **basin** itself can be deep or shallow, wide or narrow. Some are deep and wide. Others are practically non-existent; there is hardly a depression at all. Even when there's no noticeable dip, the area around the calyx is still called the basin. **The sides of the basin can be abrupt, meaning steep**. **They can be obtuse, meaning gently sloped**. The Mac basin is medium to narrow in width and medium deep to deep in depth. The sides are abrupt. You'll also note that the rim around the Mac basin is very smooth. We call that basin **regular**.

On other varieties the rim is ripply or even deeply grooved. When the rim of the basin resembles the gentle swells on Penobscot Bay, we call it **wavy**. When the waves are pronounced, the basin rim is **furrowed**. The furrows around the basin are the end points of ribs. Deeply furrowed apples are all **ribbed**. Those five bizarre points on the Red Delicious apex are called the crown. Many apple varieties have a **crown**, although only a few have one as distinct as Red Delicious.

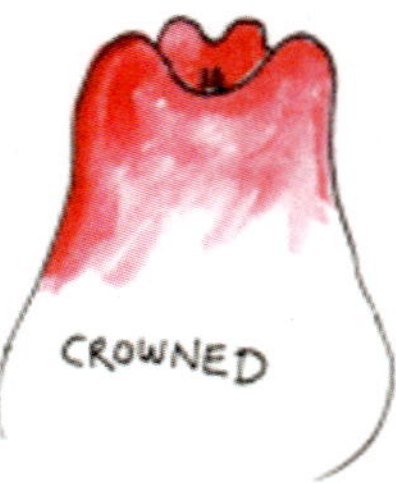


***Color***: Even if you get the stem length wrong and you call it acuminate when it should be acute, you can redeem yourself by learning to recognize the colors. The coloring on some apples is hard to miss: Black Oxford, for instance, or Winter Banana with its brilliant yellow base and fiery, red blush. Fallawater is blue-green. Starkey is rosy red. Ben Davis is a red apple, but I see black hiding behind all that red. Northern Spys always look pink to me, even though they're red.

At some point, someone will need to create a standardized book of apple colors, something akin to the work of the ornithologist Robert Ridgway. Ridgway's first effort was *A Nomenclature of Colors for Naturalists* (1886), which included hand-painted plates of a hundred and eighty-six colors. His classic *Color Standards and Color Nomenclature* (1912) expanded to over eleven hundred colors, each one labeled. Although I don't think we need eleven hundred, it would be of great value to the apple collector and identifier to have a book of a couple hundred apple colors we could all use. Let's do it!

The coloring of McIntosh is not as straight forward as many apples. Some Macs are blushed solid red, while others are striped. Many are a combination of stripes and blush. The first Macs from Dundela were said to be solid red. We'd call those red-blushed. **A blush is a patch of color**, usually covering part of the apple, but occasionally covering its entire surface. Very few of these old, solidly red-blushed trees are still around. I know of one in Kent's Hill, Maine. Once part of an orchard, it now stands alone on a lawn. It is exceedingly old, and the Macs are nearly solid red, not striped. Rumor has it the scionwood for that tree came from the original McIntosh.

By about 1900, striped Macs began to show up. These striped Macs were sports (mutations) of the original Mac. The trees that Francis Fenton's father planted in 1906 were striped Macs, only one of which remains. It is a large and beautiful tree. Despite the fact that the striped Macs were not the original strain, they remain much beloved, often referred to as "old-fashioned Macs." You can still find them now and then.

America's love of solid colors led to the rejection of the striped Mac and a resurgence of red-blushed strains beginning with the introduction of a solid red sport in 1932 called Rogers Red McIntosh, Rogers Red, or simply Mac. These Macs are a bluish red. Not purple like Black Oxford. Other sports have come along in the ensuing years, all red-blushed. When you come across a partially red Mac these days, it's probably one that didn't get very much sun. The more sunshine, the more blush.

Looking over the dozen Macs on my table, a few of them do exhibit some stripiness. But don't be fooled. The old striped Macs are really striped. These are thoroughly modern Macs, smooth, shiny, purply red, bordering on maroon.

***Stripes***: Prominent stripes are a key defining characteristic on a number of varieties. Three distinctly striped varieties in many northern districts are Duchess, Gravenstein, and St. Lawrence. Once you see a St. Lawrence, you may never forget it. It is really striped. The red stripes contrasting with a light green ground color make the St. Lawrence stripes stand out. Same is true to a lesser extent with Gravenstein and Duchess. On some varieties, the stripes are there but barely noticeable. They blend into the overall coloring. Sometimes they lay on top of the blush, visible close up and becoming invisible from a few feet away. Do we consider McIntosh a striped apple or not? Probably better to enter "variable."

***Skin qualities***: Is the skin tough, thin, thick, shiny? Is it sticky or greasy? Macs are thin-skinned and always shiny, especially if you give them a bit of a rub. Baldwin's thick skin may be responsible for its reputation as being insect resistant. Tough skin may also be one reason Baldwins keep so well in the root cellar. The moisture can't get out. Not all thin-skinned apples keep poorly, however. Northern Spy is thin-skinned and easily bruised. Spys found a way of isolating those bruises. They keep well until spring, bruises and all. Red Tip and Kola are two varieties that were selected and introduced by Niels Hanson in about 1920. Both have highly aromatic skin covered with oil. It glistens. One light touch and you'll be smelling it for hours.

Rogers Red McIntosh: bud mutation (sport); Dansville NY 1932: the dusty-looking film is called bloom

***Bloom***: The **bloom is a dusty film that covers the surface** of many red and purple apples, as well as blueberries and blue grapes. It looks like a grey dust or pesticide. It's neither. Hedrick calls it, "minute scales of wax." I've also heard it explained as yeast. It's what makes a Blue Pearmain look blue. Macs and Cortlands nearly always have bloom when you pick them. Depending on how they've been treated post-harvest, they may still have some bloom when you obtain them. Wipe it off on your jeans or not. It's harmless.

What about **scarfskin**? In a 1999 *Scaffolds* article, Dave Rosenberger calls scarfskin a "fruit finish disorder that…occurs when the epidermis and cuticle separate from the underlying tissue. The resulting air space beneath the waxy fruit surface disrupts light transmission and produces the milky or cloudy appearance [of the surface of the apple]." Earlier authors call it a roughening of the skin's surface. Apparently scarfskin is most likely triggered by weather changes and/or certain fungicide sprays when the fruit is still only about an inch in diameter. Although scarfskin is harmless and tasteless, it tarnishes the perfect look of the grocery store apple. Growers don't like it.

Submerged dots on Rhode Island Greening

***Dots***: Look at a section of the apple's skin. If it's striped, rotate the apple to the side with the darkest stripes. Now look for the small light colored points, technically called the "dots." **Dots can be prominent**, **very prominent or not prominent.** On a McIntosh, the dots are practically invisible from three feet away but obvious from nine inches. This is not always the case with other apples. Usually the dots are prominent or they're not. Each McIntosh apple probably has a thousand dots, albeit small and not particularly noticeable. We'll say that Mac has numerous dots, small and not prominent.

Apple dots possess various physical qualities as well. They can be **rough** like those on Stark or Rolfe or the modern apple, Redfield. If you close your eyes and lightly touch the surface of the apple with your finger tip, you can feel them. They are rough to the touch.

The dots can be **sunken**, **depressed** or **submerged**, three terms that mean essentially the same thing. The dots on Baldwin, Calville Blanc, and Rhode Island Greening all appear to be below the skin's surface, as though you are looking through a pool of still water at the dots swimming just below. They are submerged.

Areolar dots on Baxter

The dots can be **stellate** or star-shaped like those on the apples off an ancient tree in Belgrade at top of Smith Hill a few feet from the clubhouse at the Belgrade Lakes Golf Course. The tree predates the course by 140 years or so. The Vermont apple, Scott's Winter, and the Maine apple, Dudley Winter, both feature stellate dots.

The dots can also be **areolar**. The Canadian variety Baxter is always covered with hundreds of prominent areolar dots. Each dot consists of a dark point set in the center of a larger, pale, light yellowish-pink blob, like a halo or a tiny frog's egg floating on the surface. The dots reflect the light in a way that makes them look yellow. The surface of the apple is a mass of pale yellow dots on a rosy red background, each dot with a tiny dark center.

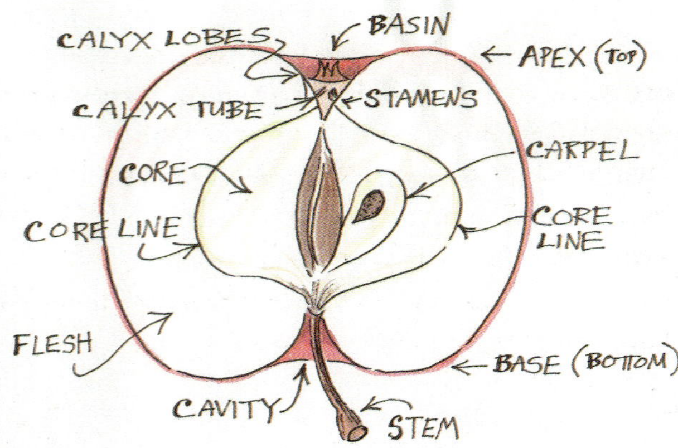


**Inside the apple**

Time to take out the knife. We'll cut three or four of our apples length-wise, from calyx to stem, being mindful to slice right through the center of the calyx and the deepest point of the cavity. If you're lucky, your slice might go right down the center of the stem.

***The calyx tube***

Look at the apple right side up, in other words, with the stem end down. The basin is now up. The calyx is sliced in half, and below it is the calyx tube. **The calyx tube is a small extension of the basin, downward towards the center of the apple**. Most calyx tubes are conic-shaped or funnel-shaped. A few are urn-shaped. **Conic-shaped** calyx tubes are shaped like a cone. Think ice-cream cone. Many apples have conical calyx tubes. **Funnel-shaped** calyx tubes are shaped like a funnel with a cone perched on top of a cylindrical tube. A few of the better known northeastern varieties with funnel-shaped calyx tubes are Bottle Greening from Vermont, Nodhead from New Hampshire, and Mother from Massachusetts. **Urn-shaped** calyx tubes are generally chubby, conic and bulge out. Gray Pearmain has a short urn-shaped calyx tube. The calyx tubes in the Mac specimens we cut in half vary somewhat in length and breadth. They are all clearly conical.

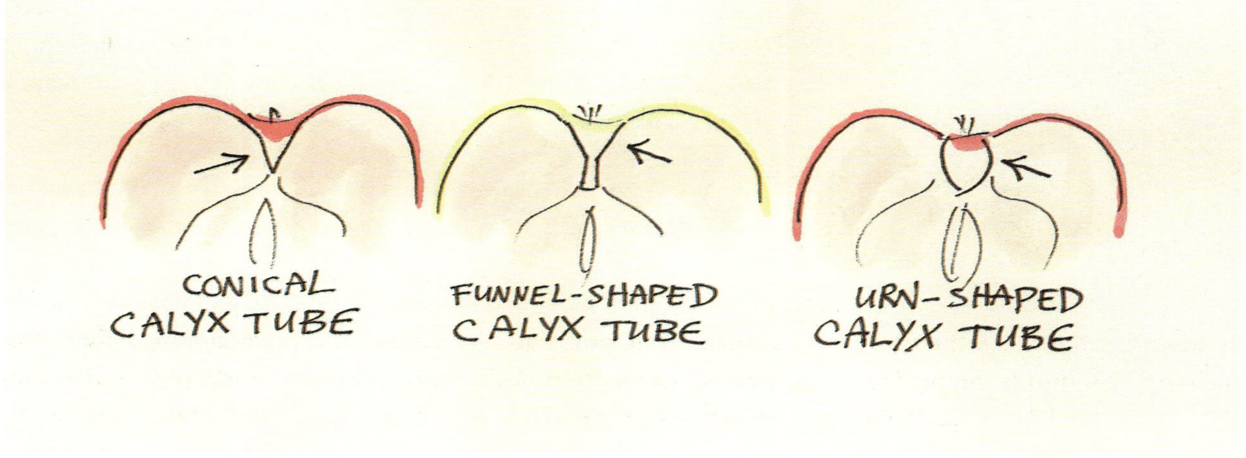

***The stamens***:

**The stamens are the short dark threads, below the calyx on the side of the calyx tube**. They are positioned part of the way down the tube towards the center of the apple. These are the remnants of the stamens. They can be marginal, median or basal. **Median stamens** are positioned at about the halfway point between the calyx and the inner end of the calyx tube. **Marginal stamens** are situated closer to the calyx. **Basal stamens** are set deeper in the cone, closest to the core. If in doubt on the importance of any one factor, remember Sherlock. "The smallest point may be the most essential." Looking at the four Macs I have before me, I'm going to say the stamens are median.

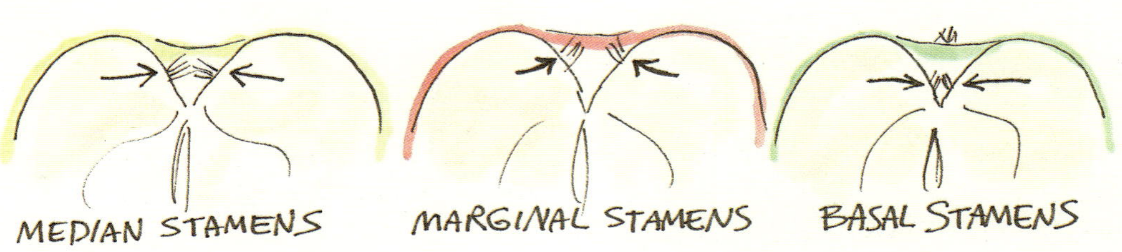


***The core***:

As you look at the apple cut in half you'll see the **core** In the center of the apple we have. **The core line is a curved line that extends from the inner end of the stem around the core to the calyx tube.** The core line can be thought of as two lines. They begin at the same spot. They circle around the core and then the two converge at the calyx tube. When the two core lines end on either side of the calyx tube, they are **clasping**. (They are "clasping" the calyx tube.) When the two core lines meet together at the tip of the calyx tube, they are **meeting**. Some apples always meet; others always clasp. Some meet or clasp. It's an important feature but one that can be variable. Our Mac's core lines meet or just barely clasp.

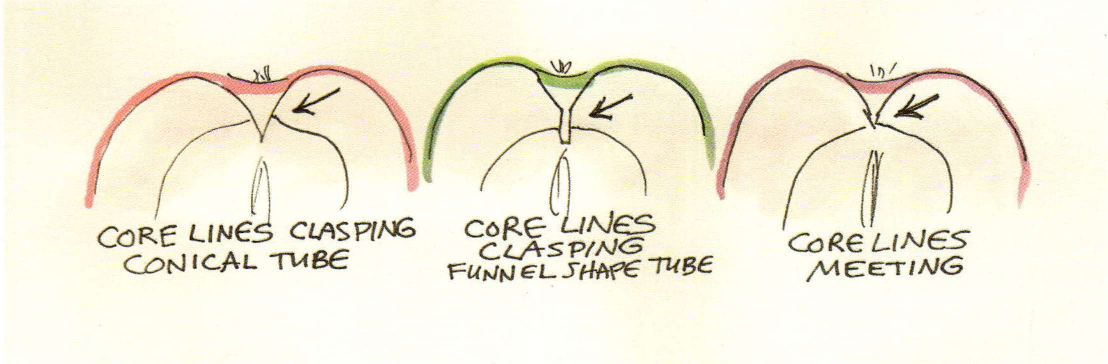


It's useful to know that **cores come in different sizes and locations**. Most cores are medium in size, including McIntosh. Some are small, like Alexander, Grimes Golden, Hurlbut, and Stark. Others are large, like Blue Pearmain, Esopus Spitzenburg, Winter Banana, and Yellow Bellflower. The Maine variety,

Brock, was selected for the apple sauce industry because of its small core. Then there's the relative position of the core. Most apple cores are centered evenly between the stem and calyx (**median**). They can also be closer to the stem (**sessile**) or closer to the calyx (**distant**). Pound Sweet and Red Canada are both sessile. Pewaukee and Porter are both distant. "I knew it had to be a Red Canada, look at that sessile core!"

***Carpels***:

The **carpels are the walls of the seed cells**. They are located inside the core lines. Sometimes they are difficult to make out when you cut the apple in half. To get a good look you may need to slice up a few apples. If you're lucky, along the way the carpels will be revealed. With a good slice, there will be two carpels, one on either side of the core. Now take the stem of your half-apple between your fingers, stem-side down. Use your imagination, and forget the apple for a moment. Pretend you're holding a stem and leaf. Together the two carpels look like the leaf, and they can be described using leaf terminology. **Cordate** is spade-shaped. **Obcordate** is heart-shaped, and so forth. McIntosh carpels aren't really any of these. They are roundish.

All carpels are also emarginate or mucronate. **Emarginate** carpel tips turn in and point towards the center of the fruit. **Mucronate** carpel tips point towards the calyx. I would say that the carpels of the Mac are slightly mucronate.

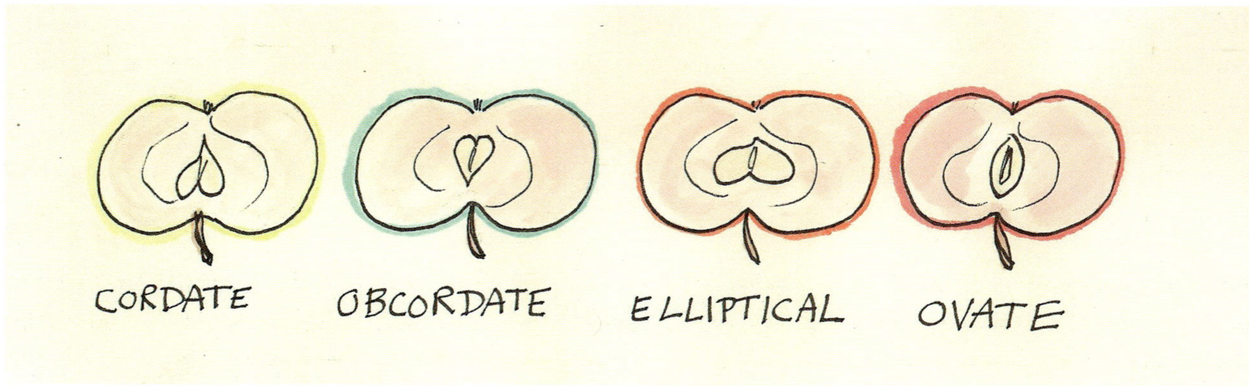


Lastly, the carpels can be tufted or not. **Tufted** carpels look as though they have a coating or rim of white fluffy stuff that resembles some form of white fungus. (It's not.) Our Macs are not tufted.

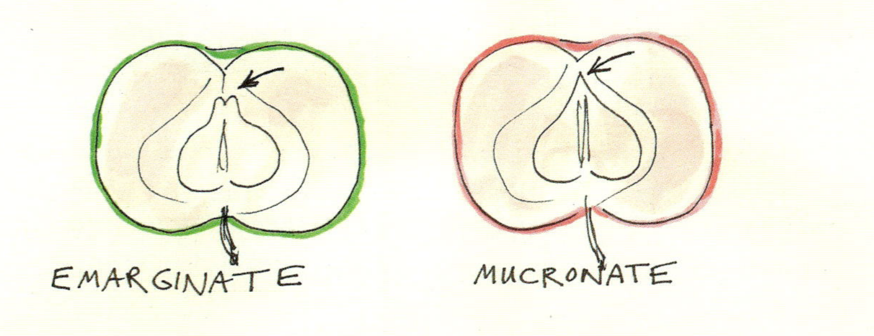

***Axile and Abaxile***: Now slice three or four Macs in half the other way, across the equator. For this step, it's best to use some of your uncut specimens. In a pinch, if you're short on apples, you can do this using the ones you've already cut in half end to end. Somehow you'll have to hold them together as you re-slice them in the other direction. This is one reason why it's good to have plenty of specimens on hand.

You can see how regular the apple is by slicing it this way. Had McIntosh been an irregular apple, it would be angular or oval. If it were a ribbed apple, the ribs would be apparent. But there are none. In the center of the slice there are normally five seed cells that together make up the core.

When the core is **axile**, the seed cells extend all the way into the center of the apple. The star center is a point. Blue Pearmain, Rhode Island Greening, and Roxbury Russet are all axile. The core is **abaxile** when the when the seed cell's walls do not extend into the center of the apple, and the center of the star is an open void. The center of the star is a space rather than a point. Northern Spy, Yellow Bellflower, and McIntosh are abaxile. Although I doubt that anyone has ever done a survey, I would guess that axile and abaxile varieties are roughly evenly split.

***Closed and open***: Axile apples can also be open or closed. The seed cells in a **closed axile** core all join at the very center. The cells extend to a point. The seed cells in an **open axile** core don't extend all the way to the center. Just most of the way. They are open. Abaxile cores are always open. As far as I know, there is no such thing as a closed abaxile core.

***Flesh color***: While we have the apples cut open and before we taste, we need to take note of the flesh color. McIntosh is an easy one. It's white. Some varieties are closer to yellow. Others are greenish even when ripe. Some are pink veined or have red bleeding just under the skin. Some are nearly solid red or pink. One of my favorites is an apple Francis Fenton collected in Mercer and topworked in his orchard many years ago. It's called Red Blaze. I never thought much about the name until one day I sliced it open long-wise and there, emerging from the calyx were two little wings of red, like flares from an oil well, one on each side of the calyx tube: the red blaze!

## Now we taste

Is the apple sweet or tart? **Sweet** means low in acidity relative to the amount of sweetness. Tart means the reverse. For description purposes, you should be able to tell from the taste. The sweet apples will taste weird and insipid. No acid. **Tart** apples will taste tart. Tart apples are also sometimes called **sub-acid**. Where does our McIntosh fit into all this? Mac is a tart apple. It makes good sauce. When you cook a Mac, it's one desire in life is to become sauce. It does not want to be in your pie. It does not want to be cider or molasses. In a pie, its internal structure falls apart, and you wind up with a collapsed crust and a pan full of apple soup. You could have better luck blending your Macs. Combine a few with dry, tart apples such as Ben Davis, and you might come up with something you're happy with.

Each apple also has its own unique flavor. Some are pronounced; others are subtle. McIntosh is one of the tart apples with a distinctive flavor. Tolman Sweet is one of the sweet varieties with a different but equally pronounced and distinctive flavor. Once you get to know that Mac flavor, it becomes the easiest way to tell a Mac. It might be small or large or striped or blushed, but it will taste like a Mac.

All apples vary in their crispness and juiciness. We always note both as we key out each variety. Is it crisp, moderately crisp, or not at all? Is it juicy, moderately so, or not at all? Not only does the fresh-eating public want their apples a little on the tart side, they also want a lot of crisp and a lot of juice. The next big-money dessert star has got to be crisp, and juicy.

Chefs and cider makers are not so impressed by crisp and juicy. They are concerned with flavors, acidity levels and tannins. They want complexity. Many of the best cider apples are not very juicy at all. Some of my favorite sauce apples are also rather dry. Drier apples have more flavor relative to the amount of juice. You don't need a crisp apple to make an apple crisp.

Awareness of these basic flavor and texture components can be a huge help in identifying an apple. If it's a sweet apple, you've eliminated about ninety percent of the possibilities. It is also extremely useful to quantify the crispness and juiciness of your fruit. McIntosh is tart, moderately crisp and moderately juicy.

## And the final question

What makes this apple distinctive? In a few words, create a thumbnail sketch of McIntosh. When you come across an apple, what tells you it's a Mac? What is the essence of McIntosh-ness? All Macs have the same unique McIntosh flavor. In some it's more pronounced than others, but they all have it. All Macs are roundish and between two and three inches in diameter. They all have a narrow, round, regular, medium-deep basin with abrupt sides. They might be purply-red striped, or they might be purply-red blushed, but they all have thin, shiny skin covered with bloom.

"Hey, that's a Mac."

Every year, on a dark night in mid-January, we gather in the orchard, drink cider, sing, and celebrate the apple tree: Wassail!

# Brief List of Works Consulted And Recommended Reading

With a couple of exceptions, this list includes only older publications. There are also many excellent recent books on apple history, apple varieties, apple growing and cider.


Beach, S. A. 1905. *The Apples of New York*. Albany: J. B. Lyon Co. A good overview of American apple history. Good technical descriptions of the parts of the apple. Detailed descriptions of hundreds of varieties. Good color plates. Two volume set.

Bradford, Frederick C. 1911. *Apple Varieties in Maine*. Master's of Science in Agriculture thesis: Orono, ME. No longer in print, but available for download at https://digitalcommons.library.umaine.edu/etd/2384/. Useful for New York and New England fruit explorers. *Apples of Maine* by George Stilphen is a reprint of Bradford. See below.

Bussey, Daniel J. 2016. *The Illustrated History of Apples in the United States and Canada*. Mount Horeb, WI: JAK KAW Press. 7 volumes. Detailed descriptions of thousands of varieties as well as thousands of full size classic watercolors. If you can only afford to own one apple book, this is it. There is nothing else like it.

Calhoun, Creighton Lee. 2010. *Old Southern Apples: A comprehensive History and Description of Varieties for Collectors, Growers and Fruit Enthusiasts*. White River Junction VT: Chelsea Green. The title tells it all. The focus is on the southern states of the U.S. Beautifully done.

Cole, S. W. 1849. *The American Fruit Book*. New York: Orange Judd and Company. Descriptions of many apples and other fruit varieties with a New England focus. Especially useful for its history and uses of the fruit.

Cox, William. 1817. *A View of the Cultivation of Fruit Trees*. Philadelphia: M. Carey and Son. Excellent for early history of American cider. Descriptions of fruit varieties, primarily apples. Reprint was published by H. Frederick Janson in 1976 (Rockton, Canada: Pomona Books).

Downing, Andrew J. 1886. *The Fruits and Fruit-Trees of America*. New York: John Wiley & Sons. Descriptions of hundreds of apples and other fruits, including many obscure varieties useful to the explorer. Multiple editions beginning in 1845. They are all good.

Elliott, F. R. 1854. *Elliott's Fruit Book; or the American Fruit-Growers Guide in Orchard and Garden*. New York: C.M. Saxton. General history, cultivation and descriptions of many apples and other fruits.

Fiala, John L. 1994. *Flowering Crabapples: the Genus Malus*. Portland, Oregon: Timber Press. Excellent botanical overview of the apple, including species from around the world. Descriptions of hundreds of crabapples.

Hedrick, U. P. 1922. *Cyclopedia of Hardy Fruits.* New York: The Macmillan Company. Brief explanation of descriptive terminology followed by detailed descriptions of hundreds of varieties of tree fruits and small fruits.

Hedrick, U. P. 1925. *Systematic Pomology*. New York: The Macmillan Company. Essential textbook for the apple identifier. Includes searchable apple key with 114 common varieties of the day. Also includes searchable keys for other tree fruit and some small fruit.

Hogg, Robert. 5th edition, 1884. *The Fruit Manual: A Guide to the Fruits and Fruit Trees of Great Britain.* London: Journal of Horticulture Office. Descriptions of hundreds of varieties of major tree fruits and small fruits. His explanations of pomological terminology are excellent. He is brilliant.

Kenrick, William. 1833. *The New American Orchardist.* Boston: Carter, Hendee, and Co. Excellent cultural history. Descriptions of many apple and other fruit varieties.

Stilphen, George A. 1993. *The Apples of Maine*. Independently published. Stilphen edited, type-set and re-printed Frederick C. Bradford's 1911 University of Maine thesis. Out of print and hard to find. See Bradford above.

Thatcher, James. 1825. *The American Orchardist.* Plymouth MA: Erza Collier. Descriptions of many American apples originating before 1800. Of particular value to those interested in early apple history and especially cider.

Thomas, John J. 1908. *The American Fruit Culturist.* New York: Orange Judd Company. Features a searchable apple variety key that is a must for the apple identifier. Also includes cultural and varietal information for other tree fruits. Multiple editions beginning in 1875.

Thoreau, Henry David, 1862. *Wild Apples*. Article first published in *The Atlantic Magazine*. Now available online. Published in a number of versions. Sometimes included in Thoreau anthologies. Worth reading again every few months. It's wonderful.

Apples by John Alsop

# Index

# With Thanks

"I have not lived for years with Sherlock Holmes for nothing."
Dr. Watson, *The Hound of the Baskervilles*, p. 252

There are many who have assisted me along the way, without whom I couldn't have done any of this. I've been fortunate when it comes to mentors. A few of them I got to live with. Others I saw only now and then. Some might not remember me, or even know who I am. Still I get to claim them as mentors and appreciate them every day. So, thank you all who shared your apples and trees and stories and lives. Thanks to all the apprentices who came and lived with us on our farm. And thanks to all those in my family who've loved and supported me as I climb my way up the apple ladder.

That being said, I do want to mention a few people in particular: Abbott Meader, my college art teacher, who painted the cover; John Alsop, lifelong buddy, who painted Bunk meets Sherlock; Russ French, who photographed all the paintings for the book, as well as the photo of Francis Fenton and me on the title page; and to my sister Emily Bunker, for her editing and many suggestions. Most of the photos, the apple "portraits" at the beginning of each chapter and the "cartoons" are my creations. All the photos were taken in Maine unless otherwise noted.

Thanks especially to Bob Doan. Somehow I convinced Bob to edit, format and proofread the manuscript. I shudder to think how many hours Bob put into this project. He's been hanging in there with me for the past forty-five years.

Lastly to Campbell Watts, a.k.a. Mary C., Nona, or Cammy. More than anyone else on earth, she makes this possible every day. I think she knows more about apples than I do. Wow.

John Bunker, Abbott Meader, and John Alsop in 2013

Surrounded by mentors at the USDA Agricultural Library, September 18, 2010:
Tom Burford, Ben Watson, Nick Botner, JPB, Lee Calhoun and Dan Bussey

I once was looking for a magic weed,
And found a fair young squire who sat alone,
Had carved himself a knightly shield of wood,
And then was painting on it fancied arms,
Azure, an Eagle rising or, the Sun
In dexter chief; the scroll "I follow fame."
And speaking not, but leaning over him,
I took his brush and blotted out the bird,
And made a gardener putting in a graft,
With this motto, "Rather use than fame."
You should have seen him blush; but afterwards
He made a stalwart knight.

Merlin, from *Idylls of the King*
by Alfred, Lord Tennyson